Praise for the Previous Edition:

"Jason Corey's succinct book covers territory rarely discus [. . .] Corey offers us a book chock-full of valuable information lly to the results of audio production."

Ross Feller, *Computer Music Journal*

Audio Production and Critical Listening

Audio Production and Critical Listening: Technical Ear Training develops your critical and expert listening skills, enabling you to listen to audio like an award-winning engineer.

Featuring an accessible writing style, this revised edition includes substantial new content on parametric equalization and tonal balance, spatial audio and reverberation, dynamics processing, distortion and noise, analysis of sound recordings, as well as updated sections on headphone recommendations, hearing conservation, and new everyday ear training exercises. The interactive web browser-based technical ear training software modules provide endless opportunities to practice identifying various types of signal processing and manipulations across a range of difficulties. Working alongside the clear and detailed explanations in the book, this software completes the learning package that will help you train your ears to listen and really "hear" the technical details of your recordings through each stage of production and post-production. This all-new edition has been updated to include access to integrated software that promotes listening skills development through audio examples found in actual recording and production work, suggestions for progressive equalizer training exercises, and new analyses of recordings examples in a range of diverse genres.

This is an ideal textbook for undergraduate students in music production, audio engineering, and sound design courses, as well as novice producers and engineers honing their critical listening skills for audio.

Jason Corey, Ph.D., is Professor of Performing Arts Technology and Associate Dean for Graduate Studies at the University of Michigan School of Music, Theatre & Dance, where he teaches technical ear training and critical listening, and sound recording and production. He is active as a recording, mixing, and editing engineer, and is the co-author of *Virtual Reality Audio: A Guide to Immersive Sound in Virtual Experiences*. He holds Ph.D. and M.Mus. degrees in Sound Recording from McGill University.

David H. Benson is a software developer and audio researcher with a Ph.D. in Sound Recording from McGill University. His work has been supported by the Quebec government's Fund for Research on Society and Culture (FQRSC) and the Audio Engineering Society Educational Foundation (AESEF). In 2011 he was named the AESEF's John Eargle scholar, an honor recognizing excellence in both technology and music. As a pedagogue, he was granted the Schulich School of Music of McGill University's first Outstanding Teaching Assistant Award.

Audio Engineering Society Presents . . .

www.aes.org

Modern Recording Techniques, 10th Edition
A Practical Guide to Modern Music Production
David Miles Huber and Emiliano Caballero

Recording Orchestra and Other Classical Music Ensembles, 2nd Edition
Richard King

Sound Reproduction, 4th Edition
The Acoustics and Psychoacoustics of Loudspeakers, Rooms and Headphones
Floyd E. Toole, Sean Olive and Todd Welti

Classical Recording, 2nd Edition
Practical Guide in the Decca Tradition
Caroline Haigh, John Dunkerley and Mark Rogers

Handbook for Sound Engineers, 6th Edition
Edited by Glen Ballou and Doug Jones

Audio Effects, 2nd Edition
Theory, Implementation and Application
Joshua D. Reiss and Andrew McPherson

Virtual Reality Audio
A Guide to Immersive Sound in Virtual Experiences
Anıl Çamcı and Jason Corey

Immersive Sound Volume II
The Design and Practice of Binaural and Multi-Channel Experiences
Edited by Agnieszka Roginska and Paul Geluso

Audio Production and Critical Listening, 3rd Editon
Technical Ear Training
Jason Corey

For more information about this series, please visit: www.routledge.com/Audio-Engineering-Society-Presents/book-series/AES

Audio Production and Critical Listening

Technical Ear Training

Third Edition

Jason Corey

Technical Ear Training Software

David H. Benson

Routledge
Taylor & Francis Group
NEW YORK AND LONDON

Designed cover image: Shutterstock

Third edition published 2027
by Routledge
605 Third Avenue, New York, NY 10158

and by Routledge
4 Park Square, Milton Park, Abingdon, Oxon, OX14 4RN

Routledge is an imprint of the Taylor & Francis Group, an informa business

First edition published by Routledge 2010
Second edition published by Routledge 2017

Library of Congress Cataloging-in-Publication Data
A catalog record has been requested for this book

ISBN: 978-1-032-60936-2 (hbk)
ISBN: 978-1-032-60933-1 (pbk)
ISBN: 978-1-003-46117-3 (ebk)

DOI: 10.4324/9781003461173

Typeset in Times New Roman
by Apex CoVantage, LLC

Access the Support Material: https://webtet.net/

Contents

Preface to the Third Edition

I am delighted that you have decided to invest time and energy to improve your critical listening skills. And I am honored that you have chosen this book and the accompanying software to help you on your journey of auditory discovery and training. In this book, we will discuss a range of different listening exercises, strategies to maximize progress using the accompanying technical ear training software, and relevant topics in audio theory and auditory perception. My overarching goals with this book are to help you develop and increase your critical listening skills for audio and to provide tools for ongoing technical ear training.

In this third edition of *Audio Production and Critical Listening: Technical Ear Training* I have revised and expanded many sections throughout the book to improve clarity and explain key concepts more effectively. I have also added new material not included in the first two editions. Since the publication of the second edition, I have continued to explore new ways of developing critical listening skills, and this edition includes additional exercises and approaches to technical ear training.

Outline of the Book

I organized the book around common audio processing tools. We will look at audio attributes and their associated audio signal processing devices, including:

- Timbre or tonal balance: parametric equalization and filtering
- Spatial attributes: reverberation, early reflections, delay, and panning
- Dynamic range control: compression/limiting and expansion/gating
- Signal degradation: distortion and noise
- Amplitude envelope and audio excerpt cut-off points: source-destination editing

In Chapter 1, we look at technical ear training and analytic listening, and their importance to audio engineers. I outline some goals, introduce some listening exercises, and provide relevant information about loudspeakers and headphones. Because the sense of hearing is crucial to our work as audio engineers, I include some information on sound levels and hearing health, and end with some suggestions of things we can do to protect our hearing.

In Chapter 2, I focus on spectral or tonal balance and how it relates to parametric equalization. Spectral balance is the relative level of various frequency bands within the full audio band, across the range of human hearing, ideally from 20 to 20,000 Hz. This chapter also describes the equalization ear training software module and outlines a progression of quiz parameter settings in increasing levels of difficulty.

Chapter 3 focuses primarily on reverberation, and it also covers panning, channel correlation and stereo image width, delay, and comb filters. The reverberation ear training module is described. The chapter ends with a section on mid-side processing and ways that this tool can be used in ear training.

Dynamics processing is used widely in most audio recording and live sound situations, with the exception of classical music for the most part. Compression, limiting, expansion, and gating all offer means to control dynamic range and modify timbre in unique and time-varying ways. Dynamic range compression may be one of the most difficult types of processing for a beginning engineer to learn how to hear and use. In many dynamics processing algorithms or circuits, the controllable parameters are interrelated, which affects how we use and hear them. Chapter 4 discusses dynamic processing and offers practice exercises on hearing artifacts produced by these different effects.

We can apply distortion intentionally to a recording as an effect such as with electric guitars or a vocal, but there are many situations where we want to avoid unintentional distortion such as that from an overloaded gain or input stage of an audio interface. Chapter 5 explores a few types of distortion, such as clipping, overdrive, bit rate reduction and perceptual encoding, as well as artifacts like extraneous acoustic or electronic noise, clicks, pops, buzz, and hum.

Chapter 6 focuses on audio excerpt cut-off points and introduces a type of ear training practice based on an audio editing technique. Listening for edit and cut-off points at the millisecond level can sharpen our critical listening skills. The accompanying software module mimics the process of finding an edit point by comparing the end point of one clip with the end point of a second clip of identical music.

Chapter 7 covers analysis techniques for recorded sound. Although there are established traditions in the theoretical analysis of music, there is no standardized method of analyzing recordings from a timbral, sound quality, spatial image, aesthetic, or technical point of view. This chapter discusses methods for analyzing musical recordings and provides some examples of analysis of commercially available recordings. We also discuss listening bias and ways that we can make the listening process more objective.

The book assumes that readers have completed some undergraduate-level study in sound recording theory and practice, and understand basic audio theory topics such as sound basics, decibels, equalization, dynamics, frequency response, microphones, and microphone techniques.

Throughout the book, I have included listening exercises, some of which can be done at any time or location and some of which require the use of reasonably capable headphones or loudspeakers in a relatively quiet listening environment.

This book builds on the work of other audio researchers, teachers, and practitioners with the goal to provide readers with methods to develop critical listening skills applicable to audio engineering. The first formal course in technical ear training that I am aware of began in 1977 at what was then called the Fryderyk Chopin Academy of Music, now called the Fryderyk Chopin University of Music, in Warsaw, Poland. Pioneering work by Andrzej Rabowski, Tomasz Letowski, and Krzysztof Szlifirski resulted in a course they called "Timbre Solfège" which was offered as part of their sound engineering curriculum. Publications by Letowski (1985), Letowski and Miskiewicz (2013), and Miskiewicz (1992), for example, describe this innovative and ground-breaking work that remains essential to the education of sound engineers. My work has also been highly influenced by work that René Quesnel did at McGill University, some of which is described in Quesnel (1996). I had the great fortune to study with Professor Quesnel during my graduate studies and learned an immense amount about technical ear training from him. Sean Olive's work at Harman International, some of which is described in Olive (1994), has also been influential to my work.

The Accompanying Software

While it is essential to learn about technical ear training strategies and associated audio theory through this book and other sources, critical listening skills are not developed without practice. To provide the reader with opportunities for focused listening practice, I have made technical ear training software modules available via the companion website https://webtet.net. I describe each practice module in more detail in the subsequent chapters. Practice modules are interactive, which means you can adjust parameters of each type of processing and get immediate auditory feedback. This way of working mimics what happens in audio production work. The software offers ear training exercises such as matching and absolute identification of processing parameters. Some of the modules provide sound processing examples only, such as the mid-side module, without specific question-and-answer exercises. Being able to isolate a specific type of processing allows us to learn the effect that it has on different audio signals. I encourage you to use the software to chart your own technical ear training path while focusing on those areas that are most needed at your stage of development.

You will need audio recordings for practice and training. I suggest some specific sound recordings later in the book, but there are other locations to obtain useful sound samples for focusing on different types of processing. Single instrument samples and stem mixes can be downloaded from many web sites, such as:

- https://freesound.org
- www.telefunken-elektroakustik.com/multitracks
- www.cambridge-mt.com/ms/mtk
- https://samples.landr.com/category/free
- www.musicradar.com/news/tech/free-music-samples-royalty-free-loops-hits-and-multis-to-download-sampleradar

Digital audio workstations such as Ableton Live, Apple Logic Pro, Apple GarageBand, FL Studio, PreSonus Studio One Pro, Reason, Steinberg Cubase, and Universal Audio Luna include libraries of instrument sounds and loops that can serve as sound sources in the software practice modules.

This book is the result of decades of work as a recording engineer and university professor. Each time I teach my technical ear training course at the University of Michigan, I learn a bit more about the various ways students learn the material and develop their skills, and I reflect on the best ways to help students gain these crucial abilities. I also continue to deepen my own critical listening skills. Based on students' questions, feedback, and the difficulty or ease with which they complete each stage of the course, I try to find ways to improve my teaching so that they learn more deeply and more effectively. I have also tried to think about my own learning and development process. How does one develop critical listening skills? How do we progress from easy to more challenging exercises in a logical way?

Even after we have developed a solid base in critical listening, we can continue technical ear training indefinitely through formal technical ear training exercises, while working on audio recording projects, and with everyday acoustic sounds. My hope is that I can impart methods for gaining critical and analytical listening skills for audio. In writing this book, I want to help you not only listen more deeply but also think more about what you hear, so that through the combined process of active listening and thinking, you can sharpen your auditory focus and give you more facility with critical listening.

Before we head into Chapter 1, I want to offer some words of encouragement. Assuming normal hearing, as determined by an audiologist, most people should be able to improve their critical listening skills through technical ear training. Renowned recording and mixing engineers—that is, the "Golden Ears"—are exceptional not because they have super-human hearing but because they have practiced listening day in and day out, year after year. Unfortunately, there is no easy way to gain critical listening skills. It takes great effort and energy to listen intently and actively, and to think about what we are hearing, but the effort pays off in better recordings and mixes, and an ability to work faster and more accurately. This book and accompanying software offer methods, strategies, suggestions, things to think about, exercises, theory, and practical tips to help you develop much-needed critical listening skills for work in audio. Anyone can turn knobs on a mixer or adjust plugin parameters. The difficult part is to hear the effect each parameter has on a project, to know what needs to be done in each situation to best support an artist, and to select the audio processing parameters needed for each situation. In short, we need to practice regularly with focused exercises, pay attention to and think about what we hear, and always listen actively to improve our critical listening skills.

Bibliography

Letowski, Tomasz. 1985. "Development of Technical Listening Skills: Timbre Solfeggio." *Journal of the Audio Engineering Society* 33: 240–44.

Letowski, Tomasz, and Andrzej Miskiewicz. 2013. "Timbre Solfege: A Course in Perceptual Analysis of Sound." In *Signal Processing in Sound Engineering*, edited by J.A. Adamczyk. Institute of Fundamental Technological Research, Polish Academy of Sciences.

Miskiewicz, Andrzej. 1992. "Timbre Solfege: A Course in Technical Listening for Sound Engineers." *Journal of the Audio Engineering Society* 40: 621–25.

Olive, Sean. 1994. "A Method for Training Listeners and Selecting Program Material for Listening Tests." *110th Convention of the Audio Engineering Society*, Preprint 3893.

Quesnel, Rene. 1996. "Timbral Ear-Trainer: Adaptive, Interactive Training of Listening Skills for Evaluation of Timbre." *100th Convention of the Audio Engineering Society*, Preprint 4241.

Acknowledgments

This book would not have come about without the direct and indirect assistance of many people.

Thank you to Dave Benson for programming and maintaining the ear training software as a browser-based application and for collaborating with me on the software since prior to the publication of the book's second edition. I am grateful for his continued support and development of the technical ear trainer site.

Thank you to the reviewers who provided valuable comments.

Thank you to David Gier, Dean of the School of Music, Theatre & Dance at the University of Michigan, for support of my professional activity, and for approving my request for a sabbatical leave to finish writing this third edition.

Thank you to the team at Routledge for help and support through the various stages of publication: Hannah Barnes-Rowe, Senior Editor; Emily Tagg, Editor; Danielle Zarfati, Squad Manager, Editorial; Cathy Hurren, Senior Production Editor; and Katherine Laidler, copyeditor.

My work on this book would not have been possible without the love and support of my family. Thank you to my family for encouraging and supporting my professional aspirations and for allowing me the time and space to finish this project.

Figures

Tables

Chapter 1

Critical Listening

1.1 Listening Skills for Audio Engineers

Critical listening skills are essential for audio engineers and producers. Whether working in music, film, video, virtual reality, podcasts, gaming, automotive audio, or live sound, audio engineers make decisions that involve microphone selection and placement, mix balances, and signal processor settings based on what they hear and how closely the sound they hear aligns with the goals of a project. Close and careful listening allows an audio engineer to optimize mix balances, dynamic range, auditory perspective, and timbre, and reduce any unwanted artifacts, such as distortion and noise, which may detract from a listening experience.

Some sound quality issues are evident to untrained listeners. Crackling or excessively distorted sound from a broken loudspeaker or the howling screech of feedback from a live sound system is an obvious defect in sound quality that a casual listener, trained or untrained, can likely hear. Other aspects of sound quality are more subtle, such as a delay effect added to a vocal recording, or a carefully chosen but subtle equalizer setting on a sound system. If a project involves hundreds if not thousands of choices, the combined effect of all these small changes on the finished product can be big. Listeners hear dramatically better-sounding music recordings, live sound systems, and film soundtracks when engineers optimize hundreds of obvious and not-so-obvious qualities. Good audio allows concert goers, music lovers, and movie viewers to engage deeply with music and film and to feel more connected to the artists, and is essential to produce impactful emotional experiences.

How do we, as engineers, distinguish fine details of tone quality or mix balance, as the seasoned engineers whose work we admire do? How do we develop and improve our critical listening skills to get the sounds that we want in each project? Technical ear training is the key to developing critical listening skills that help us work more accurately and consistently. When we hone our critical listening skills, we gain the ability to hear sonic details that may have been inaudible or at least unnoticed to us beforehand. Enhanced critical listening skills empower us to get the sound we want on every audio project because we have more control over sound quality.

Although musicians develop expert listening skills, also called aural skills, which focus on musical elements of melody, harmony, intervals, tuning, rhythm, and ensemble, this book and accompanying software concentrate on *audio* listening skills. Because the focus is on technical attributes of audio such as frequencies, signal levels, and timing, and their relation to subjective qualities of audio such as timbre or tone color, auditory perspective, and overall mix balances, we refer to it as *technical* ear training.

The practice of audio engineering is both an art and a science. As audio engineers, it helps to have some understanding of audio theory as well as highly developed critical listening skills. Each audio project has its own set of timbral, spatial, dynamics, and mix requirements, and these

DOI: 10.4324/9781003461173-1

depend on the genre or context. For example, each musical genre—pop, metal, jazz, hip hop, folk, classical—has a distinct *sound*. The word "sound" in this context can mean timbre or equalization, such as when we say a piano *sounds* different than a guitar. In a recording, equalization, dynamic range processing, reverberation, mix balance, and other processing can all contribute to the *sound* of the recording. Since each recording, mixing, mastering, live sound, or sound design project is unique and has its own set of challenges, we rely on a combination of technical knowledge and listening skills to guide our work and make decisions. Technical knowledge of analog electronics, digital signal processing, audio signal analysis, and audio theory is necessary to get the most out of our audio software and hardware tools. We get more out of our equipment and can work more efficiently when we understand the basic principles of the operation of microphones, equalizers, dynamics processors, mixers, reverberation, signal flow, and loudspeakers. Ultimately, no matter the depth and breadth of technical knowledge one may possess, most professional audio engineers rely heavily on what they hear rather than strictly on technical knowledge to make decisions about an audio project, whether those decisions are about microphone choice and location, mix balance, fader levels, or signal processing. Audio theory gives us a framework within which to work, and we should understand as much about our equipment as possible so that we can use it most effectively.

We can affect or alter audio signals in many ways, some of which produce obvious changes in sound quality and others which are more subtle. So it is important to have our ears attentively tuned to the smallest details of timbre and sound quality. As engineers, we make decisions about plugin or effects processor settings, mix levels, and microphone choice and placement with a project's artistic goals in mind. We determine by ear if each of these choices contributes to or detracts from a project's goals. To do this most effectively, we need to know how the technical parameters of audio hardware and software devices affect perceived sonic attributes. For example, we can learn the effect a particular equalizer setting has by listening to it repeatedly with different program material.

To complement technical expertise, professional audio engineers learn to recognize subtle timbral, dynamic, and technical details of sound by listening repeatedly day in and day out. They learn how to translate their aural impressions into choices about equipment settings which involve physical units such as decibels, hertz, and milliseconds. Sometimes referred to as "Golden Ears," highly experienced audio professionals possess the extraordinary—some might say magical—ability to focus their auditory attention to a high degree and translate what they hear into accurate control of audio signals. They likely do not possess super-human hearing from an audiological point of view, but they are expert listeners with highly developed and discriminating critical listening skills. Not only can they identify subtle details of sound, but they are also consistent in their judgments about what they hear. For example, an expert listener is someone who can correctly identify the frequency boosted by an equalizer each time it is presented in a blind test. An expert listener knows if the boost is there and can identify the parameter settings correctly, in terms of center frequency in hertz and gain in decibels, each time they hear it. Such experienced engineers can precisely identify sonic shortcomings they may want to correct in a project as well as features they want to bring out.

Audio engineers develop and improve their critical listening skills gradually over time through day-to-day work on the job, typically editing, mixing, processing, or recording. Most in the audio industry have developed their critical listening skills this way. But there are systematic methods that have been shown to shorten the time required to make significant progress in ear training or at least provide a jumpstart on the process. As my technical ear training professor, René Quesnel, formerly at McGill University, reported in his doctoral thesis, students who completed systematic technical ear training outperformed experienced audio professionals on tasks such as identifying frequency and gain settings of parametric equalization (Quesnel 2001). More recent research by

Denis Martin is consistent with this finding. He reported that 12 hours of technical ear training with compressors proved to be more beneficial than an average of seven years of professional experience (Martin 2019).

Although it was once common for beginning engineers to work with more experienced engineers and learn from them in the context of practical experience, the apprentice model has largely disappeared from the field. Despite this evolution in the audio industry, critical listening skills remain as important as ever, especially as we see varying audio quality in many consumer audio file formats and playback systems. For example, while some music streaming services now offer lossless audio coding formats, most continue to offer audio streams using lossy compression formats such as MP3, AAC, or Ogg Vorbis. These formats are lower in quality than non-data-reduced audio—that is, linear PCM digital audio, such as the WAV file format, or lossless audio formats, such as the free lossless audio codec abbreviated as FLAC, and Apple's lossless audio abbreviated as ALAC. Bluetooth is another commonly used technology in headphones and portable speakers that also unfortunately reduces audio quality through lossy compression. Chapters 2 and 5 discuss lossy compression further.

As we begin to consider critical listening skills related to sound recording and production, a few questions arise:

- What are the critical listening skills that experienced sound engineers, producers, mixers, and sound designers possess that allow them to make recordings, mix sound for film, or equalize sound systems better than novice engineers?
- What can experienced engineers and producers, who have extraordinary abilities to identify and manipulate sonic timbres, hear that beginning audio students may not?
- How do audio professionals hear and consistently identify extremely subtle features of or changes to an audio signal?
- How do expert listeners translate between their perceptions of sound and the physical parameters found in audio plugins and equipment?
- How can those starting out as audio engineers gain listening skills like those of seasoned professionals, which allow them to identify the physical parameters of an audio signal necessary to achieve a desired perceptual effect?
- What specific aspects of sound should novice audio engineers be listening for?

There are many great books that cover technical and theoretical aspects of sound, sound recording and reproduction, and auditory perception. The goal of this book, however, is to help you develop critical listening skills necessary for the successful practice of audio engineering, while also relating these skills to some relevant audio theory fundamentals.

To facilitate the training process, I designed and made available for free a set of software modules specifically for this book to allow you to practice listening to the effects of different types of audio signal processing and develop your critical listening skills. The software practice modules provide necessary practical training to develop technical listening skills and allow a self-guided progression through various levels of difficulty. The software is available at https://webtet.net.

1.2 Audio Attributes

In this book, we will explore many ways to think about and improve our critical listening skills for audio. As we will see, there are nearly endless ways to practice listening and apply technical ear training concepts. Distinct from musical aural skills training, or *solfège*, technical ear training focuses on the sonic effects of common signal processing tools used in audio engineering—namely, equalization, dynamics processing, delay, distortion, and reverberation/echo. My main objective with this book is to provide training methods and strategies to help you develop critical listening skills for audio in a time-efficient manner.

How well do we know our audio processing tools, and can we anticipate their effect on an audio project? Audio work is sometimes compared to cooking. A professional chef knows how much salt is needed to season a dish, which herbs and spices will create the optimal flavor, if a little acid such as lemon juice or vinegar is needed, and how much heat is needed to cook a dish. Likewise, as audio engineers we learn the ways each of our tools affects timbre and sound quality, and the possibilities each one can have for a project. After we develop critical listening skills, we are more likely to achieve the sounds we want and get there more quickly. Knowing our tools also means being able to discriminate between small changes in tone quality. When we know and can anticipate the effect of our signal processing on a project, and we can discriminate between small changes in tone quality, we can better control the sonic results of our projects.

Highly developed critical listening skills allow us to identify not only the effects of deliberate signal processing but also unintentional or unwanted artifacts that might creep in unexpectedly, such as noise, buzz, hum, and distortion. Once we identify such undesirable sounds, we can eliminate or reduce their presence. If we do not notice these unwanted artifacts until much later in the project or after someone else brings them to our attention, it may be too late or very difficult to remove them. Ideally, we can avoid unwanted noise and distortion, but this is not always possible. Unfortunately, they can threaten a project at any time. As audio engineers, we need to be vigilant to the presence of noise and distortion so that we can mitigate them, and critical listening skills help in this way too.

1.3 What Is Technical Ear Training?

Just as musical ear training or *solfège* is an integral part of musical training, technical ear training or *timbre solfège* is essential for audio engineers because it has applications in sound recording, live sound reinforcement, mixing, mastering, sound design, and audio hardware and software development. In his article "Development of Technical Listening Skills: Timbre Solfeggio," Letowski (1985) coined the term "*timbre solfeggio*" to describe training that has similarities to musical aural training but is focused on spectral balance or timbre.

Technical ear training is a type of perceptual learning. Because this type of training is pertinent to audio engineering, it is focused on attributes such as tone quality and timbre, dynamics and loudness, reverberation and sense of space, noise and distortion, and panning. Perceptual psychologist Eleanor Gibson wrote that perceptual learning refers to "an increase in the ability to extract information from the environment, as a result of experience and practice with stimulation coming from it" (Gibson 1969). Because of their professional experience, audio engineers can extract information about the sound they hear. In other words, they know that certain timbral qualities are the result of specific pieces of equipment or plugins and parameter settings. The most proficient of engineers can identify equalizer and compressor settings, console makes and models, and loudspeaker makes and models, to name a few examples, simply by listening to them. Through years of practice, they have made a range of associations between specific timbres or tone qualities and the associated signal processor settings or equipment models which are responsible for the timbres.

We can develop more reliable and accurate listening skills through focused training or through on-the-job experience, which allows us to rely on our auditory perceptions in a more concrete and consistent way. When we increase our attention to specific types of sounds and sound processing, and compare successively smaller differences between sounds, we can learn to differentiate among a variety of sound features and attributes. If we compare the critical listening skills of two listeners, one expert and one novice, with identical hearing ability, the expert listener will likely be able to recognize specific features of audio, such as equalization, dynamics, and reverberation or delay, that a novice listener will not. The expert listener has greater critical or technical listening skills than the novice. Through focused practice, a novice engineer can eventually learn to identify sounds and sound qualities that were originally indistinguishable prior to training.

We also tend to recognize more easily those timbral qualities with which we are most familiar. Here is a story from the early days of development of the MP3, a lossy audio compression file format. The story goes that when engineers at Fraunhofer, the applied research organization in Germany that developed the MP3 perceptual encoder or codec, were testing early versions of the encoding software, they used expert listener panels to evaluate their MP3 coding algorithms. With any audio codec, the two main goals are to reduce the amount of data used to represent an audio signal as much as possible and minimize any audio signal degradation as much as possible. Codec developers found that listening panel participants could more easily identify familiar distortions than unfamiliar distortions. When the developers presented listeners with a new version of the codec that produced new artifacts, the listeners had trouble identifying the artifacts at first. After listening for some time, the listeners began to recognize sonic artifacts that were initially inaudible to them. Once they could hear the new artifacts, it became impossible to not hear them. The listeners had, in essence, "unlocked" the new artifacts aurally. The same phenomenon is true in any listening situation. If we know what the artifacts of an MP3 codec or any audio processor sound like, it's easier for us to hear the artifacts because we know what to listen for, and it becomes difficult not to hear them when they are present. As a result, it becomes difficult to listen to audio with lossy compression because the artifacts are distracting, once we know what they sound like.

1.3.1 What Is Timbre?

Technical ear training focuses on the tone quality or timbre of sound with the goal to help audio engineers become more adept at distinguishing and analyzing a variety of tone qualities or timbres. The question then is: what is timbre? As it turns out, the word "timbre" is not easy to define, and scientists and musicologists continue to debate its definition and expand our understanding of it.

One of the most quoted yet unsatisfactory definitions of timbre comes from the American National Standards Institute (ANSI, 1994):

> Timbre is that attribute of auditory sensation in terms of which a listener can judge that two sounds similarly presented and having the same loudness and pitch are dissimilar.

We can distinguish between two different musical instruments playing the same pitch at the same loudness because the instruments have different tone colors or timbres. We can tell the instruments apart because they *sound* different.

Is the timbre of a musical instrument always the same? McAdams and Siedenburg write that a musical instrument does not have a fixed timbre. For example, the timbre of a clarinet when played in the lower-pitched chalumeau register is much different than its timbre when played in the higher-pitched clarion register. However, a musical instrument may have acoustic attributes that are common across a range of pitches and registers that allows us to identify an instrument by sound even if its timbre varies (McAdams and Siedenburg 2019).

What about the timbre of unpitched sounds? The late cognitive scientist Albert Bregman pointed out that for the ANSI definition of timbre to apply, the two sounds need to produce the same pitch. The problem is that many sounds do not have any pitch at all, and he gave the example of the sound of a shovel scraping gravel (Bregman 1990), of which there can be many different timbres produced depending on the shovel, the gravel, and the speed with which the shovel moves against the gravel. Another example is drums, which are unpitched. We know that hi-hat, snare drum, and kick drum all have their own distinct timbres. Anyone can distinguish a kick drum from a hi-hat due to their unique timbres even though neither has a definite pitch.

For our purposes as audio engineers, we will say that timbre is a multidimensional attribute of sound and is determined by spectral, temporal, and spectrotemporal qualities.

- Spectral—frequency content:
    - Across the range of human hearing from 20 Hz to 20,000 Hz, which frequencies, including fundamental frequencies and overtones, are present in a sound?
    - What is the relative balance of the frequencies that are present? In other words, which frequencies or frequency ranges are loudest, and which are quietest?
- Temporal—time:
    - How does the overall amplitude of a sound change over time?
    - What is the amplitude envelope of a sound? What are the attack and decay times of a sound?
- Spectrotemporal—frequency content over time:
    - How do the frequencies that are present in a sound change or fluctuate over time? If a musical instrument is played louder, usually high frequencies become more prominent as it gets louder.
    - Do the amplitudes or frequencies of any spectral components change or fluctuate over time?
    - Are different frequencies present at different times?

In normal everyday speech, we use timbre discrimination to identify vowels in speech. We recognize vowels due to the specific peaks in the frequency spectrum for each vowel sound. These peaks or resonances are called formants, and they are produced by acoustic resonances in the vocal tract. When we change the shape of the mouth and throat, we can change the frequencies of the formats. We distinguish one vowel from another using the first three formants, and we give these various vowel sounds or timbres names that correspond to specific letters. Since we map timbres—that is, spectral balance—to labels—that is, vowels—seemingly automatically when we speak or listen to someone else speaking, we may not realize that we are already doing something that corresponds directly to technical ear training. With technical ear training, we are simply adding a new set of timbres—for example, equalization applied to an audio signal—and new labels to our knowledge base.

Musicians who are not necessarily audio engineers also become timbre experts. Classical music specialists can name the instruments present in a recording based on the sound alone, even going so far as to distinguish subclasses of instruments, such as a C trumpet from a B-flat trumpet, or an E-flat clarinet from a B-flat clarinet because of the way each instrument sounds. Classical music composers write music that combines instruments in specific ways to achieve desired timbres. Electric guitar players can identify the sounds of single coil or humbucker guitar pickups. Techno and house music producers can identify various models of drum machines simply by the quality of the timbres they produce.

Sound recording has had a profound effect on the enjoyment and evolution of music since the early twentieth century. In some cases, sound recordings may simply document musical performances and engineers record microphone signals as clean as possible with no processing or mixing. More commonly, though, based on trends for pop music at least, recordings undergo extensive processing, editing, and mixing. Sound engineers play active roles in guiding listeners' attention through intentional and dramatic signal processing, editing, dynamic mixing, and panning to recordings.

1.3.2 Analytic Listening

In an article describing their *Timbre Solfège* course, Letowski and Miskiewicz (2013) describe two modes of listening: holistic and analytic. In holistic mode, listeners attend to the overall sound and listen for enjoyment. In this mode, they are not paying attention to the individual components of a mix. Sound engineers, on the other hand, listen in analytic mode and focus on individual elements that contribute to the overall sound image.

Analytic listening mode is essential when an engineer works with subtle changes in timbre that may be inaudible to a casual listener. For instance, an engineer may compare the sound of two different microphones to decide which one to use or may adjust the level of a vocal track by 0.25 dB to help it fit most optimally within a mix. Novice listeners are unlikely to hear differences in either of these situations simply because they are not attuned to the details that the engineer is paying attention to. It is an experienced engineer's responsibility to hear such subtle details and make decisions based on them so that the finished product can be much more enjoyable for all those who listen in holistic mode.

Professional recording engineers and expert listeners can focus their attention on specific auditory attributes and separate or discriminate them from the rest of a mix. Some musicians and composers have also trained their ears through experience to hear subtle attributes of audio mixes. Here is an anecdote about such a person. Many years ago, I was mixing a wind symphony recording, and the composer of the piece was present in the control room. This composer has extensive recording experience and has developed a high level of critical listening abilities for audio and sound quality. As we were listening back to an edited version of his piece, he happened to notice a misplaced clave hit buried deep in the texture of the wind instrument and percussion sounds. It took me several listens to hear this very quiet clave hit. I was a little embarrassed and frustrated that I could not hear it immediately. Once I heard it for myself, I could pick it out on each subsequent replay. This is exactly the kind of situation that requires us as engineers to develop and maintain the highest level of critical listening abilities.

Technical ear training focuses on the features, characteristics, and sonic artifacts that result from signal processing commonly used in sound recording and production, including:

- equalization and filtering,
- reverberation and delay,
- dynamics processing,
- characteristics of the stereo image.

Technical ear training also focuses on unwanted or unintended features, characteristics, and sonic artifacts that may be produced through faulty equipment, particular equipment connections, or parameter settings on equipment such as noise, hum or buzz, and unintentional nonlinear distortion. Through concentrated and focused listening, an engineer should be able to identify sonic features that can positively or negatively impact a final audio mix and know how subjective impressions of timbre relate to physical control parameters. The ability to quickly focus on subtle details of sound and make decisions about them is the primary goal of an engineer. Sometimes a component in a signal chain suddenly starts malfunctioning. Sometimes we make an inadvertent mistake in signal routing, or we don't notice that a plugin we have carefully adjusted is bypassed. Careful listening can help us identify issues like these while we still have an opportunity to correct them, rather than noticing them after a project is released.

In technical ear training, we focus not only on hearing specific features of sound but also on identifying the types of processing that cause a characteristic to be audible. To hear a difference between an equalizer engaged and bypassed is an important step in the ear training process, but it is even more helpful to know the specific settings on the equalizer. Just as experts in visual art and graphic design can identify subtle shades and hues of color by name, audio professionals should be able to do the same in the auditory domain.

Sound engineers, audio hardware and software designers, and developers of the latest versions of perceptual audio encoders such as MP3 or AAC all rely on critical listening skills to characterize audio signal processing and make final design decisions. Audio measurement tools are helpful in many situations, but objective measures do not always tell us if something will sound "good" to human ears.

1.3.3 Objective Measures of Audio

One objective measure of audio equipment quality is total harmonic distortion, abbreviated as THD. Usually, equipment designers aim to have THD levels as low as possible, but the measured THD level does not always correspond to the audibility of the distortion. For example, there are many different types of distortion, and a THD level for one type of distortion may be much more audible than for another type of distortion. Loudspeaker designers and acoustics experts Earl Geddes and Lidia Lee (2003) have pointed out that high levels of measured nonlinear distortion can be *less* perceptible than low distortion levels, depending on the nature of the distortion and the testing methods employed. When a signal is distorted, new overtones are produced, but it is possible that spectral components in the original may mask these new overtones, making them inaudible. If the distortion produces overtones that are harmonics of the signal, these new harmonics will blend with any existing harmonics present in the signal. If the distortion produces non-harmonic overtones, these may be more audible because they will not match existing harmonics.

Another objective measure of audio equipment is frequency response. In some cases, such as in audio power amplifiers, we want the frequency response to be as flat as possible. In other cases, such as in headphones, the goal may be to design them with a specific frequency response that is not flat but is based on many tests of listener preference. Sean Olive at Harman has done extensive research on headphone frequency response preferences and has developed the Harman target curve based on listening tests he has conducted. He describes some of this work in an article in *Acoustics Today* (Olive 2022). As he mentions in the article, the site RTINGS.com has an extensive list of headphone frequency measurements (Kerr 2025).

There are similar issues with loudspeakers. Listeners may prefer a loudspeaker that does not have a flat on-axis frequency response as measured in an anechoic chamber over one that does, because on-axis frequency response in an anechoic chamber is only one objective measurement of the total sound produced by a loudspeaker. Low-frequency sound is radiated out the sides and back of a loudspeaker. In a typical listening room with reflective surfaces nearby, sound that is radiated from the back of a loudspeaker will reflect off the wall behind the loudspeaker and return to the listening area, affecting a listener's overall perception of the loudspeaker. Likewise, a loudspeaker will sound different in different rooms because reflected sound from a room's walls, floor, and ceiling contribute to the sound that we hear.

Automotive audio system engineers report that listeners prefer car audio with more bass than is typical in home loudspeaker systems. With listening tests and feedback from consumers, car audio engineers have determined that consumers may not prefer car audio systems with flat frequency responses. In audio product design, the final tuning of software algorithms and hardware designs is often done by ear by expert listeners. Thus, physical measurements, although essential for equipment design and development, are usually supplemented with subjective listening tests.

It is still not possible to measure all perceived audio qualities with physical measurement tools currently available. For example, the development of perceptual audio coding schemes such as MPEG-1 Layer 3, more commonly known as MP3, required the use of expert listening panels to identify sonic artifacts and deficiencies produced by data reduction processes. Because perceptual audio coding relies on psychoacoustic models to remove components of a sound recording that are deemed inaudible, the only reliable test for this type of processing is the human ear. Small panels of trained listeners are more effective than large samples of the general population because they can provide consistent judgments about sound and can focus on the subtlest aspects of a sound recording.

Of course, with music recording and production, there is no way to measure or quantify the quality of a mix, so we rely on our ears. The work is more art and less science, but we want to be consistent from day to day, and that's why technical ear training is essential.

1.4 Tools and Training

Although this book focuses on types of signal processors used for professional audio work—namely, equalization, compression, and reverberation—it does not focus on specific models of commercially available audio processing software or hardware. It is impossible to have a one-size-fits-all approach to audio production, so my goal is to help you improve your critical listening skills generally so that you can adapt to the specific details of each new audio project. Therefore, I do not think it is helpful to recommend specific plugin settings or microphone techniques for different instruments or recording setups, since they are context dependent. While stereo and multichannel microphone techniques exist and are especially useful for classical music large ensemble recording, exact microphone placement is determined through an iterative process of careful listening and adjustment. Recording engineers estimate microphone placement that they think will provide the best sound pickup and then adjust microphone positions based on what they hear from them. Plugin presets provided by manufacturers may be helpful but again may be best thought of as starting points that we can tweak further by ear. Use the skills you gain from this book and technical ear training software to improve your ability to get any sound you want.

There seems to be a natural human tendency to want to buy more plugins and gear when the tools we already have might provide all the functionality we need. This is true in many areas of life and not just in audio. Commercial advertisements encourage us to buy the latest and greatest new product, whatever it may be. Companies exist to make profits, after all. Their goal is to convince us to buy more stuff, and they are very effective in this regard. But before we buy that fancy-looking new plugin, microphone, or audio interface, we might consider whether we have explored the capabilities and functionality of the plugins that we already own, such as those that came with our digital audio workstation (DAW), for example. You may have heard of something called "gear acquisition syndrome" or GAS. The term refers to the intense desire to buy more gear and plugins even when one does not technically need them. I can't deny the fact that it's fun to buy stuff, but it also seems as though we are driven to purchase new gear partly because, at some level, we believe it will help us become as successful as the professional engineers who endorse the products. The main problem is that we can buy all the plugins and audio gear that our budgets allow, but at the end of the day, we must know how to use them and we must know what to listen for to get the most out of them. There is a huge market for audio plugins, but simply owning a plugin on its own does not make our projects sound better or give us better critical listening skills. There are some truly excellent audio products available now that were not available in the past, which can help us be more effective audio engineers. And there are talented audio hardware and software engineers working hard to bring us extraordinary new tools. But simply owning a piece of audio hardware or software does not automatically lead to more success in the field of audio engineering. Furthermore, switching out some plugins or pieces of hardware will make little to no difference in audio quality.

In theory, any parametric equalizer with fully independent and sweepable control of frequency, boost/cut, and Q or bandwidth can replicate the frequency response curve of any other equalizer. Usually, the built-in parametric equalizer in a DAW will work as well as a third-party equalizer that costs money. Compressors and reverberation are a little different in this respect and can vary more substantially from one model to the next in terms of their control parameters and timbre-shaping capabilities. These are two areas in which it can make sense to purchase third-party plugins or hardware. Despite the differences we find among compressor and reverberation models, technical ear training helps us develop listening skills that can be transferred to any plugin or hardware device we may use. Since we are focused on increasing our ability to hear specific details of tone quality and timbre, we can apply these critical listening skills to any plugin, hardware device, or recording or mixing situation.

Besides listening to the effects of plugins or hardware choices, it is also important to remember that a musician's performance can produce greater timbral differences than those we might hear by switching out a microphone, audio interface, or microphone preamp. For example, the force with which a drummer hits the drums and cymbals, the tuning of a drum head, the way a guitarist picks the strings or the type of pick used, a guitarist's use of palm muting or not, whether a bass player uses a pick or not, the age of a guitar's strings, whether a singer is belting or singing lightly, microphone placement, or the acoustics of the recording space, among many other things, can have a significant effect on the timbre of a recording. If we want to change the timbre during a recording session, it may make sense to focus on the technique a musician may be using or microphone position before changing out any of the equipment.

Major hit recordings have been made with inexpensive equipment. For example, Billie Eilish recorded the vocals on her debut single "Ocean Eyes" with an Audio-Technica AT2020 microphone (Tingen 2019), which currently has a retail price of $119 but can be found on sale for less than that. An expensive microphone was not needed to record the vocals of a song that has been streamed billions of times and that helped launch a pop singer's astoundingly successful career. Vocal quality, performance, songwriting, production choices such as reverberation and layering, and a song's arrangement can have much more of an impact than the choice of microphone.

Technical ear training is like musical training or fitness training in the sense that it requires time and effort to make progress. As much as someone might want it to, simply owning a Steinway concert grand piano does not make them an accomplished pianist, just like owning a treadmill and a set of barbells does not improve cardiovascular fitness, physical endurance, or strength. Similarly, owning expensive audio plugins and hardware does not say anything about one's audio engineering skills. Training, in whatever skill we are trying to gain, requires struggle to make progress and improve our skills. The most in-demand professional audio engineers may own the most expensive pieces of equipment because they like them, which is perfectly fine. But they are not great engineers just because they own expensive equipment. Usually, the specific equipment models matter less than an engineer's critical listening skills and their expansive technical and theoretical knowledge, which allows them to get the most out of the tools they have. The abilities of the musical artists with whom an engineer works can also affect sound quality significantly as well. Get to know your gear by using it regularly and listening carefully to it.

1.4.1 Exercise: Limit Plugin Options

To be effective as audio engineers, we need to know our tools and their specific capabilities, but we cannot learn them all at once, so it makes sense to limit our options during the learning process so that we can understand our tools more fully. Besides being an effective learning strategy, limiting our plugin options can promote creativity since it removes the need to think about choosing the "best" plugin for each application.

In recording, mixing, and live sound work, the three most common types of audio processors are equalizers, compressors, and reverberation, which is why these are normally included with every DAW. But third-party companies provide a nearly endless array of audio signal processing options in the form of digital plugins and analog and digital hardware. To reduce decision fatigue and to get to know our tools, try the following exercise:

- Pick one equalizer, one compressor, and one reverb plugin. It doesn't matter which models or brands. The ones that are included with your DAW are a great place to start. Challenge yourself to use only these specific models for a couple of weeks or a month. For example, if you use Cockos REAPER, pick the ReaEQ equalizer, ReaComp compressor, and ReaVerbate reverberation plugins and get to know them.

By doing this exercise, you can learn more about the capabilities of the specific plugins you chose across a range of applications. This can be difficult work because maybe the plugin model does not give you the exact control you want, or the graphic design is not inspiring. In the beginning, you might need to work harder to get the sound you want with the plugin you've chosen if you don't know it very well.

After you've worked with one model of each type of processor for some time, how much more do you know about the strengths and weaknesses of each? Do you feel more confident using these plugin models now? Have you saved time and energy because you did not have to make the decision of which equalizer, compressor, or reverberation plugin to use? What were the benefits and drawbacks of this exercise? Did a plugin's shortcomings force you to find other ways to get the timbre you wanted, such as through microphone placement?

After doing the exercise once, try it again with a new set of plugin models. What similarities and what differences do you notice between the first plugin models you chose and the second set?

You can do a similar exercise with microphones or any other piece of equipment. If you have access to multiple microphones, pick one microphone and use it across a range of recording situations to really get to know how it responds in different applications.

1.5 The Importance of Timbre in Recorded Music

Listeners can recognize music recordings not only by their musical melodies, harmonies, and structure, but also by the timbres of the instruments created in the recording process. In recorded music, engineers and producers shape sounds to best suit a musical idea and artistic intention. Popular music, although often relatively straightforward in terms of melody and harmony, can use complex layers of signal processing to produce tension and release. One of the innovations in pop music production has been the development of new timbres and sonic textures using tube and solid-state electronics, analog tape recorders and DAWs, analog and digital synthesizers, and an assortment of sophisticated audio signal processing effects like equalization, compression, and reverberation. In his book *The Producer as Composer: Shaping the Sounds of Popular Music*, Virgil Moorefield (2005) describes in detail how recording and sound processing equipment contribute to the compositional process by shaping timbre. Because of pop music's focus on timbre, people can identify a pop recording by hearing only a fraction of a second of it. An excerpt of such a short duration does not allow us to hear a melody or musical tonality, but it is long enough to hear the characteristic timbre of a song. As Brian Eno said, "A fact of almost any successful pop record is that its sound is more of a characteristic than its melody or its chord structure or anything else. The sound is the thing that you recognize" (Korner 1986). This was confirmed in a study at the University of Toronto, which found that listeners could correctly identify pieces of music when presented with excerpts of only a tenth of a second (i.e., 0.1 seconds or 100 milliseconds) in length (Schellenberg et al. 1999). Popular music radio stations are known to challenge listeners by playing a short excerpt, typically less than half a second, from hit songs and inviting listeners to call in to identify the song title and artist. Such excerpts are too short to indicate the harmonic, melodic, or rhythmic progression of the music, so listeners must rely on the timbre or combination of sonic features to make a correct identification. Daniel Levitin, in his book *This Is Your Brain on Music* (2006), also illustrates the importance of timbre in recorded sound and reports that "Paul Simon thinks in terms of timbre; it is the first thing he listens for in his music and the music of others" (p.152). With all the audio processing options available, it's important for audio engineers to be aware of fine differences in timbral quality derived from the many possible sound colors one can derive.

One effect recording studios have had on music is to help musicians and composers create sonic landscapes that are impossible to realize acoustically. Purely non-acoustic sounds and sound images are most evident in pop and electronic music, in which sounds originate from purely electronic sources (whether they are analog, digital, or a hybrid) rather than through a conventional acoustic musical instrument's vibrating strings (e.g., piano), membrane (e.g., drum), bar (e.g., vibraphone), or column of air (e.g., clarinet). Electronic music often combines recordings or samples of acoustic sounds with electronically generated sound (such as from a synthesizer). The distinction between purely electronic and purely acoustic sounds is not always clear, especially since we can use acoustic recordings as sources for digitally synthesizing completely new sounds.

There is a range of acoustic and electronic sound sources, and we might think about the continuum from acoustic to electronic sound sources within four broad categories:

- Purely acoustic sounds are generated by conventional acoustic musical instruments or voices.
- Amplified acoustic sounds, such an electric guitar, electric bass, and electric keyboard, such as a Fender Rhodes or Wurlitzer, start with a vibrating mechanism, typically a string, tine, or reed, which gets amplified electronically and is then heard from a loudspeaker.
- Extensively modified and digitally manipulated acoustic sounds are recorded sounds from which new sounds are created that may have no resemblance to their original acoustic sounds. For example, if we repeat a recording or sample of a single snare drum hit fast enough (i.e., faster than 20 times a second) we begin to hear it as a pitched, sustained sound rather than the transient sound that it originally was. The more quickly it is repeated, the higher the pitch. The possibilities to

create new timbres through sonic manipulation are endless, such as with granular and wavetable synthesis techniques.

- Purely electronic sounds are those that are generated in the analog or digital domains. A voltage whose amplitude varies over time according to a mathematical sine function can generate a sine tone. Multiple sine tones at the appropriate frequencies and amplitudes can produce the standard square and triangle waves, as well as any other imaginable timbre. If we control a tone's attack time—that is, the note onset time—and decay time—that is, the speed with which a note fades out—we can transform it from a steady-state continuous sound to a more musically oriented, time-varying sound. Synthesizers can also modulate a sine tone's frequency or its amplitude at rates faster than 20 Hz which creates new timbres through what's known as frequency modulation (FM) and amplitude modulation (AM) synthesis, respectively.

Purely acoustic musical instrument recordings can also be altered with common audio signal processors and plugins (such as equalizers, dynamic range controllers, and reverberation), to create new sounds that cannot be produced acoustically. In the process of recording and mixing, we can manipulate any number of parameters, depending on the complexity of a mix and the musical goals of the project. Many of the parameters that we adjust during a mix are interrelated, so that when we alter one track, it influences our perception of other tracks. The level of each instrument can affect the entire feel or focus of a mix, and an engineer and producer may spend countless hours adjusting levels—down to fractions of a decibel—to create the right balance. For example, a slight increase in the level of an electric bass might impact the perception and musical feel of a kick drum or even an entire mix. Each parameter change applied to an audio track, whether it is level (gain) change, compressor, reverberation, or equalization parameter change, can influence the perception of other individual instruments and the entire track.

1.6 AI and Audio Production

Artificial intelligence is touching nearly every aspect of life, and audio production is no exception. AI-enabled audio tools have been available for some time now, and typically these tools analyze (listen to) audio signals to determine characteristics such as frequency content, spectral balance, stereo width, and dynamic range. Mastering tools are available in two main forms: online and plugin.

Based on an AI tool's analyses and the models it uses, a plugin or online service applies processing that it deems most appropriate for a given song or album. Typically, processing takes the form of a mastering chain with proposed settings for equalization, compression/limiting, and stereo width control, and sometimes also de-essing, distortion, and noise reduction or gating. Plugins usually offer users the ability to adjust settings as well.

LANDR was one of the first online mastering services and is perhaps the biggest name in this area. Here is a list of AI-enabled mastering tools currently available:

- Apple Mastering Assistant, plugin included with Logic Pro 11 (Mac only).
- Brainworx bx_mastering studio, standalone app (free).
- eMastered, online (paid).
- iZotope Neutron 4 and Ozone 11, DAW plugin (paid).
- LANDR Mastering Pro plugin and LANDR Online Mastering (paid).
- Maastr, online (paid).
- Waves Online Mastering (paid).

These tools seem to be useful for quick mastering at relatively low cost. The quality and appropriateness of the results from AI mastering services varies considerably from model to model, which means trained humans are still needed to decide which mastered versions are most suitable or which mastering settings need to be adjusted further. At this point, however, it does not seem that AI can replace audio engineers. For example, AI cannot adjust microphone locations to optimize timbres, respond in real time to make live sound mixing decisions, nor make mixing and mastering decisions that most closely reflect the tastes of audiences at any given point in time and for a specific genre. AI tools can be helpful to generate ideas for signal processing, even if an artist or producer decides not to release AI-mastered versions of songs or albums. At this point, AI tools cannot replace the work of competent engineers. Successful recording and mixing engineers understand the need for detailed mixing choices that target spectral balances, dynamics processing, and spatial treatments most appropriate for the genres in which they work. And ultimately, this helps artists connect most strongly with their fans.

1.7 Goals of Technical Ear Training

The main goals of technical ear training are:

- To develop mental connections between timbre perceptions and physical properties of sound, and to develop our memory for these connections. We can learn to associate audio equipment parameters—for example, an equalizer's center frequency in hertz and amount of boost or cut in decibels—with our perceptions of the resulting audio signal's timbre and we can subsequently develop a memory for these associations.
- To increase our sensitivity to subtle features and attributes of sound so that we can perceive differences in tone quality and effects processing when comparing nearly identical sounds.
- To increase our ability to focus on specific details of sound quality or signal processing. For example, professional audio engineers can focus on the timbre/dynamics/reverberation/delay effects of a single instrument or voice within the context of a full mix.
- To increase the speed with which we can identify features of sound, translate between auditory perceptions and signal processing control parameters, and decide which equipment parameters need to be changed in each situation.
- To maintain consistency in our critical listening skills.

1.7.1 Link Perception with Physical Properties of Sound

The first goal is to link our perception of timbre with physical properties of an audio signal modified by a signal processor. For example, a 12 dB boost at 1,000 Hz with an equalizer creates a change in timbre that we can perceive. The goal is to link that perception of timbre with the frequency that was boosted on the equalizer. One challenge with critical listening for audio is that there is no obvious connection between sound qualities we perceive and the hardware/software equipment parameters we can control. How do we link the perception with the equipment parameters? If a mix sounds "muddy," there is no "mud" parameter to adjust. Instead, we need to determine by ear which frequency range seems to be causing the issue we wish to correct—in this case, the perceived muddiness—and reduce it.

In this problem-solving listening process, it is helpful to determine if the issue we wish to correct arises from a single instrument or if it is across an entire mix. With experience using audio equipment, we connect our perceptions of timbre to effects processor settings, which allows us to anticipate what, for example, an equalizer boost at 1 kHz or a reverb predelay time of 125 milliseconds will sound like. The more experience we have, the more accurate we become in anticipating and estimating the changes to sonic characteristics that result from effects processor parameter settings.

Audio equipment parameter settings correspond to physical attributes of an audio signal, but what do these objective parameters sound like? A parametric equalizer, for instance, allows control of center frequency, gain (boost or cut), and Q (or bandwidth). These physical attributes as they are labeled on a device have no natural or obvious correlation to an audio signal's perceptual attributes, and yet engineers engage them to affect listeners' perceptions of a mix. How does one know what a 6 dB boost at 315 Hz with a Q of 2 sounds like? Without experience using equalizers, how could anyone predict the timbral change resulting from this equalizer setting? A novice audio engineer may understand the term "compression ratio" conceptually, but may not know how to adjust the parameter effectively or may not understand how sound is changed when that parameter is adjusted.

Then there is the issue of describing timbre and sound quality with words. When we talk to collaborators about a project, we try to describe the timbres we hear using words that make sense to us with the hope that our collaborators will know what aspect of a mix or what timbre we are talking about. Subjective descriptions can be unreliable because they are often imprecise and sometimes ambiguous, have different meanings from person to person or across situations, and do not have direct relationships to the physical properties of sound. As scientists in music technology and psychoacoustics have confirmed, timbre lacks a vocabulary that is specific to the sense of hearing. Saitis and Weinzieri (2019) wrote that we communicate our perception of timbre with:

- words borrowed from other senses, such as bright, warm, sweet,
- onomatopoeias—that is, words that imitate or suggest a source of a timbre, such as ringing, buzzing, shrill, or
- words that are unrelated to any sensory modality, such as rich, complex, harsh.

If a track or a mix is described as bright, harsh, or boomy, how do we know which physical characteristics are responsible for any of these qualities? Are they the result of an equalizer, a compressor, artificial reverberation, distortion, some combination of all of these, or perhaps something else altogether? A bright snare drum sound may have resulted from an equalizer boost around 4 kHz to 8 kHz, a cut below 500 Hz, or something else. Even if the brightness of a snare drum is due entirely to an equalizer setting, there is no direct connection between the word "bright" and an equalizer frequency response curve.

Although most equalizers do not have subjective quality labels, members of the Interactive Audio Lab at Northwestern University developed equalizer and reverberator interfaces that map language-based descriptors to audio processing parameters (Sabin et al. 2011; Pardo et al. 2012; Cartwright and Pardo 2013). Based on user studies, they have linked equalizer descriptors such as "warm" and "tinny" and reverberator descriptors such as "bathroom-like" and "church-like" to objective signal processing parameters.

Another challenge in linking perception with physical properties is that sound is ephemeral and intangible, and yet as audio engineers we are tasked with shaping it to obtain specific results. We may rely on visual aids such as level meters, waveform displays, and signal processing parameters to help with the intangibility of sound, and these kinds of visuals can be helpful, but they do not represent the full complexity of auditory perception.

Psychophysicists have known for many years that our senses are attuned to detect changes in our environment. They are most sensitive to abrupt changes in our environment, such as a flash of light or a clap of thunder. We tend to ignore stimuli that are continuous and do not change or change very slowly, such as ambient light levels and air temperatures, an air handler's continuous rumble, or the hum of a refrigerator. We tend to notice continuous sounds only when they stop. If you have ever had your eyes tested, you know that optometrists rely on their patients' ability to notice differences in visual clarity when they switch quickly from one lens to another to find the correct prescription. It's the same idea with sound. We notice differences in sound quality more easily immediately following a change. For example, when we switch from one headphone or loudspeaker to another, the differences in spectral balance are most apparent in the first few seconds after the change. One model may have more low-frequency power, more high-frequency power, or more mid-frequency power, but as we continue to listen to it, we get used to its frequency response, and the tonal coloration that we noticed immediately after the switch gradually becomes less apparent. After we get used to the sound, we may begin to hear the sound as being relatively balanced spectrally, assuming the monitors are relatively full range and not laptop or smartphone speakers. If we adapt to one specific monitor, which is normally what happens after listening for some time, it is easy to make mixing decisions to compensate for that monitor's deficiencies. If a monitor lacks low-frequency extension, a mixing engineer might compensate by boosting low-frequencies in their mixes. But there are ways to use this sensitivity to timbral change to our advantage. For example, we can switch from one loudspeaker or headphone model to another while mixing to help reduce our inclination to adapt to one monitor's qualities (or deficiencies). We can also use reference recordings to help make mixing more objective and consistent, as Mike Senior clearly describes in his excellent book *Mixing Secrets for the Small Studio* (2nd ed. 2018).

Subjective descriptions of sound can be vague, but ambiguity is reduced if we know the exact meaning of the adjectives we use. Although we can develop our own vocabulary to describe various qualities of sound, these descriptors may not match the words that other engineers or collaborators may use to describe the same qualities. Why not use labels that already exist on parametric equalizers? We are most precise when we describe tonal balance in terms of equalizer settings: center frequency in hertz, Q or bandwidth, and amount of boost or cut in decibels.

It is critical to develop a memory for the ways specific frequencies are associated with perceptual attributes of a signal. With practice, we can learn to estimate the frequency of a peak or of a dip in the power spectrum of an audio signal. Professional audio engineers develop their own personal methods to translate between perceived auditory sensations and technical parameters that they can control with the equipment available to them. They also develop a highly tuned awareness of subtle details present in sound recordings through analytical listening every day. They may be listening for the delay setting on a snare drum. How many times does the delay repeat? Does the electric bass sit within a mix in the right way? Or does it need more compression or less?

A recording engineer can have as much command of a recording studio and its associated signal processing capability as a professional musician has command of an instrument. Accomplished violinists know precisely when and where to place their fingers on the strings and can anticipate what effect each bow movement will have on the sound produced. An audio engineer should have this same level of knowledge and sensitivity of sound processing and shaping before reaching for an effects processor parameter, a mixer fader, or a microphone model. It is important to know what a 3 dB boost at 4 kHz or an increase in compression ratio is going to sound like even before it is applied to an audio signal. There will always be times when we will not be able to identify a unique combination of signal processing and equipment, but the work progresses more quickly if we are not continually guessing what an effects processor setting will sound like.

By knowing ahead of time what a particular parameter change will have on the sound quality of a recorded signal, we can work more efficiently and effectively. Working at such a level, we can respond to sound quality quickly, in the same way musicians can respond nearly instantaneously to each other in an ensemble.

Perhaps with enough facility and confidence for experimentation, the recording studio can become a musical instrument that we "play," as producer Brian Eno has written and spoken about. As engineers, we have direct input and influence on the artistic outcome of any music recording in which we are involved. By adjusting balances, mixing sonic layers, and shaping spectra, we can focus the auditory scene for listeners, guiding them aurally to a musically satisfying and moving experience that expresses the intentions of the musical artist.

1.7.2 Increase Auditory Sensitivity

The second goal of technical ear training is to increase our awareness of subtle details so that we can discriminate minute changes in physical parameters of sound. Experienced audio engineers can make small changes in sound that may not be apparent to an untrained listener, and they understand the need to hear subtle changes in sound. They also know the sources of these changes and ways to remedy problems using audio processing and recording techniques. Over the course of a complex recording project, an engineer might make hundreds if not thousands of decisions about sound quality and timbre that contribute to the finished product. Some decisions have a more audible effect than others, but here are a few of the things that engineers make decisions about during a recording project:

- **Microphones:** The make/model, polar pattern, position, and orientation for each instrument being recorded.
- **Preamplifier:** The make/model and gain settings for each microphone, usually optimizing microphone signal levels to avoid clipping.
- **Signal levels:** The gain stages for each track (mic preamp level, fader, auxiliary send, etc.) that maximize signal-to-noise ratio and minimize analog-to-digital conversion quantization error and distortion/clipping through to the recording medium.
- **Spectral/tonal balance and timbral quality:** Equalizer and filter settings. Microphone choice and placement also play a role in spectral balance. Coloration or warmth, usually the result of distortion from analog tape, transformers, and vacuum tubes, is generally subtler but contributes audibly nonetheless to the tonal balance.
- **Dynamic range and dynamics processing:** Audio signals have a range from loud (*fortissimo*) to soft (*pianissimo*), and this range can be altered through dynamics processors, such as compressors and expanders, and fader adjustment and automation.
- **Spatial characteristics:** Recording room/hall acoustics and microphone placement within a room, and parameter settings on artificial reverberation, delays, as well as panning and positioning of sound sources within the stereo or surround image.
- **Balance:** The relative levels of tracks or microphones as controlled by the individual faders on a multitrack/multi-microphone mix.
- **Processing effects settings and mix levels:** Flanger, phaser, chorus, tremolo, distortion/overdrive.
- **Noise:** It takes many forms, but in general it is any sound that is not intended to be part of a recording. There are two main categories:
  - Electronic—clicks/pops, tape and electronics hiss, quantization error, 50- or 60-Hz power supply or ground loop hum/buzz.
  - Acoustic—air handling noise (which can be in the form of a low rumble and therefore not immediately apparent), external and environmental sounds such as traffic noise, horns, and subways, talking, foot tapping, etc.

These are broad categories of technical parameters that affect the perceived audio quality and timbre of an audio signal. Each of these items can have numerous levels of detail. For example, digital reverberation plugins, depending on the model, may provide control over parameters such as decay time, predelay time, early reflections, modulation, diffusion, room size, filtering, and decay time multipliers for each frequency band, among other parameters.

Some parameter setting decisions have an insignificant sonic effect, but when added together to form a coherent whole, the cumulative effect makes each processing stage critical to a finished project. Choices made early in a project that degrade sound quality may not be able to be reversed later in a project. A common refrain is "We'll fix it in the mix," but audio problems cannot always be fixed during the mixing stage. It's best to address a problem when we notice it and there is still an opportunity to fix it, rather than try to fix it during the mixing or post-production stage. This means we must listen intently and actively to the result of each signal path and processing decision that we make and for changes to sound quality that may be due to equipment malfunction. A low-level hum on a microphone signal might seem insignificant until we compress it and raise its level by 12 dB. Likewise, we may inadvertently introduce comb filtering into a track with a signal routing or microphone placement choice. If a signal has been recorded with comb filtering, it is impossible to undo that effect. We will look at comb filtering more in Chapter 3. When listening at a focused, analytical level, we can respond to sound quality and timbre quickly and identify potential problems in the moment that may come back to haunt a project at a later stage. To use a visual analogy, artists use specific paint colors and brush strokes in subtle ways that combine to produce powerful finished paintings. Viewers may not see the individual brush strokes of a painting; rather, they see it as a whole. Artists know how to combine brush strokes and paint colors so that the human brain sees a coherent image. In a related way, recording engineers focus on specific sonic characteristics and signal processing that, when heard as a whole, can combine, blend, and support one another to create more powerful, meaningful final mixtures of sounds.

We hope that increased sensitivity in one area of critical listening, such as equalization, will promote increased awareness and sensitivity in other areas, such as compression and reverberation, and support an overall improvement in listening skills. I have no evidence to support the idea, but I can offer an analogy in the visual domain. Years ago, I became interested in font types and basic graphic design principles. Once I started learning about these concepts, I began to look at websites and print much differently than before, noticing alignment within images and text, color scheme, smoothness of images and text, and differences in fonts. I am certainly no graphic designer, but I noticed visual design elements differently after only a brief introduction to some of the principles in that field. I think that once introduced to issues of sound quality, we begin to listen differently and notice things that we hadn't noticed earlier.

1.7.3 Increase Focus on Specific Details of Sound

Audio engineering work requires analytical listening that focuses on specific details of sound. There are almost endless numbers of details that engineers can pay attention to but here are a few examples:

- The timbre of a vocal or instrument. Timbre can be shaped and controlled by any one of a variety of things including equalization, microphone type and placement, compression, room modes and reflections originating from a recording space, the instrument or instrument amplifier. If we use an equalizer to adjust the timbre of a track, which frequency should be adjusted? Would the track benefit with a cut at 300 Hz? By how much and with what Q? Would a boost at 2,000 Hz help? What other frequency ranges need attention?

- The level of each track in a multitrack mix. How well are voices and instruments balanced? Does a track need to be louder or quieter and by how much? To get more specific, does a vocal level need to come up or down for a phrase or for a single word? By how much? What about the next word or phrase? In addition to applying compression to a vocal track, some mixers adjust vocal track levels word by word using fader automation. This may seem like a tedious process, but it can really make a vocal sound more polished and clear overall.
- How are compressor settings affecting the attack of a snare drum, a kick drum, or cymbals? Does the attack time, release time, ratio, or threshold need to be changed?
- Is the reverberation predelay time on the drum kit affecting the feel of the tempo? Is it helping create a sense of space while not cluttering the mix?
- How well does the reverberation suit the vocal or the overall mix? Should the decay time be longer or shorter?
- Which components in a mix are masking or covering other elements?
- Are the processing and mix level choices allowing the musical elements of a piece to feel coherent, as if they are working together?
- Does changing the level of the bass or the kick drum change the perception of the rhythm or groove? Does a high-frequency boost on the kick drum help with the perception of its low-end clarity?
- At any moment over the course of a song or piece of music, where is the sonic spotlight aimed? In other words, what is most important in the mix? Is a mix allowing listeners to hear what they should be hearing at each moment?

There are many more details when it comes to sound quality, of course, and any one of these can improve a final mix or detract from it. When we can concentrate our attention on subtle details, it allows us to correct problems, enhance and bring out essential details, and ensure all the sonic elements work together to provide listeners with the best experience.

1.7.4 Increase Speed of Detection

The fourth goal is to increase the speed with which we can identify and decide on appropriate engineering parameters to change. Recording and mixing sessions can occupy large amounts of time, within which hundreds of subtle and not-so-subtle adjustments are made. The faster we can zero in on any sonic characteristics that we decide need to be changed, the more effective our time will be. During a recording session, valuable time can be consumed while comparing and changing microphones, and the quicker we can recognize either a deficiency or an optimal sound, the faster we can move on to other tasks.

1.7.5 Increase Consistency

Some audio equipment companies employ panels of expert listeners to evaluate their products and compare them to competitors' products. Expert listeners are deemed to be expert in part because their judgments about sound quality and audio attributes are generally very consistent. When expert listeners participate in a blind listening test to compare and rate loudspeaker sound quality, for example, they tend to rate a given loudspeaker the same way each time they take the listening test, from one listening session to another. It may seem remarkable that a listener can give a loudspeaker the same rating day after day, especially since they do not know anything about it other than what it sounds like. Perhaps you are so familiar with the loudspeakers or headphones that you use on a regular basis that you could identify them in a blind listening test compared to another unknown model. You are an expert listener for the monitors you use on a regular basis because you are so familiar with their timbral characteristics. Expert listeners can identify features in the sound of a loudspeaker quickly and accurately, even for models that they have not heard prior to a listening test. Maybe the frequency response of a loudspeaker has a peak around 100 Hz, a dip at 1250 Hz, and another peak around 8 kHz. An expert listener can identify and remember these qualities. Each time they hear these peaks and dips in the spectrum, they know which loudspeaker they are listening to. Because they can evaluate timbre with a high level of accuracy and they have developed a robust memory for timbre, expert listeners know when they hear a particular speaker again because they recognize those characteristics by ear. We want to maintain consistency in our critical listening from day to day and from project to project, just as expert listeners do.

Being consistent also helps us work faster and more confidently. If we can always identify a peak at 125 Hz and not confuse it with a peak at 250 Hz or 63 Hz, then this consistency will help us do our jobs each day. Be aware, though, that during the learning-listening process you may not always feel as though you are making progress. In fact, you may feel as if you are getting worse at times. This is normal. My students have reported that a few weeks after beginning equalizer technical ear training exercises, their consistency deteriorates briefly. It is not clear why this happens, but it seems to be a typical stage of the learning and memorization process. Awareness increases as we train, and perhaps we become overconfident too quickly before we have really solidified our memory for frequencies. Or perhaps we become more aware and more critical of our abilities, which makes us think we are not making progress even when we are. Also, be aware that insufficient sleep, food, or hydration can affect this as well, as can alcohol consumption. Critical listening skills do improve with continued practice overall, even if it can sometimes feel as though we're not as good as we once were. Keep pushing through this stage because it is temporary.

Because there is a significant amount of creativity involved in audio recording, mixing, mastering, and sound design, there are no "correct" ways to do things. What may be the perfect equalization settings for an instrument in one situation may not be suitable for other instruments or other situations. My hope is that this book can give you a guide so that you can develop listening skills to identify not only when you have the sound you want but also when you need to fix up a problem area. A novice engineer working on a mix before developing critical and analytical skills may simply have the vague feeling that something is not right with a mix or with the sound of an instrument, or that some type of improvement could be made, but may not know what the specific problem is or how to fix it. When we first start out, we may be able to recognize that something doesn't sound right but not be able to identify exactly what's wrong or why a mix doesn't sound the way we want it to. An experienced engineer with highly developed critical listening skills can listen to a mix and know specifically where the problems lie and how they can be corrected. For example, maybe 250 Hz is boosted by too much on the kick drum, the piano needs more 1 kHz, the voice has too much 4 kHz, the bass needs more compression and to be brought up by 2 dB, and the reverb decay time needs to be 0.5 seconds longer. Technical ear training gives us the skills to help solve specific problems such as these.

Within each of the main categories of signal processing—that is, equalization, dynamics, and reverberation—numerous makes and models are available at various price ranges and levels of quality. Take compressors, for example. We know that most compressors perform the same basic function: they adjust the level of a signal automatically and provide control over dynamic range, which alters the amplitude envelopes of our tracks. Most compressor models have many of the same parameters—that is, threshold, ratio, attack time, and release time—but a compressor's effect on sound quality can vary enormously from one model to another. Differences in the analog electronics or digital signal processing algorithms among compressors create a variety of sonic results, and each model will have a unique sound. Through the experience of listening, we learn the variations in sound quality between different makes and models. This allows us to choose a certain model because of the specific sound quality it imparts on a track, the control it affords, or the potential artifacts it adds. Many of the most famous analog signal processors have software plugin versions in which the screen image of the plugin looks nearly identical to the faceplate of the hardware device. Sometimes, because the two devices *look* identical, it may be tempting to think that they also *sound* identical. Unfortunately, they do not always sound alike, but it is possible to be fooled into thinking the sound is replicated as perfectly as the visual representation of the device, especially if we do not have the analog version for comparison. Suffice to say, there is not always a direct translation between analog electronics and the computer code that performs the equivalent digital signal processing, and there are various ways to create digital models of analog circuits; thus, we have differences in sound quality between the analog versions and their digital emulations. That may be fine because we can treat a plugin for what it is, rather than as a model of some historically important piece of gear. But we do need to keep in mind that if we are used to a particular analog compressor, the digital emulation may not behave the same way. You may have also noticed that more than one plugin emulation of a well-known vintage analog compressor exists. If you compare different plugins produced by different companies but all emulating the same analog compressor, they can sound different and can produce different sounding artifacts.

Although each compressor model, for example, has a unique sound, we can transfer knowledge of one model to another and be able to use an unknown model effectively after a short period of listening. Just as pianists must adjust to each new piano that they encounter, engineers adjust to the subtle and not-so-subtle differences between pieces of equipment that perform a given function.

Studies, such as those by René Quesnel (2001) and Sean Olive (1994, 2001), provide strong evidence that training people to hear specific attributes of reproduced sound makes a significant difference in their ability to consistently and reliably recognize features of sound, and it also increases the speed with which they can correctly identify these features. Listeners who have completed systematic timbral ear training are able to work with audio more productively and effectively.

1.8 Sound Reproduction System Configurations

Before examining critical listening techniques and philosophies more closely, let's outline and define some common sound reproduction systems.

1.8.1 Monaural: Single-Channel Sound Reproduction

A single channel of audio reproduced over a single loudspeaker is typically called monaural or simply "mono" (see Figure 1.1). Even if there is more than one loudspeaker, it is still considered monaural if all loudspeakers reproduce the same audio signal. The earliest sound recording, reproduction, and broadcast systems used only one channel of audio, and mono was the main audio format for many decades. Although two-channel stereo and multichannel and two-channel spatial audio are now common, the prevalence of mono Bluetooth and smartphone speakers means that mono sound reproduction is still widespread. Mono sound reproduction creates some restrictions

Figure 1.1 Monaural or single-channel listening with a single loudspeaker shown at the top and a listener's head shown at the bottom.

for recording engineers since panning and sound image width are lost. Loudspeaker manufacturers often use mono sound reproduction for subjective evaluation and testing of their products, perhaps because of the lack of stereo and spatial images.

1.8.2 Stereo: Two-Channel Sound Reproduction

Two-channel sound reproduction systems, more commonly called stereo, allow sound engineers greater freedom than they have with mono to pan sound sources, to control the width of sound images, and to experiment with a sense of spaciousness. Stereo is still the primary format for sound reproduction whether it is over loudspeakers or headphones. Figure 1.2 shows the ideal listener and loudspeaker locations for two-channel stereo.

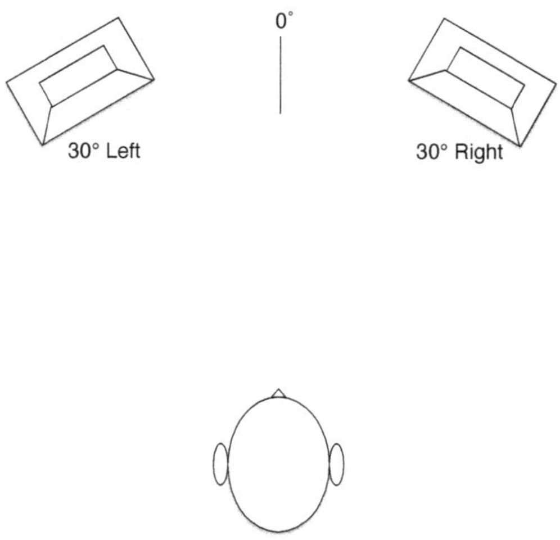

Figure 1.2 This diagram shows the ideal two-channel stereo loudspeaker and listener placement. Left and right stereo loudspeakers form an equilateral triangle with the listener.

1.8.3 Headphones

Headphone listening with two-channel audio has advantages and disadvantages with respect to loudspeakers. With modestly priced headphones, relative to the price of equivalent-quality loudspeakers, we can experience high-quality sound reproduction. Good-quality headphones can offer more clarity and detail than loudspeakers, because headphones are not subject to the acoustical effects of listening rooms such as early reflections, room modes, and reverberation. With headphone listening, we are always in the perfect listening location, unlike with loudspeakers. Headphones are also portable and can be easily taken to other locations where loudspeaker characteristics and room acoustics may be unfamiliar to an engineer.

One disadvantage of headphones is that for mono sound sources, we perceive the sound to be located inside the head (intracranial) when we listen using headphones, an effect also known as lateralization. In-head localization with headphones contrasts with the way we localize natural acoustic sound sources or even loudspeakers, which are localized externally. We perceive natural acoustic sounds to originate from outside the head because the sound bends around or reflects off the head, shoulders, and visible portion of the outer ears. The path that a sound takes when it bounces off the head, shoulders, and outer ears causes it to be acoustically filtered, and this filtering causes us to externalize sound sources. Because headphone sound does not bounce off the head, shoulders, and outer ears before entering the ear canals, it does not contain the acoustic filtering normally present in natural listening, and thus we localize headphone sound inside the head.

We can avoid or minimize in-head localization with headphone reproduction if we filter audio signals with what is known as a head-related transfer function or HRTF. Simply put, an HRTF is a mathematical representation of the transformation or filtering that sound undergoes after it leaves a sound source and bounces off the head, shoulders, and outer ears and enters the ear canals. An HRTF also includes the interaural time differences (ITD) and interaural level differences (ILD) for a given sound source location. Sound sources located in different angular locations and elevations have different HRTFs, since each location in space (elevation and azimuth) has its own interaural differences and filtering due to the way sound bounces off the head, shoulders, and outer ears. Not only do HRTFs vary with source location, but they also vary from person to person because of the differences in the dimensions of heads, shoulders, and outer ears.

If an audio signal is filtered with HRTFs for a sound source located at 30° to the left at ear level (0° elevation) and we listen to it over headphones, we should hear the sound as though it is coming from 30° to the left at ear level. In practice, not all listeners will hear the source at that precise location because everyone has their own HRTF, and the HRTFs used to filter the sound may differ from a listener's own HRTFs. Because everyone has their own HRTF, there is no universal HRTF that works equally well for everyone; every pinna is unique, and the shape of the pinna along with ear spacing determines the filtering that occurs. HRTFs are more widely used now than ever for 3D or spatial audio for music and movies over headphones, virtual reality, and loudspeaker listening simulations. Music streaming services such as Apple Music, Amazon Music, and TIDAL offer spatial audio versions of portions of their catalogs in Dolby Atmos or Sony 360 Reality Audio format.

Headphone listening differs from loudspeaker listening also since there is no interaural crosstalk with headphones. That is, the left and right channels on headphones are completely isolated from one another. With loudspeaker listening, sound from both loudspeakers reaches both ears, a characteristic known as interaural crosstalk, as illustrated in Figure 1.3. While it may not seem intuitively correct, loudspeaker listening with its inherent crosstalk can sound more natural than

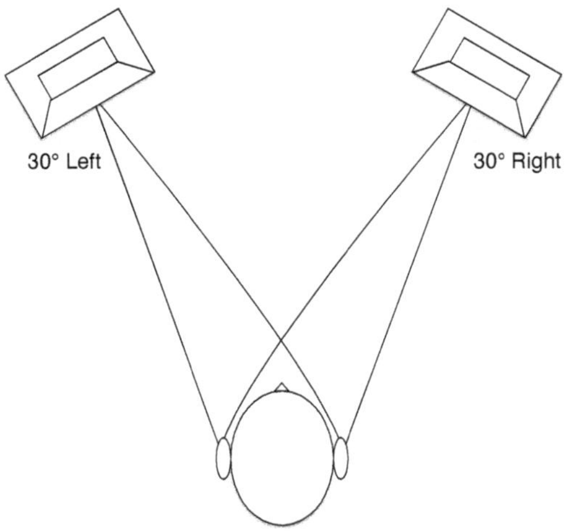

Figure 1.3 Stereo loudspeaker listening with crosstalk. The lines from each loudspeaker to each of the two ears represents the paths that sound can take.

headphone listening. Although phantom images like those produced by a pair of loudspeakers do not generally occur in natural acoustic listening, headphone listening with hard panning to one side, in which only one ear hears a sound, is also not found in natural acoustic listening; except in unusual circumstances, sounds normally reach both ears in natural listening.

If you are monitoring a project using headphones and want to simulate the crosstalk of loudspeaker listening, software plugins and hardware devices are available that cross feed audio between the left and right channels. Some options include the Goodhertz CanOpener-Studio plugin, Grace Design m900 headphone amp, and RME ADI-2 DAC FS digital-to-analog converter.

1.8.3.1 Open-Back and Closed-Back Headphones

Open-back headphones are more generally natural-sounding than closed-back headphones. With open earcups, reflections and standing waves are minimized so there is less coloration of the sound. Open-back headphones have better transient response, which means they respond more quickly to fluctuations in signal levels because there is less pressure against the back of the driver which otherwise impedes its movement. Open-back headphones are also lighter in weight than closed-back headphones, which makes them easier to wear for long stretches of time. The main disadvantage of open-back headphones is that they do not block outside sound. Similarly, others may be able to hear sound coming from your headphones.

Closed-back headphones provide more sound isolation than open-back headphones since they block background noise passively. Likewise, less sound from closed-back headphones leaks out, so others around you may not hear what you are listening to, assuming your listening levels are low or moderate. Because of their ability to reduce the amount of sound that leaks into microphones, closed-back headphones are better for studio musicians while recording. Closed-back headphones usually provide stronger response in the bass frequency range since the earcup enclosure can be tuned to emphasize low frequencies. Another potential issue that seems to affect closed-back headphones more than open-back headphones is the effect that headphone position can have on timbre or frequency response. The timbre of some closed-back headphones can vary with small adjustments in headphone placement on the ears. Open-back headphones do not seem to suffer from this issue to the same extent for some reason. Try it for yourself: when listening on closed headphones, move the headphones up or down slightly and listen for timbre changes.

Assuming you have access to a relatively quiet space, I recommend open-back headphones for critical listening and technical ear training.

1.8.3.2 Active Noise-Cancelling Headphones

Headphones with active noise cancellation (ANC) are an excellent choice for casual listening in noisy environments because of their ability to attenuate background noise. A benefit of background noise reduction is that we can listen at a quieter level, which supports hearing conservation. The main drawback of noise-cancelling headphones is that they are usually not suitable for critical listening. The signal processing that attenuates background noise can also distort the sound we want to hear in ways that make it difficult to rely on such headphones for critical audio decisions or ear training. Distortion from noise cancellation signal processing is often dependent on the background noise that may be present. Active noise-cancelling headphones are usually most effective at attenuating continuous low- to mid-frequency sound, such as that from airplane engines or air conditioners. They are less effective at attenuating sound from barking dogs, people talking, or children playing since these types of sounds are more transient and contain more high frequencies. Also, the amount of attenuation that an ANC headphone provides varies by frequency. It does not attenuate all frequencies equally like a volume knob does. Therefore, the sound that we hear when listening with ANC headphones is a mixture of the audio signal we wish to hear and the external sounds that the headphones could not attenuate. The problem is that, at any given moment, we may not know for certain if any of the sound we hear originated externally, nor can we know which parts of the frequency spectrum are being affected at any given moment. Since external sounds are often changing and unpredictable, and the ANC circuit does not always provide perfect attenuation for all frequencies equally, the ways that our audio can be affected change from moment to moment in unknown ways. Furthermore, some ANC headphones adjust the level of noise cancellation depending on the surrounding noise. They may also apply equalization that changes with playback level to compensate for the way the frequency response of the human ear changes with listening level, also known as equal-loudness contours. (See Section 2.4 for a longer discussion of equal-loudness contours.) For these reasons, ANC headphones are not my first choice for critical listening and technical ear training assuming a quiet working environment. If you need to work on an audio project on a plane or in another noisy environment, ANC headphones are a better choice for hearing health.

1.8.3.3 Wireless Headphones

Active noise-cancelling headphones or earbuds are most often found in wireless configurations, which means they use Bluetooth technology to transmit from the device to the headphones. Older Bluetooth wireless transmission protocols do not have the bandwidth to support data rates of linear, uncompressed PCM audio, even for standard digital audio quality at 44,100 samples/second and 16 bits/sample. To make real-time Bluetooth transmission possible, audio signal data is compressed by the music-playing device before transmission to wireless headphones. The problem is that most Bluetooth transmissions used by headphones and earbuds degrade audio quality since they use "lossy" codecs—that is, the codecs throw away audio data in a similar way as MP3 and AAC codecs which are also lossy. Even if you select lossless audio from your music streaming service, but you listen through wireless Bluetooth headphones or speakers, the audio that you hear goes through an audio codec and is thus not lossless. Audio quality can be even worse if the audio on your device is lossy, since audio from your device, which is encoded with one codec such as MP3, AAC, or Ogg Vorbis, might get encoded again—that is, transcoded—using a Bluetooth codec.

Fortunately, there are now some earbud and headphone models that provide higher-quality wireless lossless audio when paired with compatible devices. Qualcomm's Bluetooth codec called aptX Lossless is one such codec, but both the headphones and the device need to be capable of using it. Currently, only a few manufacturers support aptX Lossless on their earbud/headphone models, such as Beyerdynamic, Bose, Bowers & Wilkins, Denon, and Sennheiser. Audio quality can fluctuate dynamically depending on the headphones' proximity to their transmitting device, with audio quality being reduced if headphones are too far away.

Besides the aptX Lossless codec, there are a couple of options for uncompressed audio reproduction over headphones. One is to use wired headphones and loudspeakers. Another option comes from Sennheiser who announced in fall 2025 a new audiophile headphone (the HBD-630) that includes a Bluetooth USB-C dongle to ensure lossless audio over a wireless connection.

Earbuds have additional issues over headphones since the fit and seal in the ear affects the frequency response. A loose fit usually results in attenuated low frequencies. While the seal might be good when we first put them on, they can gradually loosen as we move around which brings down the low-frequency level.

Because of the issues with codec distortion in wireless Bluetooth headphones, I recommend using wired headphones for critical listening and technical ear training.

1.8.3.4 Headphone Models to Consider

The number of headphones on the market has exploded over the past several years, and there are many high-quality models suited for technical ear training. Because few retail stores stock high-quality headphones where we can just walk in and listen to them, I include some suggestions at varying price points below. The list includes circumaural headphones that surround the ear, supra-aural headphones that rest on the ear, and three types of enclosure: open, semi-open, and closed. All except one model on this list are circumaural, and all are wired.

- AKG K240 Studio (circumaural, semi-open back): This model has been a popular studio-monitoring headphone for many years and is the least expensive one on this list.
- AKG K371 (circumaural, closed back).
- AKG K701 (circumaural, open back).
- Audio-Technica ATH-M50x (circumaural, closed back): A popular studio-monitoring headphone.
- Beyerdynamic DT 770 PRO (circumaural, closed back).
- Beyerdynamic DT 900 PRO X (circumaural, open back).
- Grado (supra-aural, open back): There are several Grado headphone models, and all are supra-aural designs, meaning that they rest on top of the ear. The less expensive Grado headphones, the SR60X and SR80X models, are excellent value for money, but some may find them uncomfortable to wear for extended periods.
- Sennheiser HD 560S (circumaural, open back).

- Sennheiser HD 569 (circumaural, closed back).
- Sennheiser HD 600, HD 650, HD 660S2 (circumaural, open back): These models are in the same product family and are listed in order of increasing price.
- Sony MDR-7506 (circumaural, closed back): An industry standard for recording studios, this model is widely used in broadcast and film productions, and is relatively inexpensive.

In general, circumaural headphones are more comfortable to wear for extended periods of time since they surround the ears and do not press on them.

My personal favorites are the Sennheiser HD 650[1] headphones. These headphones have been lauded by recording engineers and audio reviewers for their accurate and refined sound. They have allowed me to hear detail in recordings that I had not heard previously, presumably due to their even frequency response and low levels of distortion. To get a similar level of detail from a pair of loudspeakers would require spending significantly more money. The HD 650s are also comfortable to wear, which makes them easy to use for hours on end.

One takeaway is that it's important to know your tools. Another takeaway is that no headphones are perfect. Therefore, the better you know your headphones, the more useful they can be since you are aware of their strengths and weaknesses. Grammy Award-winning mix engineer, recording engineer, and producer Andrew Scheps[2] (2021) said that he has mixed on Sony MDR-7506 headphones for more than 20 years. These headphones work well for him not because they have a perfectly flat frequency response (they don't) but because he knows what to expect from them, they are a constant for him, and he knows them well enough to get the most out of them.

1.8.3.5 Things to Listen for When Auditioning Headphones

It can be helpful to try headphones before purchasing, if possible. When you audition new headphones, use familiar recordings in an uncompressed or lossless audio format such as FLAC (free lossless audio codec), ALAC (Apple lossless audio codec), or linear PCM, such as the WAV, BWF, or AIFF file formats. Try to avoid using audio that has been data compressed with a lossy codec such as MP3, AAC, or Ogg Vorbis since these formats degrade sound quality. If compressed audio is all that you have available, set the bit rate to the highest quality. For Spotify, this is 320 kbit/s using their app. Note that the Spotify web player has a slightly lower maximum bit rate of 256 kbit/s.

As you audition new headphones, here are some things to listen for:

- **Stereo width:** Do different headphone models present the left–right spread differently? Are some wider or narrower than others?
- **Center image:** Sounds panned center, such as a vocal, should be tight, focused, and located precisely in the center of the stereo image. If the center image on one headphone model is wide or blurry, it may indicate mismatched drivers.
- **Frequency response:** Do all frequencies seem to be equally represented? Or does it sound muddy, bright, hollow, edgy, dark?
- **Detail:** Is one headphone model providing more high-frequency clarity or detail in the reverberation? If so, this could indicate a frequency response that favors the mid-range or high end.
- **Low-frequency extension:** Are you hearing the lowest fundamental frequencies from the bass or kick drum?
- **High-frequency extension:** Are you hearing the highest overtones, such as from cymbals?

By comparing the sound of different headphones using familiar music recordings, we can get a better sense of the strengths and weaknesses of each model. There are no perfect headphones, and each model will have a slightly different sound.

To make meaningful headphone comparisons, playback levels should be matched as closely as possible. This minimizes the influence of loudness since, when it comes to music recordings, louder is often perceived as better. But matching headphone levels is not as straightforward as we might wish. For one thing, headphone playback level can vary from one headphone model to another, even if each model is plugged into the same jack with the same the headphone output volume level. Playback levels vary mainly due to differences in headphone impedance and sensitivity from one model to another. In general, high-impedance headphones will not be as loud as those with lower impedance when connected to a phone, laptop, or other portable device. For example, Sennheiser HD 650 headphones have an impedance of 300 ohms, which is considered high, whereas AKG K371 headphones have an impedance of only 32 ohms, which is considered low. When plugged into most portable devices, low-impedance headphones like the K371s will sound much louder than high-impedance models like the HD 650 due to the difference in headphone impedance. But there is at least one exception found in some Apple devices. The 3.5 mm headphone jack on compatible Mac computers3 automatically detects headphone impedance and adjusts the output level for low- and high-impedance headphones, making an external headphone amplifier unnecessary in most cases.

Another challenge with playback levels is setting and recalling precise playback levels for each headphone. A headphone distribution amp with multiple jacks makes this possible but may not be available. If you need to switch headphones with different impedances to and from a single jack, you may find that it is impossible to recall different playback levels precisely with a headphone amp that uses a variable (non-stepped) analog potentiometer. One possible option for precise level recall if you have a Mac computer is to use the Audio MIDI Setup app, since it allows you type in the output level to a resolution of 0.01 dB. Another option is to use a DAW for playback and use the mix bus fader to recall precise output levels for each headphone models. Once you match the loudness of two different headphones, you can recall the exact output level for each one. This level of precision is also helpful for mixing and mastering to ensure that listening levels are consistent from day to day.

Be aware also that the sound of a pair of headphones can depend on the specific way that the ear pads are positioned on the ears. Sometimes small changes in ear pad positions can change the frequency response of the sound we hear from the headphones. While I have not done measurements, this seems to be more of an issue for closed-back than for open-back headphones. When you put a pair of headphones on, adjust them so that they fit your head and are positioned over your ears.

Finally, it's usually best to ignore celebrity endorsements of headphones. These kinds of advertisements are not usually aimed at audio professionals anyway.

1.8.4 Loudspeakers

One advantage loudspeaker listening has over headphone listening is that there is no in-head localization with the ideal listener/loudspeaker configuration. As with headphones, however, no two loudspeaker models will sound identical, mainly because of differences in objective characteristics such as frequency response, power response, distortion, and crossover frequency. Manufacturers make compromises in the design of loudspeakers due to the laws of physics and their price-point target. As such, it is difficult to give specific loudspeaker recommendations, but we can talk about some relevant loudspeaker characteristics.

In general, active studio monitors designed for professional audio or home recording markets are probably going to serve you better than passive loudspeakers designed for the consumer market. There are some excellent passive loudspeakers, but from my experience, active loudspeakers are going to offer better value for sound quality at a given price. For one thing, active monitors, by definition, have active—that is, powered as opposed to passive—crossover filters and contain a dedicated power amplifier for each loudspeaker driver. A two-way loudspeaker will have a dedicated power amplifier for the low-frequency woofer and a separate power amplifier for the high-frequency tweeter. As such, each power amplifier in an active loudspeaker can be optimized for its respective driver and frequency range. Furthermore, in active loudspeakers, distortion can be reduced, and the frequency and phase response of the crossover filters can be controlled better than with passive circuits. These features result in better sound quality than what we might find with a passive loudspeaker system at a similar price point. Also, passive crossover filters can absorb up to 20% of an amplifier's output since signals are amplified before they are filtered. Passive consumer-market speakers often have beautiful wood finishes that contribute to a higher cost, and while they may be aesthetically pleasing for our eyes (natural wood grain finish certainly is beautiful), a wood exterior does nothing to improve the sound quality over a plain matte black finish that we find in most studio monitors aimed at the professional audio market. Most loudspeaker cabinets are constructed with medium density fiberboard (MDF) and then, if there is a wood finish, the wood layer is attached to the MDF. Professional studio monitors are designed for optimal sound quality rather than looks for a given price range. It's best to spend money on sound quality rather than looks.

To improve the low-frequency extension of smaller studio monitors, manufacturers usually design them with a ported, also known as bass reflex, cabinet which simply means that there is an opening or port in the cabinet. A port allows the cabinet to resonate like a Helmholtz resonator which acoustically amplifies low frequencies, usually at or below the cut-off frequency of the loudspeaker without a port. The most common example of a Helmholtz resonator can be created by blowing across the open top of an empty bottle. If you've ever done this, you know that you can produce a tone, the frequency of which is based on the air volume of the bottle, the bottle's neck length, and the diameter of the neck opening. The advantage of ported cabinets for loudspeakers is that designers can achieve a better low-frequency response from small speakers with small woofers. The disadvantage is that the low end might not be as controlled as we need. In other words, the resonant frequency of the port might ring slightly longer than other frequencies, causing muddiness in the low end. Most studio monitors available are ported and can perform very well, but some have issues in the low-frequency range.

Some loudspeakers have sealed instead of ported cabinets. Sealed cabinets do not have the same kinds of low-frequency resonances that we find with ported cabinets. One tradeoff with some sealed cabinets is that they have a reduced low-frequency extension. Two classic studio monitors often pictured sitting on console meter bridges in recording studios—the Yamaha NS-10, a passive loudspeaker with distinctive white woofer cones, and the Auratone 5C, a single driver loudspeaker in a cubic cabinet—both have sealed cabinets. The NS-10 has not been manufactured for many years, but the Auratone is still manufactured. Other companies such as Avantone Pro also offer copies of the Auratone and NS-10 sealed cabinet loudspeakers in active and passive versions. Another notable example of a sealed cabinet is the Neumann KH 310, an excellent active studio monitor.

In general, of the loudspeakers I have heard, I tend to prefer active studio monitors marketed to the professional audio market from companies such as ATC, Dynaudio, Focal, Genelec, Meyer Sound, and Neumann. Although marketed more to consumer audiophiles, models by Bowers & Wilkins are also well respected in the professional audio world.

1.8.4.1 Surround Sound and Multichannel Audio Reproduction

We refer to loudspeaker systems with more than two channels of audio as multichannel, surround, or Ambisonic. Surround soundtracks for film and television are common in cinemas and are becoming more common for home systems. There are many suggestions and philosophies on the exact number and layout of loudspeakers for surround reproduction systems. One of the early standardized recommendations was the five-channel loudspeaker layout from the International Telecommunications Union (2022) shown in Figure 1.4, which was widely accepted among audio researchers. Users of the ITU-recommended configuration generally also make use of an optional subwoofer or low-frequency effects (LFE) channel known as the ".1" channel (pronounced "point one"), which reproduces only low frequencies, typically below 120 Hz. There are now several multichannel configurations that include overhead or height channels to support playback of Dolby Atmos, Audo 3D, NHK 22.2, or other three-dimensional audio formats.

Multichannel systems are denoted by the number of channels they contain, such as:

- 5.1
- 7.1
- 7.1.4 – seven ear-level, one subwoofer, and four overhead loudspeakers
- 9.1
- 9.1.6
- 10.2
- 22.2
- 3/2 channel
- quadraphonic.

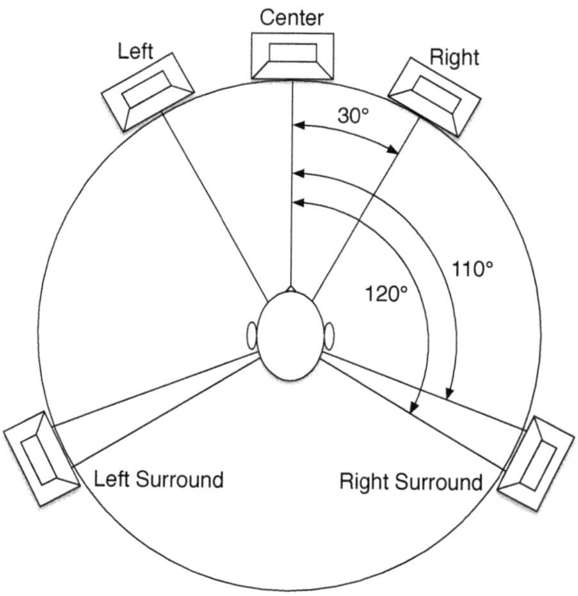

Figure 1.4 The ideal five-channel surround listening placement according to the ITU-R BS.775 recommendations (International Telecommunication Union 2022), showing three front loudspeakers at 30° left, 0° center, 30° right, 110 to 120° left surround, and 110 to 120° right surround, all equidistant from the listening position.

Multi-loudspeaker surround sound formats for music-only applications have had limited popularity among listeners despite enthusiasm among recording engineers. These formats are still not as popular as traditional stereo reproduction. The exceptions to this are the spatial or 3D audio formats such as Dolby Atmos and Sony 360 Reality Audio available from music streaming services. On music streaming services, the spatial audio formats are primarily targeted at headphone listening.

Multichannel sound systems allow much more freedom for sound source placement within the 360° horizontal plane than two-channel stereo reproduction does. Multichannel systems also provide more possibilities to create a sense of spaciousness, envelopment, and immersion within a virtual acoustic space, especially when we feed the appropriate signals to the appropriate channels. As Bradley and Soulodre (1995) and others have demonstrated, listener envelopment (LEV) in a concert hall, a component of spatial impression, is primarily dependent on having strong lateral reflections arriving at the listener 80 milliseconds or more after the direct sound. If early reflections and reverberation are reproduced from the sides and rear of a loudspeaker system, listeners feel more enveloped in sound.

There are some challenges with respect to sound localization for certain areas within a five-channel listening area. Panning sources to the side, such as between the front left speaker at 30° and the surround left speaker at 110°, produces sound images that are unstable and difficult to accurately localize. Side-panned sound sources are much more stable when side loudspeakers are included in the system, such as with seven- and nine-channel systems.

The presence of a center channel allows sounds to be locked into the center of the front sound image, no matter where a listener is located. This is a possible strength over two-channel stereo, which relies on phantom images for sounds panned to locations between the left and right loudspeakers. You may have noticed that the perceived location of sound sources panned to the center in a two-channel stereo setup tend to move when you move. Seated to the left of the ideal listening location with stereo loudspeakers, center-panned sounds seem as though they are coming from the left side, and vice versa when seated to the right of the sweet spot.

1.9 Audio Ear Training on Every Project

No matter your role on an audio project, whether you are a producer, head engineer, loudspeaker designer, assistant engineer, or student, there are always opportunities to practice active and analytical listening. Practice thinking about and listening for these qualities and characteristics on each audio project:

- **Timbre:** Listen for and evaluate the tonal balance of each instrument, voice, or sound, as well as the overall mix. Do any tracks need to be equalized? Are the microphones in the right place to give the timbre needed for the project? Is there any proximity effect (low-frequency boost) from close microphone placement? Think about general tonal balances first—low-, mid-, high-frequency bands—and then try to get more specific.
- **Dynamics:** Are sound levels varying too much, or not enough? Can each sound source be heard throughout? Are there any moments when a sound source gets lost or masked by other sounds? Is there any sound source that is overpowering others? Is every word of the singer clearly audible?
- **Overall balance:** Does the balance of musical instruments and other sound sources make sense for the music? Or is there too much of one component and not enough of another?
- **Distortion/clipping:** Is any signal level too high, causing distortion?
- **Extraneous noise:** Is there a buzz or hum from a bad cable or ground problem? Are there other distracting noises, unrelated to the recording?

- **Space:** Is the reverb/delay/echo right for the music, film/video, or game project? Is it consistent with the visual component?
- **Panning:** How is the left/right balance of the mix, as you hear it from the loudspeakers? Is it too narrow or too wide? Is it balanced from left to right?

1.10 Everyday Listening

We are exposed to a variety of sounds every day, whether we pay attention to them or not. Acoustic sounds in everyday life may be loud, quiet, near, far, sustained, transient, or some combination of all these qualities. Some sounds may have a lot of bass or low frequencies, like a freight train or distant thunder, while others may have more treble or high frequencies, such as bird songs or emergency vehicle sirens. We can use acoustic sounds as opportunities to practice critical listening.

As an exercise, take a moment right now and listen to the sounds around you. What do you hear? What qualities do the sounds have? Where are the sounds located around you? How distant are they?

Our physical environment plays an important role in the sound that we hear. Objects, walls, and other physical structures may reflect, absorb, or diffuse sound in our surroundings. Reflected sound in our environment can tell us about the physical properties of our location. Our surrounding environment becomes audible in a sense, even if it is not creating sound itself, through patterns of sound reflection and absorption. Just as a light source illuminates objects around it, sound sources allow us to hear the general shape and size of our physical environment. Echoes or strong reflections give us clues about sound source distance. The more we listen to everyday sounds, the more we become aware of acoustic effects that we might have ignored previously such as subtle echoes, reflections, reverberation, low-frequency rumble, flutter echoes, and so on. As our awareness of acoustic sound increases, our critical listening skills improve, and we can bring these skills to our audio projects.

Sound that reaches our ears tells us about its source. The frequency content and tonal balance of sound in our environment gives us some clues about the sound sources and our proximity to them. Outside of live music concerts, sound with a lot of low-frequency power can originate from large mechanical devices, such as diesel engines, airplanes, or helicopters, or they can originate from natural sources, such as waterfalls, waves crashing onto a seashore, or thunder.

Humans without visual impairments are often more visually oriented, so it may take some dedicated and consistent effort to focus our auditory awareness. As professional audio engineers know, the effort it takes to focus our aural awareness is well worth the satisfaction when we notice that our critical listening skills have improved. The hard work of focused, active listening pays off.

Although the concept of critical listening is relatively simple, the challenge lies in the practical application: focusing our attention consistently hour after hour, day after day on an audio project. It takes energy and discipline to listen with focus, but regular, targeted practice hones our awareness and gives us the ability to work more efficiently when tackling any audio project.

We can develop critical listening skills everywhere we go, even when we are not actively working on an audio project. For instance, near a construction site, we may hear sharp bursts of sounds, known as impulse noise, from jackhammers or nail guns. And we may hear echoes of that direct sound being reflected from nearby building exteriors, arriving fractions of a second after the direct sounds. The timing, location, and amplitude of echoes provide us with information about nearby buildings, including approximate distances to them. The timbre of direct and reflected sound may be different, especially in the case of long echoes. This indicates that air, the medium through which sound travels, as well as the surfaces from which it has reflected absorb some frequency content from the reflected sound.

An acoustic music performance—that is, without amplification—is another ideal place to practice critical listening. The next time you attend a music performance in a large acoustic space, listen for the reverberation, sound that continues to linger on and slowly fade out in the seconds after the ensemble has stopped playing. You may feel enveloped by or immersed in sound produced in a concert hall or other large space, especially if the sound seems to be coming from all directions. Try to focus only on the reverb and reflected sound. Is it coming from a specific direction or is it coming from all directions? Try to be aware not only of sound coming from performers on stage but also of the reflected sound coming from all around the venue.

In contrast to a large space, music performed in a carpeted living room will sound noticeably different than it does in a large public space such as a concert hall. There are a few reasons for this difference. Physical characteristics such as room dimensions and surface treatments, such as carpeting, curtains, and furniture, affect the sound we hear. As you may have experienced, a living room's acoustical characteristics will be markedly different than those of a concert hall. The reverberation time will be significantly shorter, and early reflections will arrive much sooner in a living room because the volume of a typical living room is much less than that of a concert hall. Carpeted floor, curtains, and fabric furniture upholstery will also influence spectral balance of reflected sound, since these materials absorb more high frequencies than low frequencies, making the reflections sound duller. A wooden floor and other hard surfaces will reflect high frequencies and therefore make the reflections sound brighter.

The relatively close walls in a living room will reflect sound back within milliseconds of the arrival of direct sound and at nearly the same amplitude. This small difference in time of arrival and near-equal amplitude of direct and reflected sound will create constructive and destructive interference at our ears. In other words, some frequencies will be cancelled (completely attenuated), and other frequencies will be boosted. An extreme example of this effect is comb filtering (which I discuss more in Chapter 3), where a sound is mixed with a single delayed version of itself. The effect is most apparent when we mix a sound with a delayed version of itself electrically or digitally in a mixer. We hear comb filtering all the time in everyday life, but because reflected sound delay times change as we move and because there are so many delayed reflections arriving at our ears, the effect is smoothed out, and we do not get the same deep notches in the spectrum that we do with only a single delayed reflection mixed with the original non-delayed sound.

Active listening is crucial to our work in audio engineering, and we can take advantage of times when we are not specifically working on an audio project to heighten our awareness of the auditory landscape and practice our critical listening skills. Walking down the street, sitting in a restaurant, and attending a live music concert all offer opportunities for us to hone our listening skills and thus improve our work with audio. Barry Blesser and Linda Salter's book *Spaces Speak, Are You Listening?* (2007) expands upon the idea of listening to acoustic spaces and offers a detailed exploration of aural architecture.

As audio engineers, we are concerned with capturing, mixing, shaping, and modifying sound. Whether we work with recorded acoustic sounds or those generated entirely within an electronic medium, one of our goals is to ensure that the audio communicates the intentions of the musical artist, podcast maker, film director, game producer, car audio company, or whomever we happen to be working for or collaborating with.

An important aspect of audio that engineers seek to control is the relative balance of instruments or sound sources, whether through manipulation of recorded audio signals or through microphone placement around instruments and ensembles. Sound source balances in a recording can have a tremendous effect on the musical feel of a composition. Musical and spectral balance is critical to the overall impact of a recording. Through the process of shaping sound, no matter what equipment is being used or what the end goal is, our focus is simply to listen. We need to constantly analyze what we hear to assess a track or a mix and to help make decisions about further adjustments to balance and processing. Listening is an active process, challenging us to remain continuously aware of any subtle and not-so-subtle perceived characteristics, changes, and defects in an audio signal.

Our ears capture sound from all directions all the time, so we can find endless opportunities for ear training exercises wherever we go. Below I describe three exercise that you can try when you are not working on an audio project or a technical ear training exercise. The idea is to be mindful that ear training can happen almost anywhere. The main feature of these exercises is that they require attention to sound, and this is consistent with technical ear training which seeks to expand our capacity to sustain our attention on sound.

1.10.1 Everyday Listening Exercise 1—Everyday Sounds

As you go about your everyday activities, try this exercise a few times each day, specifically when you are not working on an audio project, to help train your ability to focus on sound quality.

What sounds do you hear right now? Describe the timbre and qualities of individual sounds as well as the overall soundscape, using these points as guidelines:

- **Frequency content:** Are sounds wide band—that is, do they extend across the full audible frequency range? Or are they narrow band and contain only low frequencies, only mid-range frequencies, or only high frequencies? Are there recognizable pitches or tones, or are the sounds mostly noise-based—that is, without any definite pitch? If you turn your head, does the timbre of sound change? Can you hear the timbre gradually change while you are moving your head? For noisy sounds like machinery or power tools, are all frequencies represented equally or are there certain frequencies that stand out?
- **Temporal qualities:** Are there repeating, periodic, or rhythmic sounds? Are there transient sounds? Are there steady-state or continuous sounds?
- **Spatial characteristics:** Where are the sounds located relative to you? How close are they? What is their angular location or azimuth? How high or low are the sounds' positions—what is their elevation? Are their locations clear and distinct, or diffuse and ambiguous? If there are recognizable echoes, where are they originating? How wide is the soundscape? How wide are the sound sources?
- **Background sounds:** Besides the more obvious sounds, are there any continuous, quiet background sounds, such as those from air handling systems (HVAC), lights, or a refrigerator that we might otherwise ignore?
- **Loudness:** How loud are the sounds, overall and relative to one another?
- **Room acoustics and reverberation:** Is the sound reverberant or dry? Are there any distinct echoes or reflections? Do you hear any resonances or flutter echoes? Can you estimate the reverberation decay time? How big does the room sound and does the sound match the way it looks? While we can use visual cues to judge the size of a room, try to use only your ears to characterize the environment.
- **Moving sound sources:** Do you notice the Doppler effect on any car horns or sirens or other moving sound sources? Have you ever listened closely to the buzz of an insect flying around your head?

Sometimes we come across odd or unusual acoustic features by accident. For example, I discovered by chance that I can make one frequency resonate at a specific location near my bathroom sink. One day when I was talking while I was standing in front of the sink with my head over the sink, I noticed that one frequency of my voice was louder than the other frequencies. I had discovered a resonant frequency. After I experimented a little with my head position and humming different pitches, I figured out that my head position above the sink affects the level of the resonant frequency.

Listen in your own spaces. You may be able to find resonant frequencies in unexpected places. If you notice an acoustical resonant frequency within the range of your voice, try to hum the pitch. Then change the pitch of your voice and notice how your voice seems to get louder at the resonant frequency and quieter at other frequencies. Bathrooms in general can be interesting acoustical spaces because they often have hard reflective surfaces everywhere and can have relatively long reverberation times for their volumes. That's probably why singing in the shower is so popular.

If you've been in a vacant house or apartment, you may have noticed prominent reverberation and echoes. Furniture, curtains, and rugs/carpet are all sound absorbers, and when they are removed from a room, it's easier to hear reflections and reverberation. Likewise, building lobbies or entrances with hard, reflective surfaces are more reverberant. Parking garages may have even longer reverberation times. Offices often have short reverberation times because of the sound-absorbing effects of ceiling tiles, carpet, and furniture. When you walk from one room or space into another, listen to the way the acoustics change.

1.10.2 Everyday Listening Exercise 2—Recorded Music

When you find yourself in an environment where recorded music is playing, but not when you are working on an audio project, try analyzing the timbre and sound quality that you hear. Maybe you are in a store or restaurant, someone in your house is playing some music, or you are walking down the street and hear music. Take the opportunity to analyze the sound you hear for a moment.

- What is the tonal or spectral balance of the recording in terms of low-, mid-, and high-frequency ranges? Is there enough bass/mid/high? Too much bass/mid/high?
- Are all the elements of the mix clearly audible? If they are not, which elements are difficult to hear? Which are most prominent?
- If you are familiar with the recording, how is the timbre of the sound affected by the system and environment through which it is presented? Does it sound different than what you remember? If so, how? Does the mix balance seem the same as what you have heard in other listening situations?

1.10.3 Everyday Listening Exercise 3—Speech

In this exercise, the task is to listen to the way the spoken voice sounds in a variety of different environments. Listen to recorded or live speech over loudspeakers or headphones from sources such as podcasts, radios, movies, video conference calls, and TV shows. Listen to acoustic speech such as when you're in the same room as someone who is speaking. Listen to speech that has been amplified over a sound system, such as at a public event in which someone is speaking over a public address system. There are endless amounts of recorded speech online, and any of it can be used in this exercise. Once you start listening more closely to spoken word audio, you'll discover that a range of different microphone placements, such as close/dry versus distant/reverberant, are used. You will also notice that dynamic range, level differences between different voices in the same recording, and timbre can all vary widely from one recording to another. Try listening for the following qualities:

- **Tone quality:** Does the voice sound deep, mellow, shrill, raspy, or have some other quality? Is there a lot of low-frequency power in the voice, as we hear on some radio announcers, or is it more even tonally, as we might hear from a news announcer? If there is more than one voice in the recording, do they all have the same timbre or are they different? If they are different, describe the ways they are different.
- **Pitch variation:** Notice the pitch contour of the speech. Unless the person is speaking in a completely monotone voice, there will likely be some variation across a range of fundamental frequencies.
- **Compression/limiting:** Does the voice sound compressed? Can you describe what quality in the audio is giving you clues that it is compressed? Some recordings and live broadcasts, such as sporting event announcers, have obvious dynamic range compression or limiting. Are there distracting artifacts from the pumping and breathing of a compressor?
- **Room acoustics:** Listen for room reflections, reverberation, echoes, and room modes. How close do you think the microphone is to the person speaking? Some podcasts are very roomy-sounding, making it seem as though they recorded two or three people sitting around a table in a room with highly reflective surfaces, with a built-in microphone on a laptop.
- **Distortion:** Is there any distortion or clipping on the voice?
- **Mix:** If there is background music mixed with the voice, what is the relative balance like, and can you hear the voice well enough over the music? Most professional audio mix engineers for television and radio broadcast will use ducking compression on any background music when it is mixed with speech to make sure that the music is always mixed below the speech so that the speech is clearly audible.

The next time you listen to a recording of spoken voice, pay attention to the quality of the recording. Television or radio broadcasts from studios offer relatively high sound quality for speech, but spoken word recordings such as podcasts or YouTube videos made by non-audio professionals vary widely in quality. From an analysis and critical listening point of view, these types of recordings offer some great examples. Speech recordings offer an excellent opportunity for critical listening and analysis, and there is a wealth of sound sources online to analyze.

1.11 Effort, Attention, and Rests

At various points in this chapter, I have highlighted the fact that critical listening requires real effort. Because of the energy required to listen closely with sustained attention, long training sessions can be tiring. One sign that we're getting tired and may need a break is when we start making more mistakes than usual. If we continue to train when fatigued, we can begin to feel frustrated doing technical ear training. So, how do we minimize fatigue so that we can sustain hard work and maximize our progress?

One way to minimize fatigue is to practice for shorter times but on a regular—that is, daily or near-daily—basis. Regular practice for shorter periods of time several times a week is much more productive than extended but less frequent practice sessions. Like learning to play a musical instrument or speak a new language, regular daily practice keeps our brains and bodies connected with the new skill we are trying to develop. Because of the effort needed to learn something new, it is best to limit practice sessions to 20 to 25 minutes. Pay attention to your fatigue and limit training time to shorter stretches if you need to. Five minutes a day is more effective than trying to cram in a three-hour practice session once a month. Regular and consistent repetition helps us learn these skills. Because highly focused listening requires intense energy, practicing for longer periods at a time can be counterproductive. Eventually, as you get used to this focused type of listening, you may want to increase the duration of your practice periods. Still, you may find that 30 to 45 minutes is the upper useful limit for a given practice session of intensely focused attention on listening tasks.

As your technical ear training times get longer, you may benefit from short breaks to help maintain the levels of concentration required to learn and develop these skills. It may be useful to apply findings from research in Attention Restoration Theory (ART), which suggest that breaks in natural environments can help restore mental focus, boost waning attention, and enhance concentration (Kaplan and Kaplan 1989; Kaplan 1993, 1995). Although it may not be feasible to take a walk in nature, there are other options. Even simply looking at images of natural scenes can be restorative. In one experiment, participants were assigned to view one of two images for only 40 seconds. One image was a simulated view of a city scene with a bare concrete roof, and the other was of a city scene that included a green roof planted with a flowering meadow. Participants assigned to view the flowering meadow green roof made fewer errors in a post-treatment task compared to those who viewed the concrete roof image (Lee et al. 2015). This research suggests that even micro-breaks less than a minute long looking at digital images of natural scenes are restorative and can provide valuable attention boosts.

For breaks to be effective, avoid picking up your phone so that you are not tempted to read newsfeeds, watch videos, or respond to notifications while taking a breather. When you feel that you need a break, try to look at a nature image screensaver or out the window, go for a walk in nature, do some stretches, tend to a plant, get outside for a moment, or just walk around your room. Technical ear training requires repetition and practice, but to sustain the intense focus that allows us to make progress, taking breaks to look briefly at nature scenes can help restore our attention so that we can come back to training again.

1.12 Hearing Conservation and Sound Level Measurement

Hearing conservation is essential to our discussion of technical ear training. We want to protect our hearing not only for our careers but also for general quality of life and health. Noise-induced hearing loss or NIHL, sometimes called sound-induced hearing loss or SIHL, results from exposure to loud sounds of any type. No matter whether the sound source is heavy machinery, a musical instrument, a movie, a game, a power tool, or amplified speech, it can cause hearing loss if it is loud enough. Although audio engineers may typically work with music or speech reproduced over loudspeakers and headphones rather than with noisy machines such as those found in factories or construction sites, exposure to high sound levels can cause irreversible hearing loss and associated disorders of tinnitus and hyperacusis[4] regardless of the source of the sound. Tinnitus refers to "ringing in the ears" when no other sound is present, and can sound like hissing, roaring, pulsing, whooshing, chirping, whistling, or clicking. Hyperacusis is a condition that causes a person to be unable to tolerate everyday noise levels without discomfort or pain. Hearing loss, tinnitus, and hyperacusis not only make it difficult or impossible to work in audio, but these conditions also make everyday life more difficult and unpleasant. Exposure to high noise levels is also associated with high blood pressure and cardiovascular disease. Unfortunately, hearing loss is permanent so it's important to know about sound levels and hearing protection, and take appropriate measures to reduce our exposure to loud sound.

1.12.1 How Loud is Too Loud?

Loud sounds can damage hearing permanently, but how loud is too loud? John Eichwald, an audiologist now retired from the U.S. Centers for Disease Control (CDC), says a good rule of thumb is that in a noisy environment if you must raise your voice to be heard by someone 3 feet (1 meter) away, you are at risk of causing hearing damage (Center for Hearing and Communication 2021). In the workplace, the National Institute for Occupational Safety and Health (NIOSH) in the United States recommends that exposure to sound levels of 85 dBA should be for no more than eight hours per day. The NIOSH recommended time limits for different sound levels are listed in Table 1.1.

The time limits include a 3 dB time-intensity tradeoff, which means that for every increase of 3 dB in sound level, the exposure time is reduced by half (National Institute for Occupational Safety and Health 2024).

The unit "dBA" means A-weighted decibels. Decibels of sound pressure level, or dB SPL, is a measure of sound level relative to the quietest level we can hear, also known as the threshold of hearing. Measurements using the dBA weighting filter approximate the frequency response of the human ear. A sound level meter set to dBA filters or attenuates low frequencies prior to measurement to approximate human perception of sound at normal listening levels. Many sound level meters can also measure in C-weighted decibels or dBC, which includes a weighting filter with less attenuation of low frequencies than dBA and which correlates with human perception of very loud sounds. A C-weighted measurement may be used to evaluate sounds that are particularly loud or contain very low-frequency sounds. If a C-weighted measurement is higher than an A-weighted measurement, it means that the sound has significant low-frequency content. Sound level meters and apps sometimes also include the Z-weighted or dBZ measurement unit, which means unweighted or unfiltered. For the purposes of hearing conservation, A-weighted sound level measurements are used to predict hearing risk.

The NIOSH recommended exposure times for each sound level consider only noise exposure on the job, but of course we also encounter noise outside of the workplace from a whole host of sources. Ambient noise pollution from sources such as road traffic, air traffic, lawn care machines, and construction can be significant, but we can also encounter high sound levels from activities we choose to do. Attending concerts, listening on headphone/earbud and automotive audio systems, attending sporting events, playing video games, going to bars, nightclubs, and movies, using power tools and vacuum cleaners, going to a shooting range, and using kitchen appliances like blenders are all activities that can involve high sound levels potentially putting our hearing at risk.

Table1.1 The maximum exposure times for various sound levels as recommended by the U.S. National Institute for Occupational Safety and Health.

Sound Levels	*Maximum Exposure Time*
82 dBA	No more than 16 hours a day
85 dBA	No more than 8 hours a day
88 dBA	No more than 4 hours a day
91 dBA	No more than 2 hours a day
94 dBA	No more than 1 hour a day
97 dBA	No more than 30 minutes a day
100 dBA	No more than 15 minutes a day
110 dBA	No more than 90 seconds a day
115 dBA	No more than 28 seconds a day

The World Health Organization (WHO) and International Telecommunication Union (ITU) recommend more conservative sound level exposure than NIOSH. They state that a safe sound level for leisure noise is below 80 dB for a maximum of 40 hours a week, and this includes audio on personal listening devices. At sound levels below this amount, we do not put our hearing at risk (World Health Organization and International Telecommunication Union 2019). If we consider the full 24-hour period of daily living, a U.S. Environmental Protection Agency (EPA) report (1974) states that exposure to a daily average sound level of no more than 70 dBA would protect 96% of the general population from developing hearing loss. But many people regularly encounter sound levels above 70 dBA in everyday life. Busy urban streets have higher noise levels than that. One study found that mean street level noise in New York City was 73.4 dBA and levels ranged from 55.8 to 95.0 dBA varying by location (McAlexander et al. 2015). Recreational activities also put us at risk. Sound levels at live music shows and sporting events can go above 110 dBA, putting audience members and fans at risk of hearing loss after only a few minutes of exposure. During a playoff game in 2024, the sound levels at a Detroit Lions football game reached 133.6 decibels, which was the loudest game ever at their stadium, Ford Field (Dozier 2024). As exhilarating as events such as this are for those attending in person, given the NIOSH sound level recommendations, it's not hard to imagine how hearing can be damaged without the use of hearing protection.

The WHO and ITU are also working to raise awareness about the need for safe listening levels in venues, at events, and on personal devices in an initiative called "Make Listening Safe" (2026). One of the WHO recommendations is that events with amplified music should not exceed a sound level of 100 dB $L_{\text{Aeq, 15min}}$—that is, 100 dB A-weighted sound level, averaged over 15 minutes (World Health Organization 2022). In response to research conducted by the WHO, an organization in the United Kingdom called Healthy Ears, Limited Annoyance (HELA) provides online training and certification for safe listening at live events for anyone who works at a venue, including bar tenders, security staff, and sound technicians, since they are all exposed to high sound levels (PSW Staff 2025). Hearing conservation is important for everyone who works at events or venues that produce loud sound, not just the sound engineers.

1.12.2 Measuring Sound Levels

How do we find out what sound levels we might be exposed to? Aside from hiring a qualified acoustician, there are three categories of options for casual sound level measurement that can provide a guide for listening levels:

- basic sound level meter (SLM),
- smartphone sound level meter app,
- smartwatch noise level app.

Basic sound level meters measure and display sound levels in real time. A reliable sound level meter that is popular among audio professionals is the Galaxy Audio Check Mate CM-140. With a measurement accuracy of ±1.5 dB, a frequency range of 31.5 Hz to 8 kHz, A-weighted or C-weighted decibel readings, it complies with the IEC 651 Type 2 and ANSI S1.4 Type 2 standards for sound level meters. Although the IEC 651 standard has been superseded by the newer IEC 61672 standard, meters that meet the IEC 651 standard like this one are still considered to be usable for many applications. Sound level meters that do not comply with either IEC standard may give unpredictable or inaccurate readings which can put you and others at risk for hearing loss if you are relying on the readings to determine the need to wear hearing protection. A basic handheld sound level meter like the Galaxy Audio Check Mate CM-140 is useful for measuring playback levels in the recording and mixing studio or when mixing live sound reinforcement. At the high end of the quality scale, manufacturers like Hottinger Brüel & Kjaer make sound level meters that provide highly precise measurements along with more advanced analysis and data-logging features appropriate for professional noise control and acoustical engineering applications.

Besides its important function to determine safe listening levels, a sound level meter can be used to ensure that studio monitoring levels are consistent from day to day. Since the frequency response of the human ear varies with sound level, consistent monitoring levels help ensure that we are not mixing to the equal-loudness contours of the ear. The company Blue Sky offers calibration test files and instructions on calibrating a monitoring system on their website.[5]

Handheld sound level meters are not practical in every situation, such as when attending a concert or sporting event as an audience member or fan. Fortunately, there are sound level meter smartphone apps available. The main problem with sound level meter apps is that they vary in quality and accuracy. If you have an Apple iOS device, I recommend the free NIOSH Sound Level Meter app[6] because it has been tested and validated, it offers accurate measurement of noise levels on any iOS device, and it can provide several additional measurements including noise dose percentage or dosimeter (Kardous and Celestina 2017). Sound levels usually fluctuate, so a noise dosimeter is useful to determine sound exposure over time since it shows what percentage of the maximum allowable daily dose you have been exposed to. A person continuously exposed to 85 dBA will reach 100% of their daily dose after 8 hours, per the NIOSH recommendations (listed in Section 1.12.1 above). Basic inexpensive handheld sound level meters do not include dosimeters, so this is a useful feature on the NIOSH sound level meter app. The app also includes information about noise and hearing loss prevention. Using a calibrated microphone, such as the i436, i437L, or i437C microphone, which is calibrated to meet the IEC 61672–3 Class 2 standard, with an iOS device provides more accurate measurements in the app. The NIOSH Sound Level Meter app is only available for iOS devices because, as they say on their website, the app developers can verify that it will perform as expected on any iOS device. Android devices, on the other hand, are manufactured by many different companies, and thus it is impossible to guarantee that the app will perform as expected. The NoiSee app for iOS devices has also been tested along with other apps for accuracy and appears to be as accurate as the NIOSH app (Celestina et al. 2018; Kardous and Shaw 2014).

Smartwatches provide another convenient way to monitor sound levels since they can provide notifications if a specific decibel threshold is reached. While a watch may not typically be located at the ear, it can nonetheless be useful to warn the wearer when ambient sound levels are dangerously high. Not all smartwatch wearers have noise notifications turned on, but the fact that mainstream devices like Apple Watch provide notifications at all when sound levels go above 80 dBA is an encouraging and positive sign that hearing conservation awareness could grow among the general public.

1.12.3 Hearing Protection Options

Even without a sound level meter we should remain aware of sound levels. If it sounds loud, it is loud. If we find ourselves in an environment with potentially damaging sound levels, we usually have some options. Maybe we control the level of a sound system and can turn the level down, we can use hearing protection, we can move to a quieter area, or we can take breaks from the sound.

Fortunately, there are many hearing protection options available now. Some companies are even marketing earplugs to be both fashionable and functional. It is remarkable to think that earplugs could make a fashion statement, considering that there seems to be little awareness of the need for hearing protection among the general population. At the same time, we are exposed to ever higher sound levels at concerts and sporting events, as well as in everyday life. The main problem with earplugs is that they can detract from our enjoyment of music depending on the type, and they usually don't make a fashion statement in the way that sunglasses do. Any option that encourages hearing conservation is likely to help extend the number of years we have with normal hearing.

The efficacy of earplugs and earmuffs is usually described with a specification called NRR, short for noise reduction rating. The U.S. Environmental Protection Agency, which requires that hearing protectors' packages include a noise reduction rating, defines NRR as:

> A single number noise reduction factor in decibels, determined by an empirically derived technique which takes into account performance variation of protectors in noise reducing effectiveness due to differing noise spectra, fit variability and the mean attenuation of test stimuli at the one-third octave band test frequencies.
>
> (Noise Control Act of 1972)

Many hearing protectors also include a chart with mean attenuation levels at various octave or one-third octave frequencies from 63 to 8,000 Hz, which provides more detail than a single NRR number.

The main categories of hearing protectors are foam earplugs, earmuffs, high-fidelity earplugs, and custom-molded ear plugs.

- **Foam earplugs or "foamies" (in-ear, disposable, generic):** Basic foam earplugs are cheap, widely available, and are highly effective when inserted correctly in the ear canals. Companies such as 3M, Honeywell Howard Leight, and Mack's all offer inexpensive but high-quality foam earplugs with an NRR of 30 to 33 dB. Foam earplugs provide excellent protection from loud sounds, especially in non-critical listening situations. The main tradeoff is that foam earplugs attenuate high frequencies more than low frequencies which means things sound muffled when wearing them. Therefore, people might be less likely to use them at concerts. It is also critical that they are inserted correctly. They need to be rolled and compressed with the fingers and then quickly inserted fully into the ears. Use one hand to pull the pinna (outer ear) up and back to make insertion easier. The CDC has a short video on how to wear foam earplugs.[7] Improperly inserted foam earplugs do not provide the proper amount of sound attenuation, which means a wearer could expose themselves unwittingly to high sound levels.
- **Earmuffs or ear defenders (over-the-ear, reusable, generic):** This kind of hearing protection fits over the ear like headphones, is reusable, and provides protection in very loud environments. 3M and Honeywell Howard Leight manufacture earmuffs with NRRs of 27 to 31 dB, but noise reduction rating varies by model. Both foam earplugs and earmuffs are great options for noisy environments and can be used together for maximum attenuation. When worn for long periods of time or in hot environments, foam earplugs may be more comfortable than earmuffs. In cold environments, earmuff cushions may stiffen and become ineffective. The advantage of earmuffs is that they are easier to remove and replace than foam earplugs if noise is intermittent. Some live sound engineers use earmuffs on the job anytime they are not actually mixing audio, such as during setup and strike, or between sets, to help reduce exposure to loud sound.
- **High-fidelity earplugs (in-ear, reusable, music-specific):** Also known as uniform-attenuation earplugs, high-fidelity earplugs are a better option for sound quality than foam earplugs because

they reduce sound levels approximately equally across all frequencies, which means they maintain clarity much better. Some excellent options in this category for live music, events, and travel include:

- CURVD Everyday Earplugs (NRR 15 dB),
- CURVD Live Earplugs (NRR 14 dB),
- Etymotic Research ER20XS High-Fidelity Earplugs (NRR 13 dB), and
- Etymotic Research ER20 High-Fidelity Earplugs (NRR 12 dB).

Etymotic Research is a highly respected company in hearing protection, earphones, and auditory research, but I find the CURVD earplugs to be much more comfortable than the Etymotic ones. Perhaps it's because my ear canals seem to be smaller than average. Other brands in this category include EarPeace Music PRO, which has three attenuation options: NRR 16, 13, and 9 dB, and Westone TRU Music with NRR 10 dB. Because high-fidelity earplugs sound clearer than foam earplugs, they are more likely to be used to protect hearing. Like foam earplugs, high-fidelity earplugs need to be inserted fully to ensure maximum protection from loud sound.

- **Custom-molded musicians' earplugs (in-ear, reusable, music-specific):** Custom-molded earplugs sound better than foam earplugs because they attenuate more evenly across all frequencies. They also usually fit better than high-fidelity earplugs, meaning they are more likely to be used. The downsides are that they are much more expensive than high-fidelity earplugs and require a visit to a hearing specialist such as an audiologist for fitting. Companies such as Etymotic Research, Sensaphonics, and Starkey offer custom-molded musicians' earplugs that come with an attenuating filter. If you are regularly exposed to very loud sounds, choose a filter with the most attenuation—typically 25 or 26 dB, depending on the brand.

As the American Speech-Language-Hearing Association (ASHA) website says: "Don't be fooled by thinking your ears are 'tough' or that you can 'tune it out'! Noise-induced hearing loss is usually gradual and painless but, unfortunately, permanent. Once destroyed, the hearing nerve and its sensory nerve cells do not repair" (ASHA 2026). If your ears continue to ring in the hours or days after attending a concert, you may have suffered some amount of permanent damage to the hair cells in your inner ear. If you have concerns about your hearing or just want to find out how your ears measure up, make an appointment with an audiologist, and get your hearing tested.

To practice safe listening in loud acoustic and amplified settings, Dr. Bruce Edwards, a retired University of Michigan audiologist, recommends that we:

- understand the risks of repeated exposure to high sound levels without protection,
- wear filtered ear plugs consistently,
- use active noise-canceling headphones,
- keep personal listening device volume ≤60% of maximum,
- take listening breaks,
- reduce exposure times, and
- separate from very loud sources.[8]

1.13 Summary

In this chapter we explored active listening and its importance in recording projects as well as everyday life. By defining technical ear training, we also identified some goals toward which we will work in the book and software practice modules. We finished by giving an overview of the main sound reproduction systems and principles of hearing conservation. In the next chapter, we will move on to ideas and exercises with a focus on equalization.

Notes

1 I have not received any compensation from Sennheiser or any other company to review these headphones. This recommendation is based on my own evaluation of the headphones.
2 "Mixing with Headphones—Andrew Scheps & Fab Dupont Share Expert Tips" on Puremix (August 27, 2021): www.youtube.com/watch?v=s4YuXNTCU2Y, starting at 1:55.
3 Apple Support page, "Use high-impedance headphones with your Mac": https://support.apple.com/en-us/108351.
4 The American Speech-Language-Hearing Association (ASHA) has more information about tinnitus and hyperacusis: www.asha.org/practice-portal/clinical-topics/tinnitus-and-hyperacusis.
5 Blue Sky, https://abluesky.com/support/blue-sky-calibration-test-files.
6 National Institute for Occupational Safety and Health, NIOSH Sound Level Meter App: www.cdc.gov/niosh/noise/about/app.html.
7 CDC, Roll, Pull, and Hold. How to Wear Soft Foam Earplugs, YouTube, https://m.youtube.com/watch?v=zM3R_1JceWo.
8 From unpublished workshop presentation slides which cite World Health Organization, Making listening safe: www.who.int/activities/making-listening-safe.

Bibliography

American Speech-Language-Hearing Association. 2026. "Loud Noise Dangers." www.asha.org/public/hearing/loud-noise-dangers.

ANSI. 1960/1994. *Psychoacoustic Terminology: Timbre.* New York, NY: American National Standards Institute.

Blesser, Barry, and Linda-Ruth Salter. 2007. *Spaces Speak, Are You Listening? Experiencing Aural Architecture*. The MIT Press.

Bradley, John S., and Gilbert A. Soulodre. 1995. "Objective Measures of Listener Envelopment." *Journal of the Acoustical Society of America* 98 (5): 2590–97.

Bregman, Albert S. 1990. *Auditory Scene Analysis: The Perceptual Organization of Sound.* The MIT Press.

Cartwright, Mark, and Bryan Pardo. 2013. "Social-EQ: Crowdsourcing and Equalization Descriptor Map." Paper presented at 14th International Society for Music Information Retrieval Conference, Curitiba, Brazil.

Celestina, Metod, Jan Hrovat, and Chucri A. Kardous. 2018. "Smartphone-Based Sound Level Measurement Apps: Evaluation of Compliance with International Sound Level Meter Standards." *Applied Acoustics* 139: 119–28. https://doi.org/10.1016/j.apacoust.2018.04.011.

Center for Hearing and Communication, dir. 2021. *CDC Audiologist John Eichwald's Healthy Hearing Tips.* YouTube.com. www.youtube.com/watch?v=v6DE-ml9Y8g.

Dozier, Emily. 2024. "How Loud Is Ford Field? Lions Fans Set New Record for Decibel Level during 2024 NFL Playoffs." *The Sporting News*, January 21. www.sportingnews.com/us/nfl/news/ford-field-decible-level-loud-record/8a48b33349cfff76c58eb225.

Environmental Protection Agency. 1974. *Information on Levels of Environmental Noise Requisite to Protect Public Health and Welfare with an Adequate Margin of Safety.* 550/9-74-004. Environmental Protection Agency Office of Noise Abatement and Control. www.nrc.gov/docs/ML1224/ML12241A393.pdf.

Geddes, Earl, and Lidia Lee. 2003. "Auditory Perception of Nonlinear Distortion: Theory." *115th Convention of the Audio Engineering Society*, Preprint 5890.

Gibson, Eleanor. 1969. *Principles of Perceptual Learning and Development.* Appleton-Century-Crofts.

International Telecommunication Union. 2022. *Recommendation ITU-R BS.775-4: Multichannel Stereophonic Sound System with and without Accompanying Picture*. ITU.

Kaplan, Rachel. 1993. "The Role of Nature in the Context of the Workplace." *Landscape and Urban Planning* 26: 193–201.

Kaplan, Rachel, and Stephen Kaplan. 1989. *The Experience of Nature: A Psychological Perspective*. Cambridge University Press. https://archive.org/details/experienceofnatu00kapl/mode/2up.

Kaplan, Stephen. 1995. "The Restorative Benefits of Nature: Toward an Integrative Framework." *Journal of Environmental Psychology* 15: 169–82.

Kardous, Chucri A., and Metod Celestina. 2017. "New NIOSH Sound Level Meter App." *NIOSH Science Blog*, January 17. www.cdc.gov/niosh/noise/about/app.html.

Kardous, Chucri A., and Peter B. Shaw. 2014. "So How Accurate Are These Smartphone Sound Measurement Apps?" *NIOSH Science Blog*, April 9. www.cdc.gov/niosh/bulletin/2014/sound-app.html.

Kerr, Stephen Eldon. 2025. "Our Headphone Sound Quality Tests: Raw Frequency Response." *RTINGS.Com*, June 3. www.rtings.com/headphones/tests/sound-quality/raw-frequency-response.

Korner, Anthony. 1986. "Aurora Musicalis." *Artforum*. www.artforum.com/features/aurora-musicalis-206915

Lee, Kate E., Kathryn J.H. Williams, Leisa D. Sargent, Nicholas S.G. Williams, and Katherine A. Johnson. 2015. "40-Second Green Roof Views Sustain Attention: The Role of Micro-Breaks in Attention Restoration." *Journal of Environmental Psychology* 42: 182–89.

Letowski, Tomasz. 1985. "Development of Technical Listening Skills: Timbre Solfeggio." *Journal of the Audio Engineering Society* 33: 240–44.

Letowski, Tomasz, and Andrzej Miskiewicz. 2013. "Timbre Solfege: A Course in Perceptual Analysis of Sound." In *Signal Processing in Sound Engineering*, edited by J.A. Adamczyk. Institute of Fundamental Technological Research, Polish Academy of Sciences.

Levitin, Daniel J. 2006. *This Is Your Brain on Music: The Science of a Human Obsession*. Dutton/Penguin Group.

Martin, Denis. 2019. "Dynamic Range Compressor Ear Training: Improving Critical Listening Skills Through Software-Assisted Practice." Doctoral Thesis, McGill University.

McAdams, Stephen, and Kai Siedenburg. 2019. "Perception and Cognition of Musical Timbre." In *Foundations in Music Psychology: Theory and Research*, edited by Peter Jason Rentfrow and Daniel J. Levitin. The MIT Press.

McAlexander, Tara P., Robyn R.M. Gershon, and Richard L. Neitzel. 2015. "Street-Level Noise in an Urban Setting: Assessment and Contribution to Personal Exposure." *Environmental Health* 14: 18. https://doi.org/10.1186/s12940-015-0006-y.

Moorefield, Virgil. 2005. *The Producer as Composer: Shaping the Sounds of Popular Music*. The MIT Press.

National Institute for Occupational Safety and Health. 2024. *Understand Noise Exposure*. www.cdc.gov/niosh/noise/prevent/understand.html.

Noise Control Act of 1972, Pub. L. Nos. 92-574, Title 40 Code of Federal Regulations (1972). www.ecfr.gov/current/title-40/part-211/subpart-B#p-211.203(q).

Olive, Sean. 1994. "A Method for Training Listeners and Selecting Program Material for Listening Tests." *97th Convention of the Audio Engineering Society*, Preprint 3893.

Olive, Sean. 2001. "A New Listener Training Software Application." *110th Convention of the Audio Engineering Society*, Convention Paper 5384.

Olive, Sean E. 2022. "The Perception and Measurement of Headphone Sound Quality: What Do Listeners Prefer?" *Acoustics Today*.

Pardo, Bryan, David Little, and Darren Gergle. 2012. "Building a Personalized Audio Equalizer Interface with Transfer Learning and Active Learning." Paper presented at Second International ACM Workshop on Music Information Retrieval with User-Centered and Multimodal Strategies, Nara, Japan. *MIRUM '12: Proceedings of the Second International ACM Workshop on Music Information Retrieval with User-Centered and Multimodal Strategies*.

PSW Staff. 2025. "HELA Announces Training & Certification Now Available." *ProSoundWeb* (Framingham, MA), March 3. www.prosoundweb.com/world-hearing-day-hela-announces-training-certification-now-available.

Quesnel, René. 2001. "A Computer-Assisted Method for Training and Researching Timbre Memory and Evaluation Skills." PhD Thesis, McGill University. https://escholarship.mcgill.ca/concern/theses/h702q794m.

Sabin, Andrew Todd, Zafar Rafii, and Bryan Pardo. 2011. "Weighting-Function–Based Rapid Mapping of Descriptors to Audio Processing Parameters." *Journal of the Audio Engineering Society* 59 (6): 419–30.

Saitis, Charalampos, and Stefan Weinzieri. 2019. "The Semantics of Timbre." In *Timbre: Acoustics, Perception, and Cognition*, edited by Kai Siedenburg, Charalampos Saitis, Stephen McAdams, Arthur N. Popper, and Richard R. Fay, vol. 69. Springer Handbook of Auditory Research. Springer Nature Switzerland AG. https://doi.org/10.1007/978-3-030-14832-4.

Schellenberg, E. Glenn, Paul Iverson, and Margaret C. Mckinnon. 1999. "Name That Tune: Identifying Popular Recordings from Brief Excerpts." *Psychonomic Bulletin & Review* 6: 641–46. https://doi.org/10.3758/BF03212973.

Senior, Mike. 2018. *Mixing Secrets for the Small Studio*. 2nd ed. Routledge.
Tingen, Paul. 2019. "Billie Eilish—Finneas O'Connell: Producing When We All Fall Asleep, Where Do We Go?" *Sound on Sound*, July. www.soundonsound.com/people/billie-eilish.
World Health Organization. 2022. *WHO Global Standard for Safe Listening Venues & Events*. World Health Organization. www.who.int/publications/i/item/9789240043114.
World Health Organization. 2026. Making listening safe. www.who.int/activities/making-listening-safe.
World Health Organization and International Telecommunication Union. 2019. *Safe Listening Devices and Systems: A WHO-ITU Standard*. World Health Organization and International Telecommunication Union. www.who.int/publications/i/item/9789241515276.

Chapter 2

Equalization and Tonal Balance

2.1 What Is Tonal Balance and How Do We Describe It?

"Tonal balance" is a term that describes the relative amplitudes of frequencies present in an audio signal. Also known as spectral balance, it is largely responsible for the perceived timbre of an audio signal. At a basic level, it is the relative balance of bass, middle, and treble frequencies, but as audio engineers, we want to be more specific and targeted in our approach to tonal balance. From single instruments and voices, to stems and full mixes, tonal balance at every level is a crucial aspect of audio projects, whether we are talking about recordings, loudspeaker design, sound for movies, or automotive audio systems. The overall tonal balance of an audio mix determines whether it sounds full or thin, even or uneven, warm or harsh, dark or bright, clear or muddy, and how effectively it allows an artist to communicate their art and emotion with audiences. Likewise, individual instruments each have their own tonal balances which also impact a mix. One of the keys to great-sounding audio is finding a tonal balance that is appropriate for the context in which we are working. While we may not want all frequencies at the same level in music recording, game audio, or film soundtracks, a flat frequency response from loudspeakers and headphones is an ideal that design engineers may work toward. Manufacturers of high-end loudspeakers, headphones, and car audio systems continue to tune their systems by ear to achieve an optimal tonal balance.

Our ability to hear and control fine details of tonal balance is crucial to becoming outstanding audio engineers. The good news is that we can develop these necessary critical listening skills through focused listening and practice with different types of effects processing. But it is also useful to expand our awareness of sound and to consider theoretical concepts that can guide us to discover sound quality details that may have eluded us before. There is no magic to the process. It takes consistent effort, but the work pays off.

Ideally, we should be able to hear if or how each audio tool, whether it is hardware or software, may alter the tonal balance or spectral content of our audio. If we know what is being added or removed to our audio signals by each device or plugin, we can make the most of our tools. The more familiar we are with our tools, the more we can draw on them for specific purposes in our projects since we know their capabilities and the ways in which they will affect our audio.

When collaborating with others, it's helpful to use a shared language when it comes to describing sound quality. In loudspeaker and headphone development, the process can move much more quickly if the team doing the listening tests can talk about tonal balance in terms of specific frequencies in hertz.

How do we begin to describe the tonal balance of a mix, a track, or a loudspeaker? If high frequencies are prominent, we might say that it is bright or harsh. If low frequencies dominate,

DOI: 10.4324/9781003461173-2

then maybe it sounds boomy or bassy. We can find plenty of lists of descriptors online with definitions. As we saw in Chapter 1, descriptors such as these can be imprecise, ambiguous, or have different meanings for different people. For example, when we say something sounds boomy, it usually means there is too much low-frequency content. But what specific range of low frequencies does the word "boomy" refer to? The term "low frequencies" might refer to a range of frequencies anywhere from 20 Hz to around 500 Hz or so, but the upper cutoff frequency depends on whom you ask, since there is no strict definition of low frequencies. Some people might say that low frequencies are those below about 250 Hz or so. Whatever the exact range, how do we know which frequency needs to be attenuated to reduce a subjective impression such as boominess? The context in which we are working, such as the musical genre, likely matters, but it is also the case that the word "boominess" may have different meanings for different people. My definition of boominess might correspond to a boost around 125 Hz, but your definition might correspond to 80 Hz, for example. The words we use to describe subjective impressions, such as "boomy" or "tinny," do not appear as parameters that we can control on equalizers or any other effects processors.

Some consumer sound systems offer basic tone controls such as bass and treble, and many guitar amps likewise include bass, middle, and treble controls. These labels seem intuitively understandable, but they lack precision. What specific frequency ranges are affected when we boost the treble or bass? It's usually difficult to know for sure, and frequency ranges can vary from one manufacturer to the next. Some guitar amps also have additional controls labeled as "contour" and "presence" which can affect tonal balance. Since it's not immediately apparent which part of the frequency spectrum these control, or if the specific frequencies affected are consistent from one amp model to another, a user unfamiliar with such parameters would need to change them and listen to the resulting difference in tone color to learn how they are affecting the sound. Guitarists get to know the tonal controls afforded by the amplifiers they use, but this knowledge is often gained through trial and error, and may not translate to other amps which may have different cut-off frequencies in their filters.

As audio engineers, we require more precise control than we can get with the basic bass, middle, and treble controls. We also need clear descriptors. Parametric equalizers are the natural choice for most of us and are normally included with digital audio workstations (DAWs). Graphic equalizers are also useful, especially for training, but they limit what we can control. Historically, telephone companies designed equalizers to compensate for high-frequency loss that occurred over telephone transmission lines. The goal was to selectively boost certain frequencies to equalize them with other frequencies—thus the term "equalization."

2.2 Parametric Equalizer: A Powerful Tool for Audio Engineers

Because equalizers are the primary tools for altering the spectral balance of sound, engineers often describe tonal balance using an equalizer's parameters:

- center frequency (in hertz),
- Q (no unit), or its inverse, bandwidth (in octaves),
- amount of boost or cut (in decibels).

By using equalization parameters to describe tonal balance, not only do the descriptors match equipment settings, but we also gain precision in language. We do not need to wonder how words like "boomy," "bright," or "contour" translate to an equalizer setting. Instead, we can use the unambiguous equalizer parameter settings themselves—center frequency in hertz, boost/cut gain in decibels, and Q or bandwidth—to describe sound quality. We need to know how these parameters affect the timbre of our audio, and one of the goals of this book, of course, is to help you develop a memory and ability to identify equalizer settings by ear.

Despite its straightforward simplicity, a parametric equalizer is an exceptionally powerful tool. In fact, I believe that a parametric equalizer, with fully independent and sweepable control of center frequency, gain, and Q (or bandwidth), is an audio engineer's most valuable device.

Given an equalizer's power to enhance or destroy a mix, how can we maximize its usefulness in our work? With the wide range of timbres we may want to correct or enhance, there are no hard and fast rules that apply to equalizer use across all situations. Each circumstance is unique, and many factors influence our use of equalization, including the timbres of individual instruments/voices or sound sources, the overall mix balance, the recording room acoustics, musicians' performances, musical genre, and the microphones used. Only after we can hear where problems exist can we apply appropriate corrective measures with an equalizer. That is why technical ear training is so important, so that we can quickly pinpoint problems and correct them.

If an equalizer is the best tool to correct tonal balance problems, how do we identify problems? What do we listen for? What exactly can we do with an equalizer and what should our goals be when using it? Does it make sense to aim for a flat tonal balance in a recording or mix?

Tonal balance is different from frequency response. Tonal balance is a characteristic of a signal or a mix, and it describes the relative balance of frequencies present in a sound. Frequency response, on the other hand, is a characteristic of a device. Many manufacturers measure and publish the frequency responses of their devices. Unless a device's frequency response is flat, it will impart its own equalization curve onto a signal in some constant way. On the other hand, a music signal's spectral content will vary according to the harmonies, melodies, transients, unpitched sounds, and dynamic levels of each instrument or voice. Some test signals, such as white noise, contain all audible frequencies of sound, with energy spread equally across the frequency spectrum. Even though white noise has every frequency represented equally, meaning that it is "flat" in terms of its spectrum, it doesn't sound flat. In fact, it sounds as though there is much more high-frequency content than low-frequency content. Pink noise, in contrast to white noise, does sound "flat" to the human ear. We look more closely at the characteristics of pink noise in Section 2.5.1. We might say a music signal's tonal balance is flat if its frequency range is represented appropriately, but what does "appropriately" mean? Does it mean that a musical recording should sound identical to the way it sounds when we hear a live performance? Does this apply equally to both amplified music and purely acoustic music? Is that possible or even desirable? Does it vary by genre? I have not seen any rules about tonal balance, but these questions can help guide your creative process in recording and mixing.

Classical music recording engineers often try to recreate an ideal live concert experience in their recordings. For other genres such as pop, hip-hop, or rock, recording engineers may want to recreate the feeling and energy of a live performance but without the negative aspects of stadium acoustics. A pair of audience microphones at a pop or rock stadium concert can capture the sound waves at that location in the space, but would that recording be as engaging as a recording made with close microphones and therefore less of the stadium acoustics? What about the sound quality of a concert video recorded by an audience member on their phone? The audio quality of a recording made in a stadium by an audience member can suffer from several possible deficiencies including clipping or overload distortion, excessive audience noise, spectral imbalances due to the phone microphone and location, and an overabundance of reflected and reverberant sound because the recording device is far from the stage. Sound from a phone video can be washed out and unfocused due to the strength of reflected sound relative to the direct sound. Audience microphones alone do not provide an engaging recording of an amplified concert in a stadium, even though the recording might be acoustically consistent with the sound heard at that location in the stands.

Professional engineers adjust equalization and spectral balance to best suit the situation they are working in. For instance, the kick drum in a jazz recording will undoubtedly have a different spectral balance than that of a heavy metal or hip-hop recording. Experienced recording engineers understand and can identify specific timbral differences for a given instrument across a range of recordings. (In Section 7.6, we will consider the variety of different snare drum timbres across a range of different recordings.) Well-developed listening skills help us determine the equalization or spectral balance for a given recording situation. When approaching a recording project, we should be familiar with existing recordings of comparable music, film, or game audio and know each project's timbral goals.

2.3 A Human RTA

Sometimes called a frequency analyzer or spectrum analyzer, a real-time analyzer (RTA) measures and displays an audio signal's frequencies and their respective amplitudes, and it updates the display as the audio signal changes. In other words, a real-time analyzer displays the tonal balance of an audio signal. Most equalizer plugins overlay a real-time analyzer display on the equalizer settings. Figure 2.1 shows a snapshot of the frequency content of a mix. Although it may be tempting to use a real-time analyzer to determine equalization settings for creative audio work, it is generally not as effective as just using our ears. Frequency spectrum measurements visualize and quantify audio signals which can supplement our subjective impressions, but they usually do not provide the right level of detail needed to be useful nor do they provide information about context or stylistic norms that experienced engineers know. An analyzer

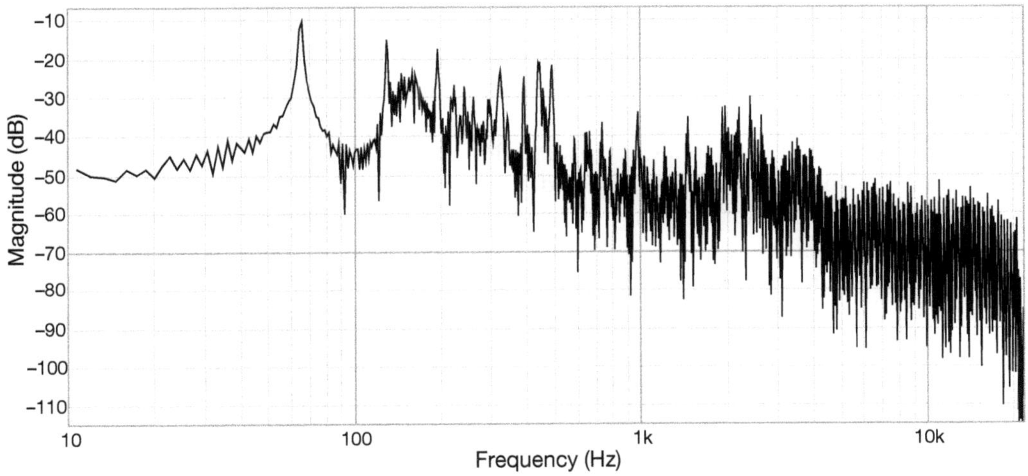

Figure 2.1 The frequency content of a mix at one moment in time. The lowest peak sits at approximately 65 Hz, which corresponds to a pitch of C_2, with subsequent peaks at higher harmonics.

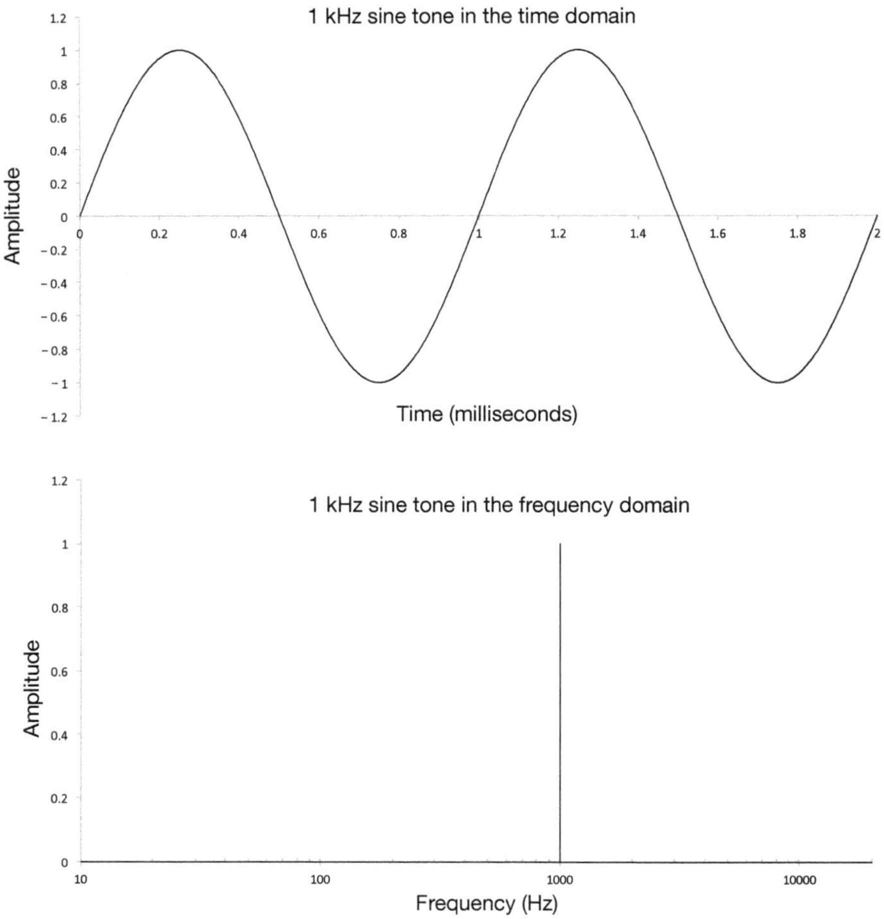

Figure 2.2 The time domain (top) and frequency domain (bottom) representations of a 1 kHz sine tone. Note the x-axis labels on the two plots.

measures the energy in each frequency band, often using a mathematical operation such as a fast Fourier transform (FFT). An FFT provides what is called a frequency domain representation of a signal, plotted with frequency on the x-axis and amplitude on the y-axis, as we see in equalizer plugins. This contrasts with the time domain or waveform representation of a signal that we see in the tracks of a DAW editor, which display tracks as time domain waveforms. Most equalizer plugins provide real-time frequency domain representations, such as Figure 2.1. Real-time analyzers provide continuously updated snapshots of an audio signal's frequency content. As an example, Figure 2.2 shows a time domain and frequency domain representation of a 1 kHz sine tone.

We might say that experienced audio professionals are human spectral analyzers because they identify and characterize tonal balance to a high degree of accuracy, information that they can then use to make mix and signal processing decisions. They monitor the spectral balance of individual microphone signals as well as the overall spectral balance of multiple, combined microphone signals at each stage in a recording project.

Audio engineers tend to rely on their ears rather than RTA power spectrum measurements for mix decisions. Instead of an RTA, they switch between different monitoring systems such as another set of loudspeakers and headphones, and they use familiar reference tracks to increase objectivity in the recording and mix process. RTA spectral measurements are largely ineffective for creative audio work for three reasons:

1. Music signals fluctuate in frequency and in amplitude, and these fluctuations make the frequency spectrum displays difficult to read and interpret with any degree of accuracy or subtlety.
2. The frequency spectrum of a short excerpt provides a static visual display that can be easily analyzed, but because it is a snapshot, the timeframe is too narrow to be useful as harmonics change over time with music.
3. If we average the spectrum over several minutes, fluctuations inherent in an audio signal are smoothed out so that the spectrum is easier to read, but the downside is that issues we might want to identify are obscured or averaged out.

Even if we make clearly audible changes to the spectrum with an equalizer, such changes are usually indiscernible to the eyes on a fluctuating spectral display. More subtle equalizer changes are completely impossible to see. Besides that, there is no objective reference for a recording's power spectrum, so it's not clear what it "should" look like. The moral of the story is that we should mix with our ears and not with our eyes.

Frequency spectrum visualizations are complicated a bit further by the tradeoff between time resolution and frequency resolution. Real-time analyzers use a fast Fourier transform or FFT to convert time-domain representations—that is, the waveforms we see in a DAW timeline—to frequency-domain representations that show the amplitudes of frequencies present in a signal. Most equalizer plugins visualize the frequency response curve they apply to an audio signal, which is the frequency domain representation since the goal is to adjust the levels of some frequencies but not all by the same amount. To convert from a time domain representation—the waveform—to a frequency domain representation—the frequency spectrum—real-time analyzers process in-coming audio in chunks or blocks of samples called windows. The size of a window is the number of samples of audio that gets processed at one time. Window size is often set to 512, 1024, 2048, or 4096 samples, but windows with smaller and larger numbers of samples exist as well. The spectrum analyzer function available with Sonic Visualiser[1] software, for example, allows users to select window sizes as low as 32 samples in length and as high as 32,768 samples in length. There is an inherent tradeoff between window size and the resulting frequency and time resolution. A smaller window—that is, fewer samples processed at a time—gives better time resolution meaning the display is updated more frequently, but the tradeoff is that smaller windows have lower frequency resolution. Longer windows give better frequency resolution but worse time resolution, since the process doesn't update as frequently. If the display updates more frequently with better time resolution, frequency resolution decreases and we see less detail across frequencies. If we increase the frequency resolution, the analyzer smears transients because the time resolution is reduced. It is because of these trade-offs that real-time analyzers are not very useful for subjective equalization decisions. So we must rely on our ears.

Live sound engineers, on the other hand, will often use real-time spectral analyzers to help them tune a sound system for live sound reinforcement. The difference in their use of an RTA is that they have a reference signal, usually pink noise or a recording. The analyzer compares the spectrum of the original audio signal, which is a known, objective reference, to the output of the loudspeakers. The goal in this kind of situation is a bit different from that of recording and mixing because a live sound engineer is adjusting the frequency response of a sound system so that the input reference signal and the system output signal spectral balances are as similar as possible.

2.4 Shaping Tonal Balance

Although an equalizer may be our primary means to control tonal balance, other tools and techniques are available which can also affect spectrum, including dynamic range compression (as we discuss in Chapter 4) and microphone selection and placement. There are also factors related to monitoring that influence our spectral balance choices, such as our choice of headphones and loudspeakers, and their placement in our room, and effects from listening room acoustics such as room modes and early reflections. In this section, we consider some ways that we can alter spectral balance directly and indirectly.

2.4.1 Monitoring Factors That Affect Spectral Balance

In addition to the specifics of recording, mixing, mastering, or other production techniques, we also need to be aware of factors that affect our work but that are easy to overlook. It may be tempting to think we can solve all our production issues with just the right plugin, but that's not the case, especially when it comes to monitoring and room acoustics. Because there is no direct connection between recorded audio signals and the auditory processing center of the brain (at least not yet), we need to keep in mind that audio signals are altered in the transmission path from our loudspeakers or headphones to our brains.

Three main factors influence our perception of the spectral balance of an audio signal in our studio control room:

- studio monitors: loudspeakers and headphones,
- room acoustics,
- sound levels.

Studio monitors are usually fixed for a studio space, until it's time to upgrade, of course, so we do not adjust their characteristics in a direct way as we do when we equalize a track, for example. We purchase a set of studio monitors or headphones, and we work in a room that has certain acoustical characteristics that remain constant across a range of projects until we change acoustical treatments or purchase a new set of loudspeakers or headphones. Without accuracy in our monitoring system—that is, the loudspeakers and room acoustics as a whole—we cannot know whether our equalizer and other processing choices are valid or not. Information about room acoustics treatments to correct for room modes and reflections is beyond the scope of this book, but we want to be aware of some issues that can affect tonal balance choices we make.

Figure 2.3 illustrates the path of an audio signal from electrical to acoustical energy, highlighting three of the main modifiers of spectral balance that affect every choice we make in spectral balance processing.

2.4.1.1 Studio Monitors and Loudspeakers

Studio monitors and loudspeakers are like windows through which we perceive and make decisions about audio signals, but they are not perfect and, as Floyd Toole has pointed out, they are part of what he calls the *circle of confusion* (Toole 2026). Recordings are made with microphones,

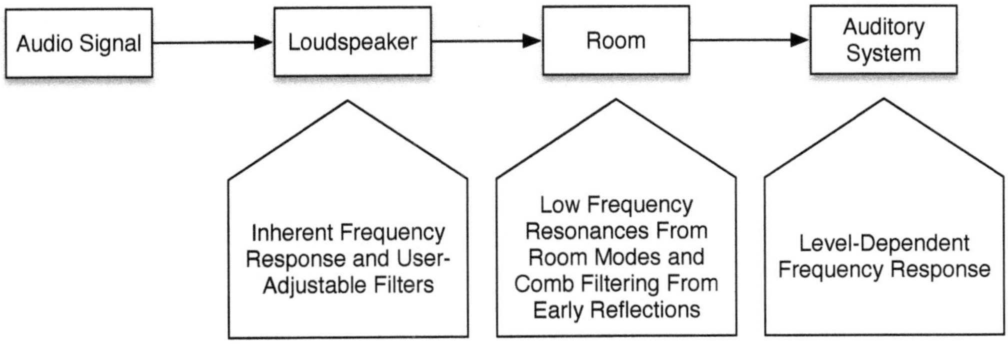

Figure 2.3 The signal path showing the transmission of an audio signal as an electrical signal to a loudspeaker where it is converted to an acoustic signal, modified by a listening room, and finally received by the ear and processed by the auditory system. Each stage highlights factors that influence a signal's spectral balance—both physical (loudspeaker and room) and perceptual (auditory system)—through the path.

equalization, and other effects, and these recordings are evaluated using monitor loudspeakers that are evaluated using recordings made using microphones . . . and so on, back around through the circle. Toole also points out that recording engineers mix on their monitors in their control rooms, which the average listener does not have access to. Listeners are always going to hear recordings differently than the way engineers originally heard them because everyone has a different playback system.

Each type and model of monitor and loudspeaker offers a unique frequency response, and because we rely on monitors to judge spectral balance, the monitors we use affect these judgments. If we monitor through loudspeakers with a weak low-frequency response, we may boost the low frequencies in the recorded audio signal to compensate.

A common way to add some objectivity to the mixing process is to audition on a second or possibly third set of speakers and headphones. That's why pictures of studio control rooms often show at least two sets of loudspeakers in the room. Experienced engineers check their mixes on several different sets of loudspeakers and headphones to hear and be able to correct any issues with spectral balance, reverberation, or dynamic range that may be more audible on one set of monitors than another. Each loudspeaker model gives a slightly different impression of a mix, and by listening to a variety of monitors, we can find the best compromise. Even inexpensive, mono loudspeakers are useful as alternatives to an expensive set of main studio loudspeakers, because they can help us hear what the average listener might hear and therefore help us make mix decisions. Each monitoring system tells us something different about sound quality and mix balance. One loudspeaker model may give the impression that the reverberation is mixed too loud. Another may sound as if there is not enough bass in a mix. We search for a compromise so that recordings will sound decently good when played back on a range of different loudspeakers and headphones, rather than sounding perfectly good only on the recording studio's monitors and quite bad on other systems. Engineers often say that a mix *translates* well if it remains relatively consistent across various types and sizes of loudspeakers. One mark of a well-made recording is that it will translate well on a wide range of sound reproduction systems, from mini systems to large-scale loudspeaker systems.

Beyond the inherent frequency response of a loudspeaker, almost all active loudspeakers include built-in user-adjustable filters—such as high- and low-frequency shelving filters—that can compensate for effects such as low-frequency buildup when monitors are placed close to a wall. Decisions that an engineer makes about spectral balance are influenced by a loudspeaker's inherent frequency response as well as by user-adjustable filtering settings on the loudspeaker. Engineers mix to what they hear, so changing the frequency response of a loudspeaker or switching to a different loudspeaker or to headphones can impact mix decisions.

Real-time analyzers and frequency response measurement tools can provide some indication of the frequency response of a loudspeaker within a room. Unless frequency response is being measured in an anechoic chamber, the measured response results from a combination of both the loudspeaker, the way it radiates sound in multiple directions, not just on axis, and any room resonances and reflections. Although some loudspeaker models offer filters for manual or automated frequency response correction, an equalizer cannot fix loudspeaker room frequency response problems resulting from room modes and reflections. As we discuss in the following section, these acoustic conditions occur because two or more waves arrive at different times and either cancel or add. Frequency response notches created by room modes or comb filtering cannot be brought back to flat with an equalizer. At frequency notches due to comb filtering, for example, there is little or no audio signal to boost because destructive interference has removed energy at those frequencies. Equalization is only effective if the frequencies we want to alter with an equalizer are present. If no signal exists at a certain frequency, there is nothing to boost.

Another reason equalization is ineffective in correcting room responses is that each point in a room will have a different frequency response. If we correct for one location, another location nearby could be made worse. As we will discuss below, frequency resonances in a room are prominent in some locations and less so in others. Equalizers are also ineffective at reducing the effects of resonances such as room modes.

A further complication is that identical frequency response curves can have different *time responses*. So even if we correct for frequency response anomalies, there may still be time response differences. Let's look at an example of this situation. To illustrate, we will look at the impulse response, in the time domain, and frequency response, in the frequency domain, of a DAW mixer channel at unity gain with no processing applied. An *impulse* is simply a very short sound at maximum amplitude. It sounds like a click. An *impulse response* is a recording of the impulse after it has propagated through a device or room, and it is a useful way to show how a device alters or does not alter a signal. In this case, the impulse response shows a perfect impulse, as short as possible and at maximum amplitude, as shown in Figure 2.4. The associated frequency response is flat, as we would expect from a perfect impulse response. As we will see below, it is possible to have a precisely flat frequency response even if the impulse is not perfect.

To illustrate that the time response can change without altering the frequency response, let's consider the effect an all-pass filter can have on an audio signal. Without getting too technical, we can simply say that an all-pass filter passes all frequencies equally—thus the name—but it does alter the phase. Sometimes called a phase rotator, an all-pass filter can be used to reduce the level of peak transients because it spreads a signal's amplitude over a longer period. Since an all-pass filter does not alter the frequency response, the effect of a single all-pass filter can be inaudible depending on how it is used. Figure 2.5 shows the impulse response of REAPER's ReaEQ[2] all-pass filter. Note that the frequency response of the filter is completely flat, consistent with Figure 2.4, but the time responses in Figures 2.4 and 2.5 are different.

We can alter the frequency response by mixing a signal with an all-pass filtered version of itself. Figure 2.6 shows the frequency response (middle) and phase response (bottom) of the REAPER ReaEQ all-pass filter center frequency (fc) set to 1 kHz and mix set to 50%. As we can see from the frequency response plot in Figure 2.6, there is a deep notch at 1 kHz. The notch results because

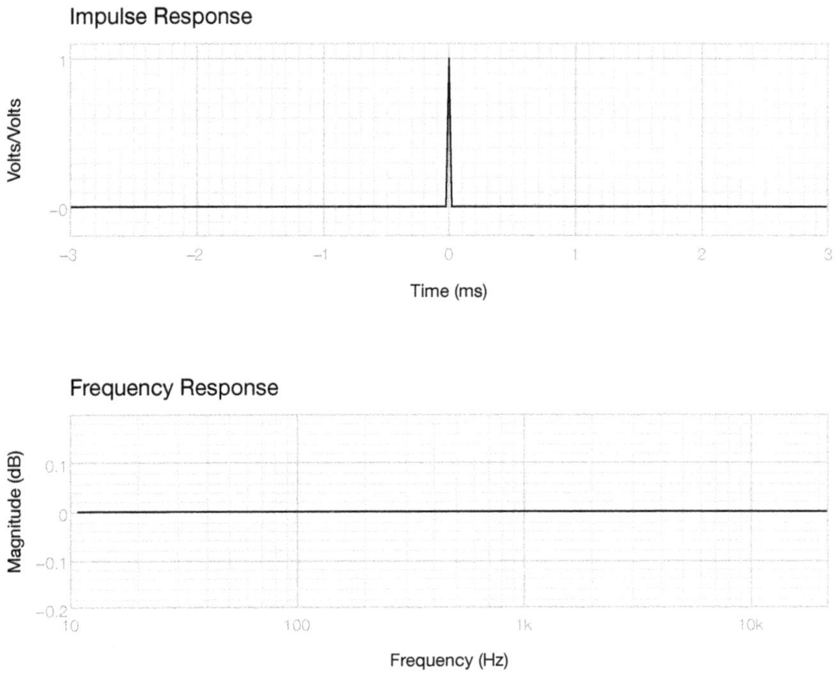

Figure 2.4 The top plot is an impulse response, in the time domain, of a single channel in the Cockos REAPER DAW at unity gain (0 dB), sampling rate 44,100 Hz, with no effects or plugins. The bottom plot is the corresponding frequency response which is perfectly flat. The impulse response appears as a perfect impulse at 0 milliseconds which is consistent with what we would expect from an audio path without any plugins or other processing. We can confirm that the impulse response contains all frequencies at equal levels from 20 Hz to 20,000 Hz, as shown in the flat frequency response in the bottom frequency domain plot.

the all-pass filtered version has a frequency-dependent phase shift with a maximum shift of 180° at 1 kHz and then a gradual decreasing phase shift above and below 1 kHz. When the original signal, with 0° phase shift at 1 kHz, mixes with an all-pass filtered version, with 180° phase shift at 1 kHz, the result is complete cancellation at that frequency. Further away from the center frequency, less cancellation occurs because the phase shift gets closer to 0° as we move further away from 1 kHz.

All-pass filters are widely used as components of parametric digital reverberation algorithms, in contrast to convolution reverberation which uses impulse responses of real spaces, to simulate the natural decay of sound in a room, which we look at more closely in Chapter 3. We can also find a net 180° phase difference between two filters in a loudspeaker crossover. A crossover is simply the combination of two filters: a low-pass filter for the woofer and a high-pass filter for the tweeter set to the same cut-off frequency so that the driver outputs combine acoustically to give a flat frequency response. Depending on the filter order and type, the phase at the crossover frequency could be shifted –90° for the woofer's low-pass filter and +90° for the tweeter's high-pass filter. The resulting phase difference between the two filters at the crossover frequency is 180° (i.e., 90° – (–90°) = 180°). The phaser guitar pedal effect also uses multiple all-pass filters mixed with the dry signal to create notch filters which sweep up and down the spectrum over time.

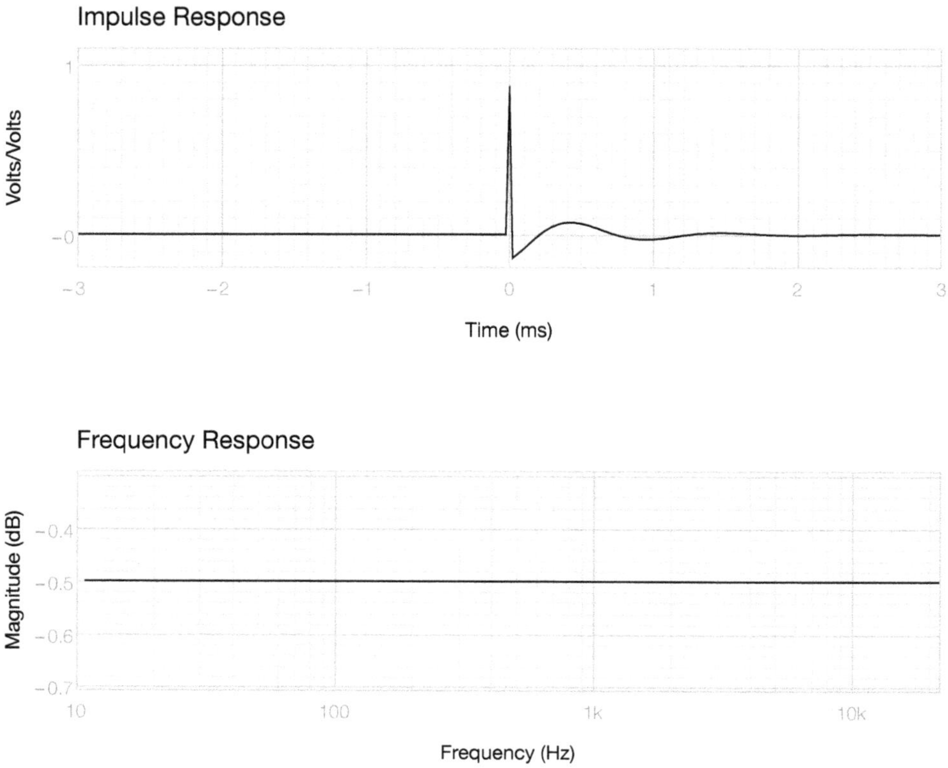

Figure 2.5 The top plot is an impulse response of the REAPER ReaEQ all-pass filter set to 1 kHz, and the bottom plot is the corresponding frequency response which is perfectly flat. In the top plot, note the ripple just after 0 milliseconds, which shows the effect of the all-pass filter in the time domain. The frequency response plot is the same as it is in Figure 2.4 even though the time response is different. The overall magnitude of the frequency response is slightly less than unity gain (it's at –0.5 dB) to avoid clipping.

2.4.1.2 Control Room and Listening Room Acoustics

Our control room dimensions, volume, and surface treatments also have a direct effect on the audio we hear. International professional societies and standards organizations, such as the International Telecommunications Union (ITU) headquartered in Switzerland, have published recommendations on listening room acoustics and characteristics. One of their recommendations, ITU-R BS.1116, defines several physical and acoustical parameters that can be applied to listening rooms such that listening tests may be somewhat comparable from one room to another (ITU-R 2015). It may be intuitive to think that an anechoic room, free of room modes, resonances, and reflections, would be ideal for listening because the room will essentially be acoustically "invisible." But a reflection-free room does not represent the acoustic conditions of a typical room so it's not appropriate for a mixing and mastering room.

Sound originating from loudspeakers propagates into a room, reflects off objects, walls, ceiling, and floor, and combines with the sound propagating directly from the loudspeaker to the listener. Sound radiates mainly from the front of a loudspeaker especially for high frequencies, but

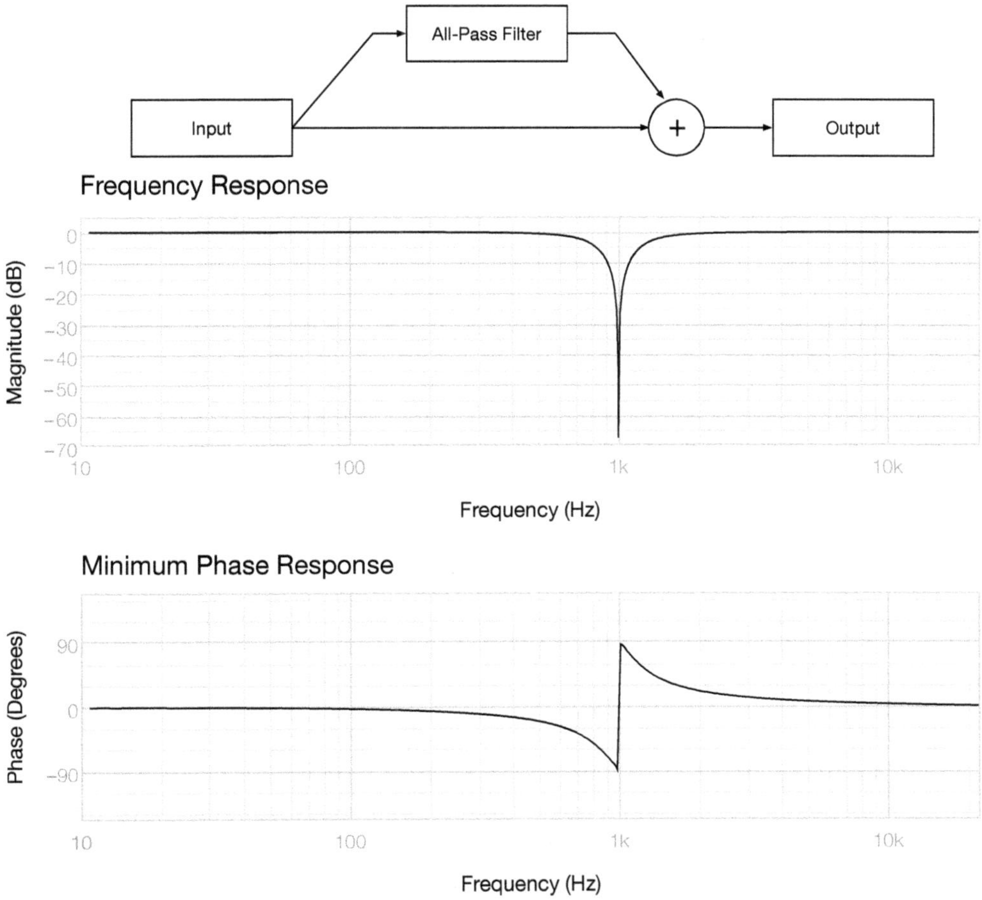

Figure 2.6 The block diagram at the top shows the signal flow in which an input signal is sent to an all-pass filter and the output of the all-pass filter is mixed with the unfiltered signal and sent to the output. The frequency response, in the middle of the figure, shows a deep notch at the center frequency of the filter. The phase response, at the bottom of the figure, shows a maximum phase shift at 1 kHz of 180° which produces the notch in the frequency response when the filtered and unfiltered signals are mixed at equal levels.

most loudspeakers become more omnidirectional at low frequencies. The primarily low-frequency sound that is radiated from the back and sides of a loudspeaker is reflected into the listening position by nearby walls and large surfaces. Walls with significant mass, such as concrete or multiple layers of drywall, reflect low frequencies better than walls made from a single layer of drywall.

Regardless of the environment in which we are listening to reproduced sound, we hear not only the sound directly from loudspeakers but also the effects that the room has on the sound. We can think of both loudspeakers and listening environments as filters that alter the sound we hear.

Room modes are one effect that a room can create. Modes or standing waves depend on a room's dimensions, and they influence the spectral balance of what we hear from loudspeakers in a room. Room modes are mostly problematic in the low-frequency range, typically below around

300 Hz. Fundamental resonant frequencies that occur in one dimension (called axial modes) have wavelengths that are two times the distance between parallel walls. For example, if our room is 3 meters (10 feet) wide and we assume sound travels at 343 meters/second (1,125 feet/second) at 20°C (68°F), there will be a room mode produced between the side walls at 56.25 Hz, with harmonics at 112.5 Hz and 168.75 Hz. The formula for calculating the fundamental room mode for a single dimension between parallel walls is:

$$f_1 = \frac{c}{2 \times L} \tag{2.1}$$

where f_1 is the fundamental frequency or first harmonic, c is the speed of sound, and L is the distance between parallel walls.

Axial or single dimension room modes are produced between any two opposing wall surfaces, that is, side to side, front/back, and floor/ceiling. Room modes also resonate in two dimensions, called tangential modes, and three dimensions, called oblique modes, within a room. Many room mode calculators exist online.[3] Although splaying or angling walls may help redirect early reflections, it does not reduce room modes. Instead, the resonant frequencies are based on the average distance between opposing walls.

Room modes are problematic not only because they amplify some frequencies and not others, but also because the magnitude of their effect varies with location in a room. Because the amplitudes of room resonances vary according to location, we need to be aware that the listening position may have a standing wave node at a particular frequency. A node is a point in a room where cancellation occurs for a given frequency due to destructive interference in the standing wave. Because the effects of room modes depend on the location within a room, it is important to listen at different locations within a room. Different frequencies will have nodes in different locations. If we are not aware of nodes in our mixing room and if we do not move around within the room, we may inadvertently boost a missing frequency with an equalizer, only to realize later when listening somewhere else in the room that the boost is too much and that we were trying to compensate for a room mode, rather than for a deficiency in the mix.

To get a new perspective on the balance in a mix, we can turn up the playback level, walk out of the room to an adjacent room while keeping the adjoining door open, and listen from the adjacent room. By taking a moment to listen from outside the mix room, we might hear balance problems that are not apparent when we listen directly in front of our monitors. Mixing is about compromises, and finding the best compromise means we want to hear our mixes under different and less-than-ideal listening conditions, not just from the sweet spot in an acoustically treated control room. We are probably not going to make subtle mix decisions listening at a distance, but this exercise can be useful when focusing on balances of vocals, lead instruments, and low end or bass especially. It is also useful to identify if any balances seem to change when we listen indirectly from outside the mixing room.

2.4.1.3 Sound Levels and Tonal Balance

Our perception of tonal balance also depends on listening levels—that is, how loud we listen. Fletcher and Munson (1933) published results of their study of human loudness perception of frequencies across the audible range at different listening levels. These equal-loudness contours show that not only does the human hearing system have wide variations in frequency response, but also its frequency response changes with loudness. In general, we are least sensitive to frequencies at the high and low ends of the spectrum, and most sensitive to mid-spectrum frequencies from about 1,000 to 4,000 Hz due to the natural resonance of the ear canal which amplifies sound in this frequency range. At higher sound levels, we become more sensitive to low and high frequencies relative to mid-frequencies. In essence, the frequency response of our ears becomes flatter as sound level increases. I hasten to add that this is not a reason to listen to high sound levels. As we saw in Chapter 1, consistent exposure to high sound levels above even 80 dBA sound pressure level (SPL) can cause permanent hearing loss.

What practical implication does this have for recording and mixing? Let's say we are monitoring at a high sound level, such as 95 dBA average SPL, and then we turn the level down to 50 dBA SPL, for example. At the lower listening level, the mix will sound as though it has much less bass at the quieter level. The spectral balance of the mix doesn't change when we change the monitoring level, but our ears respond differently to different listening levels.

Knowing that this effect occurs, professional mix engineers check their mixes at different monitoring levels to find the best tonal balance compromise. They know that listening loud makes low frequencies more audible. Furthermore, even very small differences in listening level can make differences in sound quality attributes of a mix. Since there is no standard listening level for music, we do not know how loud listeners will hear our recordings. Movie soundtracks for theatrical release are a little different in this respect since they may be mixed on systems calibrated to a reference sound level. For music, it is helpful to compare our mixes to commercial recordings to judge tonal balances at different listening levels.

2.4.2 *Equalization*

Equalizers can be used as both a correction tool to reduce particularly strong resonant frequencies and an enhancement tool to bring out or highlight characteristics of an instrument or mix. Resonances can mask other frequencies and prevent us from hearing the truest sound of an instrument, and equalizers can help mitigate that kind of issue. There is a significant amount of art in the use of equalization, whether for tuning a loudspeaker system or shaping a mix, and we rely on what we hear to make decisions about the application of equalization. The precise choice of frequency, gain, and Q is critical to the successful use of equalization, and for recording, mixing, and mastering the ear is the final judge of the appropriateness of an equalizer setting.

There are several different types of equalizers and filters including high-pass filters, low-pass filters, band-pass filters, graphic equalizers, and parametric equalizers, which provide various levels of control over spectral balance. Filters and equalizers have different functions. Filters can remove a range or band of frequencies, above or below a defined cut-off frequency. Equalizers, on the other hand, offer the ability to boost or cut selected frequencies by various amounts and with a range of shapes or Q.

2.4.2.1 *Filters: Low-Pass and High-Pass*

High-pass and low-pass filters remove frequencies below or above a defined cut-off frequency. Usually, the only adjustable parameter is the cut-off frequency, although some models do offer the ability to control the filter's slope—that is, how steeply the output drops off beyond the cut-off frequency. Figures 2.7 and 2.8 show frequency response curves for low-pass and high-pass filters, respectively. In practice, high-pass filters are generally used more often than low-pass filters. High-pass filters can be used to remove low-frequency rumble that may have been inadvertently

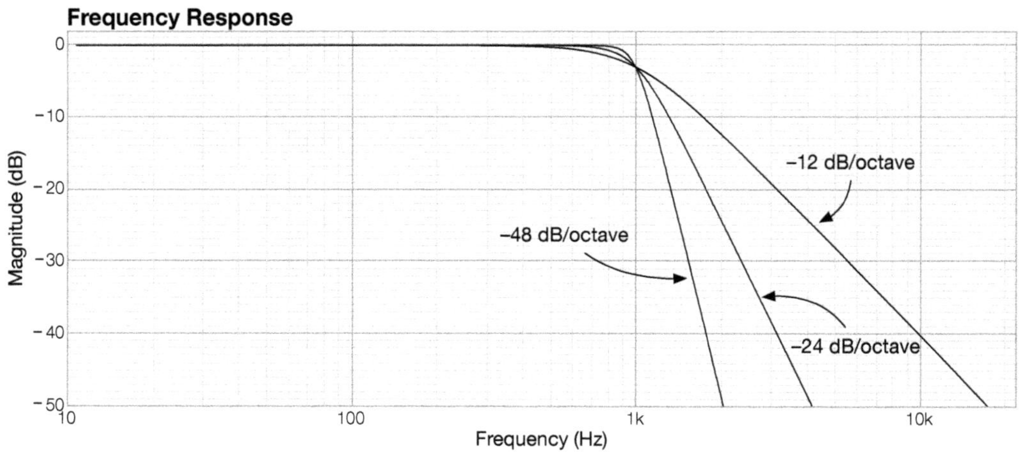

Figure 2.7 The frequency response of a low-pass filter set to 1 kHz at three different slopes: –12 dB/octave, –24 dB/octave, and –48 dB/octave.

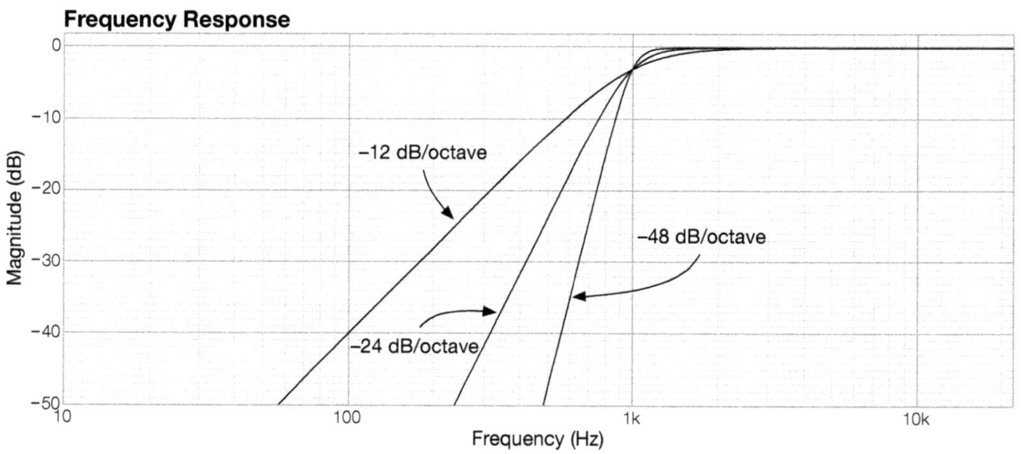

Figure 2.8 The frequency response of a high-pass filter set to 1,000 Hz at three different slopes: –12 dB/octave, –24 dB/octave, and –48 dB/octave.

picked up during a recording, but care is needed to make sure its cut-off frequency is set below the music signal's lowest frequency.

2.4.2.2 *Graphic Equalizers*

Graphic equalizers allow control of the amount of boost or cut for a predetermined set of frequencies. Usually, they have vertical sliders to control the amount of boost or cut for each frequency. The frequencies available for adjustment are typically based on the International Standards

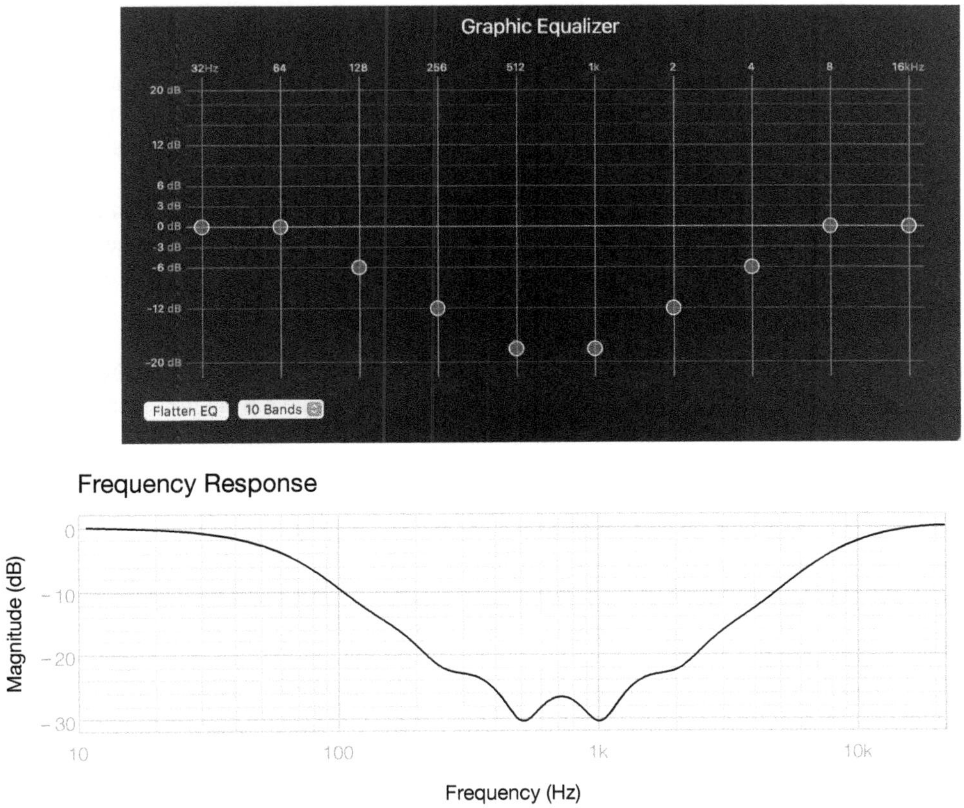

Figure 2.9 A screenshot (top) of the Apple AUGraphicEQ plugin with cuts to six frequencies, 128 Hz, 256 Hz, 512 Hz, 1 kHz, 2 kHz, and 4 kHz. Cuts are greatest for the two center frequencies (512 Hz and 1 kHz). The measured frequency response of this equalizer setting is shown at the bottom.

Organization (ISO) center frequencies, such as the octave frequencies 31.5 Hz, 63 Hz, 125 Hz, 250 Hz, 500 Hz, 1,000 Hz, 2,000 Hz, 4,000 Hz, 8,000 Hz, and 16,000 Hz. Some graphic equalizers have more frequency bands available, such as one-third-octave or one-twelfth-octave frequencies. A graphic equalizer's designer usually predetermines the Q, meaning that the user cannot adjust it. Some models, such as the API 560 Graphic EQ, have proportional Q, which means the Q is wider for small boosts/cuts and narrower for large boosts/cuts.

The graphic equalizer gets its name from the fact that the vertical sliders form the shape of the equalization curve from low frequencies on the left to high frequencies on the right, forming a graph of the frequency response. But be aware that the measured frequency response curve of a graphic equalizer does not precisely match the frequency response implied by the curve of sliders on the interface. For example, Figure 2.9 shows a graphic equalizer setting and the measured frequency response. Note that the maximum cut on the equalizer is –18 dB at 512 Hz and 1 kHz, but the measured frequency response shows –30 dB cuts at those frequencies. There is also a ripple in the frequency response in the range of the deepest cut, but this is not apparent from the equalizer interface.

Graphic equalizer advantages:

- Many possible frequencies or bands can be adjusted at once with a single hardware unit or plugin; for example, an octave band EQ has ten adjustable frequencies (32 to 16,000 Hz). A parametric equalizer usually provides the possibility to adjust only four to six frequencies, depending on the model.
- It's faster to reach for a single fader to bring down an unwanted resonance compared to parametric equalizers which require both frequency selection and boost/cut adjustment.

Graphic equalizer disadvantages:

- The frequencies are preset and cannot be changed. Only those frequencies can be adjusted.
- Q or bandwidth is usually not adjustable, and it is often not known what the actual Q values are.

2.4.2.3 Parametric Equalizers

Grammy Award–winning engineer, producer, and equipment designer George Massenburg is said to have coined the term "parametric equalizer" in his 1972 Audio Engineering Society convention paper (Massenburg 1972). A parametric equalizer allows completely independent and sweep-tunable control of center frequency (fc), gain (amount of boost or cut at that frequency), and Q or bandwidth.

Q is inversely related to the bandwidth of the boost or cut and is calculated with the following formula:

$$Q = \frac{center\ frequency\ (Hz)}{bandwidth\ (Hz)} \tag{2.2}$$

Some equalizers specify bandwidth in octaves instead of Q. To convert from bandwidth in octaves to Q, the following formula is used, where N is the bandwidth in octaves:

$$Q = \frac{\sqrt{2^N}}{2^N - 1} \tag{2.3}$$

Bandwidth is simply the difference between two frequencies, a higher frequency minus a lower frequency: f2 – f1. But it turns out that there are different ways of defining the two frequencies, and thus there are different definitions of Q or bandwidth. The classic definition of bandwidth used for most analog equalizers defines the two frequencies, f1 and f2, as the points where the frequency response is 3 dB down from the maximum boost at the center frequency (fc) or 3 dB up from the maximum cut at the center frequency (fc).

Another definition of Q, proposed by digital signal processing engineer Robert Bristow-Johnson (1994), defines the two frequencies, f1 and f2, as the points where the frequency response is the "midpoint dB gain" for any amount of boost or cut.

Figure 2.10 shows two frequency response curves of the REAPER ReaEQ parametric equalizer plugin. For both measurements, the equalizer was set to a boost of 12 dB at 1,000 Hz and a bandwidth of 0.714 octaves, which is equivalent to Q = 2.0. The top one is wider than the bottom one despite identical parameter settings. The difference between the two lies in the ReaEQ "Type" parameter setting. In the top plot, ReaEQ Type was set to "Band alt 2," and in the bottom plot, ReaEQ

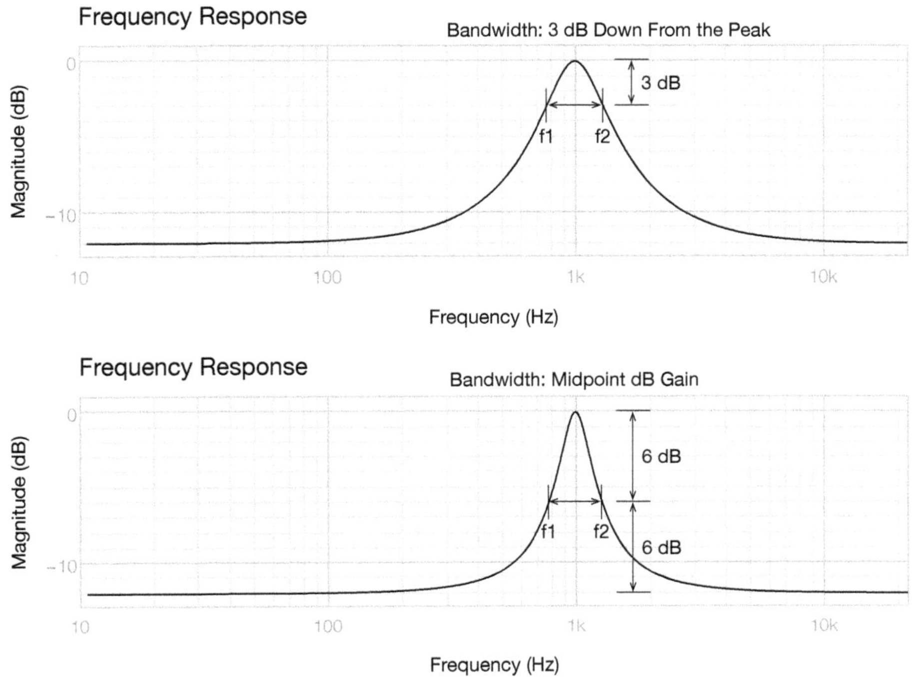

Figure 2.10 Two frequency response curves of the REAPER ReaEQ parametric equalizer plugin. For both measurements, the equalizer was set to a frequency of 1,000 Hz, gain of +12 dB, and a bandwidth of 0.714 octaves, which is equivalent to Q = 2.0. The top plot shows a wider curve than the bottom one despite the two equalizers being set to the same frequency, gain, and bandwidth. The difference lies in the ReaEQ equalizer "Type" setting. The top one is set to "Band alt 2" and the bottom one is set to "Band." The peak is at 0 dB because overall gain was reduced by 12 dB to avoid clipping in the range of boosted frequencies. From these plots, it is apparent that ReaEQ "Band alt 2" Type (top plot) uses the classic definition of bandwidth—that is, 3 dB down from the peak—but ReaEQ Band Type (bottom plot) uses the "midpoint dB gain" definition of bandwidth.

Type was set to "Band." From these measurements, we can determine that when Type is set to Band alt 2, as in Figure 2.10, top plot, the equalizer uses the classic definition of bandwidth—that is, 3 dB down from the peak. When Type is set to Band, as in Figure 2.10, bottom plot, the equalizer uses the midpoint dB gain definition of bandwidth. The bandwidth is the difference between f1 which is 781 Hz and f2 which is 1,281 Hz, giving a bandwidth of 500 Hz and a Q of 2.0. Because the bandwidth is defined differently in the two plots, the top curve is wider than the bottom curve even though both are set to the same bandwidth.

Equalizers can also have symmetrical or asymmetrical boost and cut curves. Figure 2.11 shows a symmetrical boost and cut overlaid on one plot. If we add these two curves or pass the audio through boost followed by the cut, or vice versa, there is no change in the spectral balance, and the resulting frequency response would be flat. In other words, we can undo a boost with a cut by the same amount at the same center frequency, if the equalizer has symmetrical boosts and cuts.

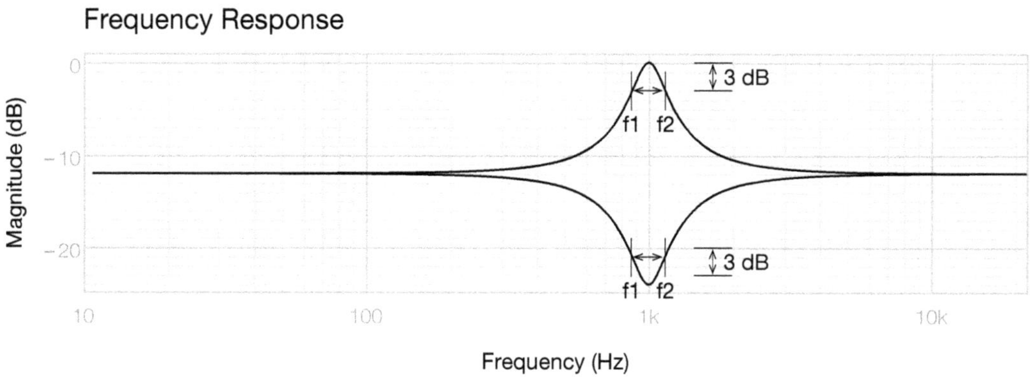

Figure 2.11 Frequency responses of the REAPER ReaEQ parametric equalizer plugin with symmetrical boost/cut. The plot shows two separate measurements, one with gain of +12 dB and the other with gain of –12 dB. For both settings, the frequency was 1,000 Hz, bandwidth 0.714 octaves, which is equivalent to Q = 2.0, and Band Type. Bandwidth is defined by the points in the curve 3 dB below the maximum for a boost, or 3 dB above the minimum for a cut.

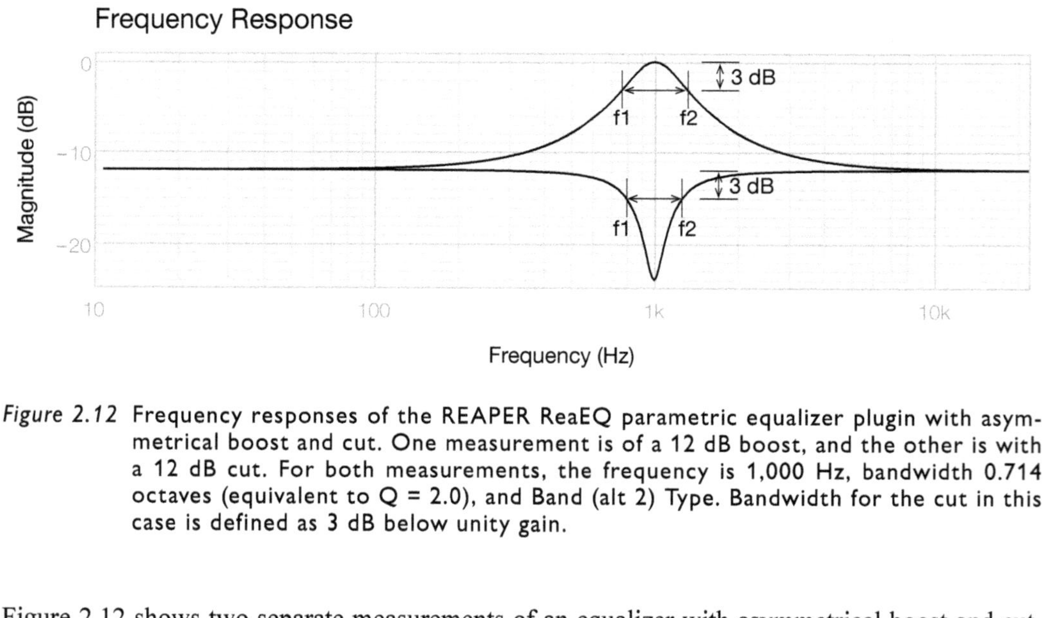

Figure 2.12 Frequency responses of the REAPER ReaEQ parametric equalizer plugin with asymmetrical boost and cut. One measurement is of a 12 dB boost, and the other is with a 12 dB cut. For both measurements, the frequency is 1,000 Hz, bandwidth 0.714 octaves (equivalent to Q = 2.0), and Band (alt 2) Type. Bandwidth for the cut in this case is defined as 3 dB below unity gain.

Figure 2.12 shows two separate measurements of an equalizer with asymmetrical boost and cut. One measurement shows a 12 dB boost, and the other shows a 12 dB cut. These curves are from the same ReaEQ equalizer plugin, measurements of which are shown in Figure 2.11, except that the equalizer Type was changed to Band (alt 2). Because the cut is narrow and deep at the center frequency, the bandwidth is determined by the –3 dB points below unity gain.

On an asymmetrical boost/cut equalizer, if we apply a boost and cut on the same audio signal at the same center frequency and bandwidth, we do not get a flat frequency response. In other words, we cannot undo a boost with a cut or vice versa as we can with a symmetrical boost/cut equalizer.

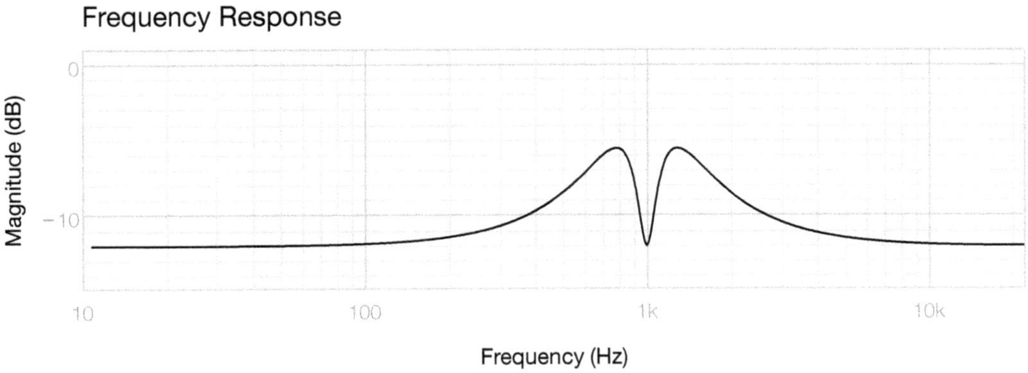

Figure 2.13 Frequency response of the REAPER ReaEQ parametric equalizer plugin set to the same parameters as shown in Figure 2.12: frequency 1,000 Hz, a simultaneous boost of +12 dB and cut of −12 dB, bandwidth 0.714 octaves (equivalent to Q = 2.0), and Band (alt 2) Type. The difference between this measurement and Figure 2.12 is that for this measurement, both the boost and cut were applied at the same time. Because the boost and cut are asymmetrical, the boost and cut do not cancel out, and the resulting frequency response is not flat.

Figure 2.13 shows the frequency response of the ReaEQ equalizer using the same settings as Figure 2.12, except that both the boost and the cut are engaged simultaneously giving a summed asymmetric boost and cut.

Among equalizer designs, some analog and digital models constrain parameter selection while others offer more flexibility. The most common constraint is the use of stepped parameter controls, which is usually found in analog models or their digital emulations. Models with stepped controls allow only specific frequencies and gain levels to be selected. Some equalizer models also do not allow independent control of bandwidth, so these are not technically true parametric equalizers.

Equalizers with fully sweepable controls offer the most control since parameters are continuously variable. Sweepable controls allow a user to select any center frequency, gain, and Q or bandwidth. There is no technical reason why digital equalizer plugins cannot be fully sweepable, so I am puzzled why plugin designers choose to limit frequency selection options in any digital equalizer plugin. An equalizer with sweepable and fully independent control of center frequency from 20 to 20,000 Hz, gain, and bandwidth, like those included for free with a DAW, can replicate the frequency response curves of any equalizer with stepped controls. Why not choose an equalizer that offers the most control? If we want to tame a resonance at 1,200 Hz, for example, an equalizer that offers only stepped center frequencies, the closest of which might be 1,000 Hz and 1,500 Hz without independent control of bandwidth, is not going to do the job. If we identify a resonant frequency that we wish to attenuate, we need to be able to cut precisely at that frequency to make the most effective use of our equalizer without affecting frequencies that are not problematic.

Although frequency in hertz and cut/boost in decibels are standard units, there appear to be more definitions of bandwidth beyond the classic 3 dB down point and the midpoint dB gain definitions. The drawback of having multiple definitions of bandwidth is that two equalizers with identical settings can produce different frequency response curves. One equalizer might have a wider measured bandwidth than another, even though the Q or bandwidth parameter value is the same on both. Since equalizers use different definitions of bandwidth, we cannot necessarily transfer parameter settings from one plugin to another to get the same sonic result. Neither can we easily compare them by ear because they will sound different if their curves are different. If we train ourselves to hear one equalizer model with its inherent bandwidth definition, we may expect to hear these characteristics with other equalizers. In practice, it may be best to set Q or bandwidth by ear rather than rely solely on the displayed value.

To investigate the differences in the ways bandwidth might be defined, I selected five different equalizers and measured their frequency responses with the same settings:

- Frequency: 1,000 Hz.
- Gain: +12 dB.
- Bandwidth: 1.39 octaves or Q = 1.00, depending on each plugin's parameter unit.

A bandwidth of 1.39 octaves is equivalent to a Q of 1.00 according to the conversion formula, Equation 2.3.

Figure 2.14 shows the frequency response of five equalizers with equivalent bandwidth settings:

- FabFilter Pro-Q 4: Q = 1.00.
- Metric Halo ChannelStrip 4 Equalizer: Bandwidth = 1.39 octaves.
- REAPER ReaEQ Type = Band: Bandwidth = 1.39 octaves.
- REAPER ReaEQ Type = Band (alt 2): Bandwidth = 1.39 octaves.
- Tokyo Dawn Records NOVA: Q = 1.00.

Only four distinct curves are visible in this plot because the REAPER ReaEQ (Type = Band) and Tokyo Dawn Records NOVA curves are identical.

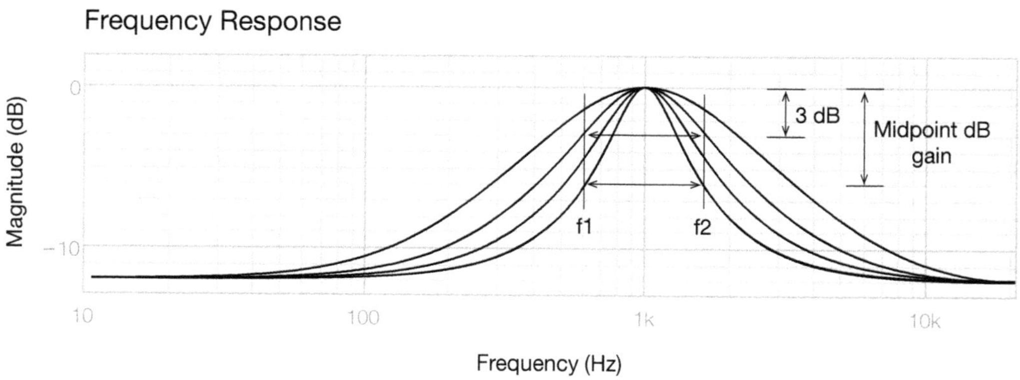

Figure 2.14 The frequency responses of five equalizers with identical parameter settings—Frequency: 1,000 Hz; Gain: +12 dB; Bandwidth: 1.39 octaves or Q = 1.00, which are equivalent values—show the variation in bandwidth that exists. The equalizers measured were FabFilter Pro-Q 4, Metric Halo ChannelStrip 4 Equalizer, REAPER ReaEQ, both Band and Band (alt 2) Types, and Tokyo Dawn Records NOVA Equalizer. Only four curves are visible because the REAPER ReaEQ Band Type and Tokyo Dawn Records NOVA curves are identical and thus overlap.

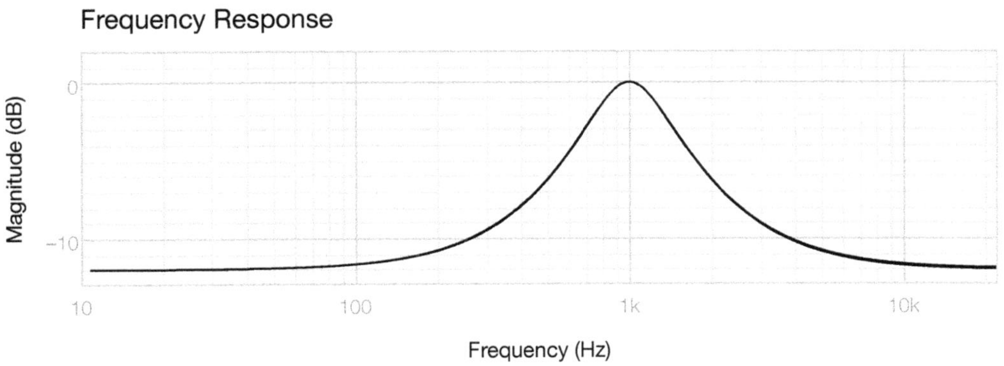

Figure 2.15 Frequency responses of five equalizers with parameters set so that they produce the same curve. Identical frequency (1,000 Hz) and gain (+12 dB) settings were used for all equalizers, but bandwidth settings differed. The equalizers measured were FabFilter Pro-Q 4, Metric Halo ChannelStrip 4 Equalizer, REAPER ReaEQ (both Band and Band (alt 2) Types), and Tokyo Dawn Records NOVA Equalizer.

The differences in frequency response curves for identical settings does not imply that one equalizer is better than another. The point is simply that we cannot rely on the absolute Q value or bandwidth in octaves to be consistent from one equalizer to another. There are evidently different ways that plugins determine bandwidth beyond the classic 3 dB down point and the midpoint dB gain definitions. One thing we do know is that, regardless of the way bandwidth is defined, a free parametric equalizer with fully sweepable parameters, like the ReaEQ included with REAPER, can produce the same frequency response curves as any other equalizer. Figure 2.15 shows the frequency responses of the same five equalizers as above, with their bandwidths set so that they produce identical frequency responses. Bandwidths were set as follows:

- FabFilter Pro-Q 4: Q = 1.00.
- Metric Halo ChannelStrip 4 Equalizer: Bandwidth = 0.64 octaves.
- REAPER ReaEQ Band Type: Bandwidth = 1.92 octaves.
- REAPER ReaEQ Band (alt 2) Type: Bandwidth = 1.01 octaves.
- Tokyo Dawn Records NOVA: Q = 0.70.

If the free parametric equalizer that comes with your DAW can replicate frequency response curves of equalizer plugins that cost money, why pay for a third-party equalizer plugin? One reason might be that the paid plugin has a better graphical interface. Or perhaps the paid one is easier to use and allows you to work faster. These are perfectly valid reasons, but they are not based on audio quality. Maybe the paid equalizer allows you to select a narrower—that is, smaller—bandwidth for more precise frequency shaping. Note, however, that the ReaEQ equalizer plugin that comes with REAPER allows extremely narrow bandwidth selection—as low as 0.01 octaves—which is narrower than many paid equalizers.

2.4.2.4 Shelving Filters

Shelving filters allow us to boost or cut a range of low frequencies or a range of high frequencies. Whereas high- and low-pass filters can completely remove a range of frequencies, shelving filters

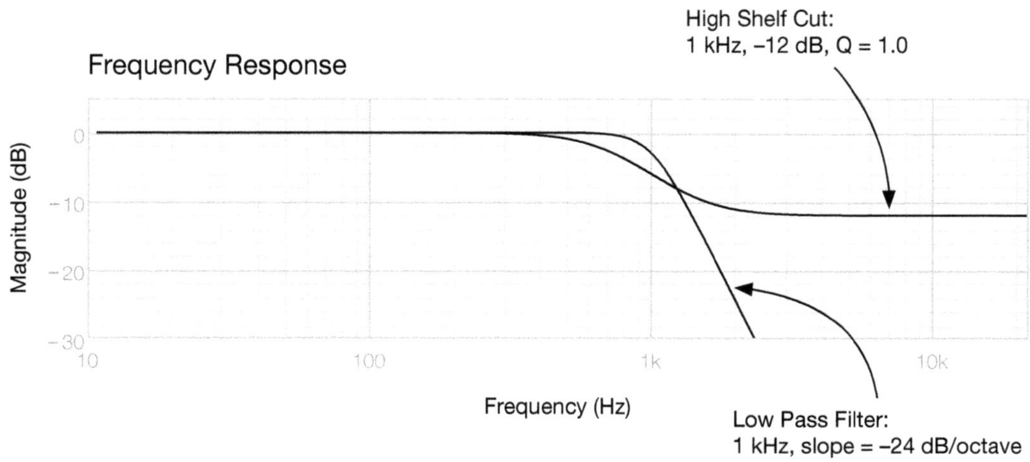

Figure 2.16 The frequency responses of a high-frequency shelf filter cut set to frequency of 1 kHz, gain of –12 dB, Q of 1.0, and a low-pass filter set to 1 kHz with a slope of –24 dB/octave.

raise or lower the level of a range of frequencies by some specified amount in decibels. High-shelf filters apply a user-specified amount of boost or cut equally to all frequencies above the cut-off frequency, and low-shelf filters apply some amount of boost or cut equally to all frequencies below the cut-off frequency.

In recording equipment and plugins, shelving filters are often found as a switchable option in the lowest and highest frequency bands in a parametric equalizer. Some equalizer models also offer high- and low-pass filters in addition to shelving filters. The treble and bass controls found on consumer audio systems typically use shelving filters. Figure 2.16 compares the frequency response of a high-shelf cut to that of a low-pass filter.

2.4.3 Microphones and Tonal Balance

Microphones provide us with important ways to affect the tonal balance of recorded or amplified acoustic sounds. Each microphone model has its own on- and off-axis frequency responses which affect the tonal balance of the captured sound. Placement is crucially important, since it shapes not just the frequency balance but also the clarity and spatial perspective of a recording. Some microphone models offer a flat frequency response for sound sources that are located on-axis (0° directly in front), whereas other models have prominent peaks and dips in their on-axis responses. Engineers often choose microphones to complement the sound sources they record. So, for example, they might choose a microphone with a peak around 2 to 4 kHz on a vocal, or one with a dip around 300 Hz for a kick drum. Classical music recording engineers often rely solely on microphone choice and placement for spectral balance and perspective and with minimal use of equalizers, if any.

While equalizers may be the most powerful tool in an engineer's toolbox, they cannot solve every problem. Some tonal balance and reverberation issues in recording can only be solved through microphone choice and placement. If we optimize microphone placement for one sound source, we may compromise the tonal balance of other nearby sound sources that bleed in, since we cannot equalize off-axis sounds separately from on-axis sounds. Any room sound that a microphone may pick up is another aspect that cannot be eliminated with equalization. If we place a microphone too far away from a source, there is no way to make it sound closer with an equalizer. Any equalization we apply to a microphone signal will affect all sounds, on- and off-axis, direct and indirect. Thus it is important to recognize sound quality issues such as these before recording while there is still an opportunity to fix them with microphone placement. Technical ear training gives us the critical listening skills to make judgments about sound quality as quickly as possible during each phase of a project and take appropriate corrective action.

During recording sessions, we place microphones next to sources and listen to the resulting audio signals. We might also compare microphones to decide which ones have the most appropriate sonic characteristics for a given situation. How do we decide which ones are most appropriate? Generally, we want to consider the characteristics of a musician's instrument or voice, the recording space, and the timbral goals for the project.

A microphone's physical orientation and location affect an audio signal's tonal balance due to:

- the microphone's on- and off-axis frequency responses,
- the microphone's proximity effect—that is, the low frequency boost from directional microphones when they are placed close to a source,
- the sound source's radiation patterns,
- the acoustic environment and any constructive and destructive interference of direct and reflected sound energy at the microphone location.

2.4.3.1 Microphone Off-Axis Response

Off-axis response is a critical, but easy-to-ignore, aspect of recording. When we aim a microphone at a sound source, it is easy to imagine that the microphone will pick up sound from only the source to which it is aimed. But if you have done any recording, you know that microphones pick up sound from other nearby sources as well. Usually, off-axis sound arriving at a microphone originates from:

- indirect or reflected sound—that is, the room reverberation and early reflections—and
- direct sound from nearby instruments or voices that we are not specifically trying to capture with a given microphone. This sound is usually referred to as bleed or leakage.

Microphones do not generally have the same frequency response for all angles of sound incidence. Sound that arrives off-axis—that is, from any direction other than directly in front of the microphone—is captured with a different tonal balance than sound that arrives on-axis because the off-axis frequency responses are usually different from the on-axis response. Even omnidirectional microphones, considered to have the best—that is, the flattest—off-axis frequency response, have frequency responses that vary with angles of sound incidence. Simply changing a microphone's orientation can alter a sound source's spectral balance. Small diaphragm, such as ¼″ diameter, condenser microphones provide the closest to a perfect omnidirectional polar pattern because a small microphone capsule is less likely to interfere with sound waves arriving off-axis.

Microphone manufacturers usually do not provide off-axis frequency responses in a Cartesian graph format—that is, a graph of frequency response on the x-axis and magnitude on the y-axis—as they do for microphone on-axis responses. Still, many manufacturers include polar plot measurements at different frequencies. It is difficult to compare the on-axis Cartesian frequency response plots with off-axis polar frequency response plots to determine the frequency response at a given angle. Some manufacturers simply claim that their microphones produce minimal off-axis coloration, without providing any measurements. We assume that a microphone with minimal off-axis coloration has a relatively flat frequency response or at least a smooth frequency response with no prominent peaks or dips for sounds arriving off-axis.

Let's discuss a practical consideration for off-axis microphone response in a recording situation. A microphone placed next to a snare drum will also pick up other elements of the drum set, such as the hi-hat and tom-toms, even its polar pattern is cardioid. For a perfect cardioid polar pattern, the output level for sounds arriving 90° off-axis is down only 6 dB relative to sounds arriving on-axis. The output level in decibels from a cardioid polar pattern can be calculated using the following formulas:

$$r(\theta) = 0.5 + 0.5 * \cos(\theta) \tag{2.4}$$

where θ is the angle of incidence in degrees, and r is the output level or sensitivity.

$$Level\,in\,dB(\theta) = 20 * log_{10} \frac{r(\theta)}{r(0°)} \tag{2.5}$$

Where the on-axis level, $r(0°)$, is equal to 1.

At 90°, the output level is 0.5, which is equivalent to –6.02 dB. In practice, some frequencies, typically high frequencies, are attenuated more than 6 dB, and others, typically low frequencies, are attenuated less than 6 dB, depending on the microphone.

Attenuation of sounds at 90° by 6 dB means that sound sources to the side of a cardioid microphone can and will audibly bleed into it. The other problem is that the tonal balance of off-axis sound is often also affected since a microphone's off-axis frequency response may not match its on-axis frequency response.

For example, with a microphone aimed at a snare drumhead, the hi-hat timbre captured by the snare microphone depends largely on the microphone's off-axis response. Drum set overhead and hi-hat spot microphones will also capture the hi-hat sound. If all these microphone signals are mixed, the hi-hat sound coming from the snare drum microphone will blend with hi-hat sound coming from these other microphones. Any significant coloration of the hi-hat from the snare microphone will have an impact on the overall sound of the hi-hat. Furthermore, any processing applied to the snare drum sound will also be applied to the hi-hat bleed in the snare microphone. Taken a step further, because these microphones are different distances from the hi-hat, the hi-hat sound will arrive at slightly different times at the three microphones and could result in comb filtering if the signals are panned to the same place.

With theoretical information such as this, we can pay particular attention to the timbral quality of indirect and adjacent sounds, knowing that timbral issues could arise due to the mix of on- and off-axis signals. When we know what to listen for, we are more adept at identifying problems when there are still opportunities to mitigate them. If we do not like what we hear, we can experiment with a different microphone location or orientation. If the microphone has a switchable polar pattern, we can see if a different pattern helps the sound quality. It is usually difficult, if not impossible, to correct the timbre of indirect sounds at the mixing stage, so we save time and effort by getting the sounds we want, directly from the microphones by adjusting their angles of orientation and location before committing to the recording. Technical ear training plays a significant role in helping us determine timbre anomalies quickly and accurately so that we can take corrective measures early in the process while they can be corrected.

2.4.3.2 *Microphone Proximity Effect*

Proximity effect is another tonal balance issue that we may need to work with. Directional microphones, such as cardioid, hypercardioid, and bidirectional polar patterns, amplify low frequencies when a sound source is close—that is, within about 1 meter or 3 feet—in a phenomenon known as proximity effect or bass tip-up. In recording sessions, listen for changes in low-frequency response if a singer is moving while singing. The proximity effect can be used to our advantage to boost low frequencies when close miking a kick drum, for instance. Some microphones, especially those marketed specifically for kick drum applications, include frequency response graphs for different source distances.

Microphone location in relation to a musical instrument, voice, or instrument amplifier can have a direct effect on the resulting spectral balance because of the sound radiation patterns of sound sources. The tonal balance from musical instruments and other sound sources varies according to the angular location relative to the source. For example, sound emanating directly out of a trumpet bell will contain more high-frequency energy than sound emanating to the side of the trumpet, assuming no reflective surface is nearby. We can affect the recorded timbre simply by changing a microphone's location relative to the instrument. A trumpet bell aimed slightly above or below a microphone will result in slightly attenuated high frequencies, compared to the sound of the trumpet when it is aimed directly at a microphone. For more detailed technical information about musical instruments' sound radiation patterns, Jürgen Meyer's book *Acoustics and the Performance of Music* (2009) is an excellent reference.

2.4.3.3 *Microphone Polar Patterns*

Microphone polar patterns also help maximize the level of one source relative to others. Cardioid polar patterns are probably the most common directional microphones, but bidirectional

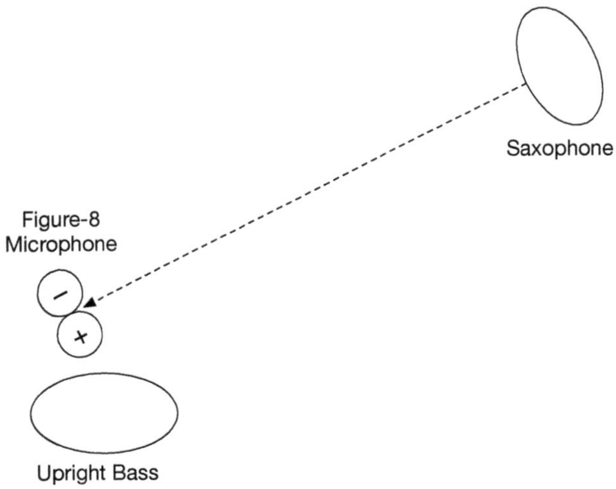

Figure 2.17 This diagram shows a figure-8 microphone polar pattern positioned close to an upright bass (double bass) such that the null of the figure-8 pattern faces the saxophone positioned further away. If we record upright bass with a figure-8 microphone in the same room as a saxophone, placing the sax 90° off-axis can help reduce its bleed in the bass microphone.

or figure-8 microphones are also highly effective in many situations but generally underutilized. To take advantage of a figure-8 polar pattern, place other sound sources so they project into the microphone's 90° off-axis null. By utilizing a figure-8 microphone's rejection of sounds coming from 90° off-axis, we can focus on sounds arriving at 0° or 180°. Strategic placement of figure-8 microphones relative to sound sources can help attenuate indirect and adjacent sound sources and give a cleaner pick-up. Figure 2.17 shows a possible placement of a figure-8 microphone to record an acoustic upright bass and maximize rejection of sound from a saxophone which is located at the 90° null of the polar pattern.

No matter how good our parametric equalization skills are, some problems like bleed cannot be fixed with equalization after the fact and must be corrected through microphone placement prior to recording. Ear training is essential so that we can identify problems in the moment and fix them.

Microphones are powerful timbral shaping devices that we can use to our advantage when we listen closely to them and take into account their frequency responses, off-axis responses, proximity effect, and polar patterns. In the recording process, microphones are the first stage of timbral shaping, but they do not automatically give us an ideal sound at just any location. Placement and orientation matter. To guide our choice for microphone placement, it can be helpful to walk around a sound source, moving our head to different locations around a source to hear tonal balance acoustically with our ears and to hear the ways that location affects tonal balance. Listening around a sound source to find a good microphone position is most effective with one ear facing the sound source and the other ear blocked with your hand. Once we decide on microphone placement based on listening acoustically, we can position the microphone and fine-tune its location based on the sound we hear directly from the microphone. The sound we hear from a microphone signal played back over our monitors may be different from the sound we heard when listening at the microphone location. Nonetheless, listening in different locations around a source with our ears can provide a useful starting point for microphone placement.

2.5 Technical Ear Training Software Module: Parametric Equalization

In this section, we look at the accompanying practice and test software module "Technical Ear Trainer – Parametric Equalization" and explore some strategies for using it. The software module is available at: https://webtet.net/apcl/#/parametric.

As audio engineers, we assess tonal balance by ear so that we can address issues that affect individual tracks, groups of tracks, and entire mixes. With well-developed critical listening skills, we can use equalization more precisely and minimize its effects on tracks other than those we wish to affect. When we can apply equalization more precisely, we are better able to blend tracks, address problematic resonances, reduce masking of one track by another, and avoid using equalization in ways that detract from a mix.

In the first half of this chapter, we explored some audio theory topics related to equalization and tonal balance. Theoretical knowledge complements technical ear training as we work to develop and improve our critical listening skills. While gaining this knowledge is indispensable to our work in audio, we also need to engage in focused listening exercises.

Using the software at https://webtet.net/apcl/#/parametric, the main task for each question is to identify by ear the equalization settings that the software has randomly selected. It randomly chooses equalization settings within some constraints that the user can specify. Figure 2.18 shows a screenshot of the user interface.

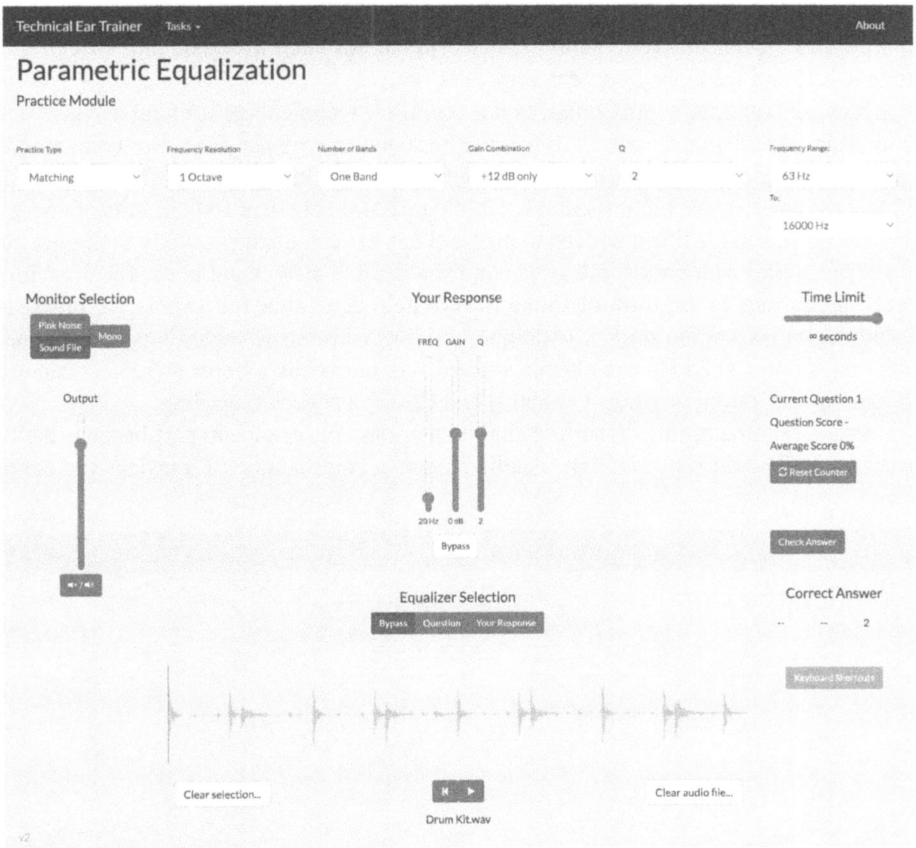

Figure 2.18 A screenshot of the software user interface for the Technical Ear Trainer—Parametric Equalization module, available at https://webtet.net/apcl/#/parametric.

We will now look at the functionality of the software and some approaches to training.

2.5.1 Monitor Selection

Perhaps the first thing to do when we are ready to begin training with the software is to choose the sound source, which can be done by selecting either "Pink Noise" or "Sound File" under Monitor Selection on the upper left side. The choice of sound source can affect the difficulty level of equalization training, and you will find that some sound sources work better than others.

Although it is not commonly used as a sound source in creative projects like music recording, live sound, film, or games, pink noise is an excellent sound source for technical ear training because its frequency spectrum sounds balanced or flat. This is because it has equal energy *per octave* and the human auditory system is sensitive to octave relationships between frequencies. An octave represents a doubling of frequency, so this means we perceive the frequency span of the lowest octave—that is, 20 to 40 Hz—to be the same as that of highest octave—that is, 10,000 to 20,000 Hz. With equal energy per octave in pink noise, it sounds as though each frequency is represented equally. Besides having an even frequency spectrum, pink noise also has a steady overall amplitude level, unlike dynamic real-world sounds such as music and speech in which amplitude levels and frequency content vary over time.

White noise, in contrast to pink noise, is not useful for technical ear training because it sounds harsh and bright. As a random signal, white noise has equal energy *per frequency* compared to pink noise which has equal energy *per octave*. In white noise, the lowest octave, 20 to 40 Hz, has 20 units of energy, whereas the highest octave, 10,000 to 20,000 Hz, has 10,000 units of energy. This is why white noise sounds like a wideband hiss and has a much harsher quality than pink noise.

Besides practicing with pink noise as a sound source, it is also essential to practice with music and speech recordings in the form of sound files to help generalize our training to practical situations. Although every audio project is unique, training with music recordings helps us hear, for example, how a boost at 2 kHz can affect a snare drum or a vocal, a boost at 125 Hz can affect an electric bass or kick drum, or a boost at 500 Hz can affect a speech recording.

Music and speech recordings are more challenging than pink noise in part because their levels change dynamically over time and they usually contain a combination of transient and continuous

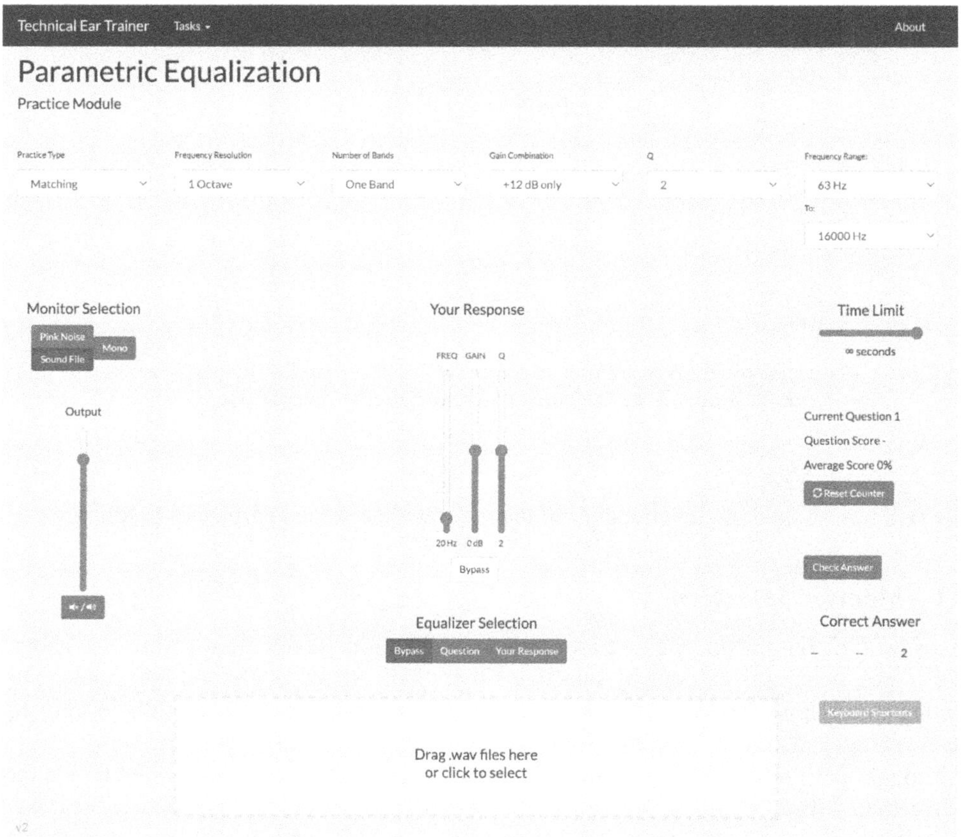

Figure 2.19 A screenshot of the lower portion of the user interface for the Technical Ear Trainer—Parametric Equalization module showing the module without a sound file loaded in.

sounds. The other reason why recordings may be more difficult as training signals is because they usually do not contain all frequencies and those frequencies that are present are not equal in level. It's difficult to imagine any sound source with a frequency spectrum that extends all the way down to 20 Hz and all the way up to 20,000 Hz. Most sound sources have a much more limited range or bandwidth. Flute, for example, does not extend below 250 Hz. Other sound sources or recordings may have gaps in the frequency spectrum. If a recording is deficient in the 500 Hz range, it's more difficult to hear the effect of an equalizer set to boost that frequency. Furthermore, the spectral content and balance of recordings vary over time, which can sometimes make it tough to hear an equalizer setting. These are all reasons why pink noise is a great sound source not only for the novice but also for experienced listeners as well.

The Monitor Selection defaults to Sound File if you do not select Pink Noise. To use a Sound File as the source, you need to load in a .WAV audio file using the rectangle at the bottom of the screen, as shown in Figure 2.19.

Under Monitor Selection, you can select mono or stereo. If a sound file contains only one track of audio as opposed to two, the audio signal will be sent out of the left output only. By pressing the mono button, the audio will be fed to both left and right output channels. Even if the source is stereo, some people prefer to do their equalization training in mono, so you may want to consider this option. The Pink Noise option is mono.

It is best to start with pink noise when beginning any new exercise or when working with a more difficult set of parameters. After getting comfortable with pink noise, challenge yourself with a variety of recordings of various instrumentation and genres. The greater the variety of sound recordings you use, the more you will be able to transfer critical listening skills to other situations. In Section 2.7, I list some recordings for training.

2.5.2 *Equalizer Selection*

In the technical ear training software module, an audio signal (pink noise or audio file signal) is routed to three places:

- **Bypass:** No equalization is applied.
- **Question:** Audio is processed with equalizer settings randomly chosen by the software.
- **Your Response:** Audio is processed with the equalizer settings chosen by the user and shown in the graphical interface equalizer.

You select one of these options to audition the audio. Depending on the Practice Type you have selected, Bypass and Question may be the only options available to audition. Figure 2.20 shows a block diagram of the practice module.

2.5.3 *Practice Types*

At the top of the equalizer technical ear training window shown in Figure 2.18 and at https://webtet.net/apcl/#/parametric, there are seven drop-down menus which are used to specify the quiz parameters or training constraints. The left-most menu, Practice Type, allows you to select one of the practice types: Matching, Matching Memory, Return to Flat, Absolute Identification, and Return to Flat Memory. The practice types provide different ways of working that vary in

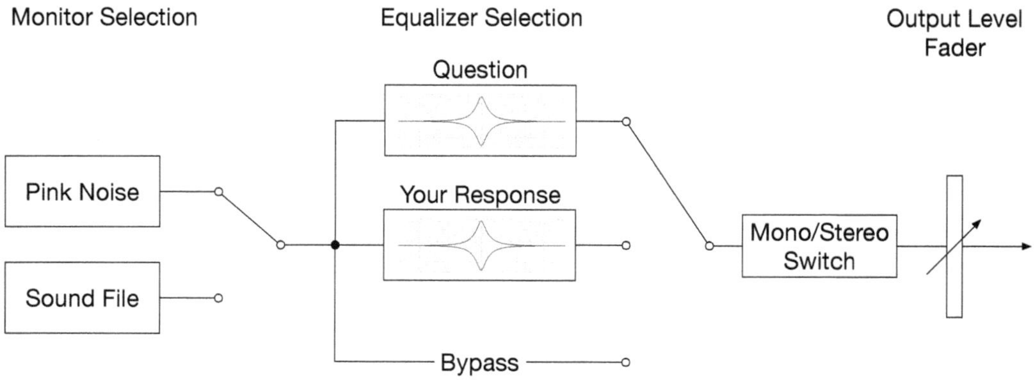

Figure 2.20 A block diagram of the signal path for the technical ear trainer practice module for parametric equalization. Sound sources enter on the left and are routed to the Question equalizer, Your Response equalizer, and Bypass, from which a user can select one at a time.

difficulty. The following bullet points describe the functionality of each Practice Type. Training approaches to each one are described in Section 2.8.

- **Matching:** This mode provides the most flexibility and is the best place to start, since it allows the freedom to switch among all three Equalizer Selection options: Bypass, Question, and Your Response. By comparing the sound of the Question equalizer to the sound of Your Response equalizer, the task is to determine if your chosen equalization setting (Your Response) matches the sound of the equalization applied by the software (Question). After setting Your Response equalizer so that it produces the same tonal balance as the Question, click "Check Answer" or hit enter or return to see the correct answer.
- **Matching Memory:** This mode is like the Matching mode with one main difference. After selecting and listening to the Question audio and then switching to either Bypass or Your Response, the equalized audio in the Question is no longer available for selection until you move to the next question.
- **Return to Flat:** In this mode, the task is to reverse or cancel the Question equalization by selecting the correct frequency and applying equal but opposite gain level. It is similar in difficulty to Matching but requires thinking in the opposite way since the goal is to remove the equalization and return the sound to its original spectral balance. For instance, if you hear a boost of 12 dB at 2000 Hz, the correct response would be to apply a cut of 12 dB at 2000 Hz, thus returning the audio signal to its original state and sounding identical to the Bypass option. Because the equalization used is symmetrical boost/cut, it is possible to eliminate any frequency boosts or cuts by applying equal but opposite boosts or cuts to the matching frequencies.
- **Return to Flat Memory:** As the name implies, this mode is a combination of Matching Memory and Return to Flat. The task is to bring the equalized version back to flat to match the sound of Bypass mode by applying gain that is opposite to the question at the correct frequency. Each new question starts in Bypass mode, but after you switch to Question or Your Response, you can no longer listen to Bypass. The goal is to remember the timbre of the flat version so you can match it.

- **Absolute Identification:** This practice mode is the most difficult. The task is to identify the applied equalization without having the option to listen to Your Response. The only equalizer selections available in this mode are Bypass and Question.

2.5.4 Frequency Resolution

The next drop-down menu to the right of Practice Type is Frequency Resolution. You can choose from two frequency resolutions. Table 2.1 lists the octave frequencies and Table 2.2 lists the one-third-octave frequencies.

Table 2.1 Frequencies available when "1 Octave" frequency resolution is chosen. This is the easiest of the two options and has nine possible frequencies.

16,000 Hz
8,000 Hz
4,000 Hz
2,000 Hz
1,000 Hz
500 Hz
250 Hz
125 Hz
63 Hz

Table 2.2 Frequencies available when "1/3 Octave" frequency resolution is chosen. This is the most difficult of the two options and has 25 possible frequencies. Note that the nine octave frequencies are included in the list.

16,000 Hz	800 Hz
12,500 Hz	630 Hz
10,000 Hz	500 Hz
8,000 Hz	400 Hz
6,300 Hz	315 Hz
5,000 Hz	250 Hz
4,000 Hz	200 Hz
3,150 Hz	160 Hz
2,500 Hz	125 Hz
2,000 Hz	100 Hz
1,600 Hz	80 Hz
1,250 Hz	63 Hz
1,000 Hz	

The octave and one-third-octave frequencies in this module correspond to the standard frequencies commonly found on graphic equalizers. The specific frequencies are determined by the International Standards Organization (ISO).

For each question, the software randomly chooses from among the list of either octave or one-third-octave frequencies.

2.5.5 Number of Bands

In the context of spectral analysis, a band is simply a range of frequencies. The number of equalizer bands determines the number of frequency ranges we can affect with an equalizer. For example, a nine-band graphic equalizer allows nine different frequency ranges to be boosted or cut. A band is a frequency range rather than simply just one frequency, because when we boost or cut with an equalizer, we affect a range of frequencies around the center frequency, not just the center frequency. In the software module, you can choose one, two, or three simultaneous bands for each question.

2.5.6 Frequency Range

The Frequency Range drop-down menus allow you to limit the range of possible frequencies from the full range of 63 Hz to 16,000 Hz to a range as small as three octave frequencies.

2.5.7 Gain Combination

The Gain Combination option refers to the possible gains that can be applied to a given frequency. The default gain is +12 dB, which is the recommended starting point for training. Possible gains include: +12 dB, –12 dB, +9 dB, –9 dB, +6 dB, –6 dB, +3 dB, and –3 dB. Combinations of different gains can be selected as well such as +12/–12 dB, or +12/–12/+6/–6 dB.

When you set the Gain Combination to include more than one possible gain, the software randomly chooses between the possible gain options for each question, which means you need to adjust the gain to the appropriate amount for each question as well. When there is only one possible gain option, you do not need to adjust the gain since it automatically jumps to the correct amount when you select a frequency.

2.5.8 Q

The Q is a static parameter for any exercise and is not a test parameter. The default setting, a Q of 2, is the best starting point for all exercises. Equalizer settings with higher Q—that is, narrower peak shape or smaller bandwidth—are more difficult to identify because they affect a smaller range of frequencies. Likewise, settings with lower Q—that is, wider peak shape or larger bandwidth—are also difficult to identify because they affect so many frequencies around the center frequency.

2.5.9 Sound File Control

The Sound File Control area of the interface is the transport section and includes a waveform display of the audio signal. After loading a sound file and clicking the play button, the audio file automatically repeats once it reaches the end of the file or the end of the selected section. You can select excerpts of the full audio file by clicking and dragging on the waveform. You can also jump to any point in the audio by clicking in the waveform.

2.5.10 Time Limit

The default time limit is unlimited, as shown by the infinity symbol (∞). You can select no time limit, that is, ∞, or a time limit of 30 seconds or less.

2.5.11 Keyboard Shortcuts

The keyboard shortcuts included in the software are meant to help you focus more on listening and less on pointing and clicking with a mouse or trackpad. They also speed up the process, allowing you to switch between Equalizer Selection options and indicate your response more quickly. Table 2.3 lists the general keyboard shortcuts and Table 2.4 lists the frequency selection keyboard shortcuts.

Pressing the space bar toggles among the three Equalizer Selection options depending on the Practice Type in these ways:

- **Matching:** Each question starts on Bypass. Pressing the space bar toggles between Question and Your Response.
- **Matching Memory:** Each question starts on Bypass.
    - o The first space bar press selects Your Response.
    - o The second press selects Question.
    - o The third press selects Your Response and makes Question unavailable, which is indicated by the grayed-out Question button.
    - o Subsequent space bar presses toggle between Bypass and Your Response.
    - o After checking your answer, but before moving to the next question, the space bar can be used to toggle between Question and Your Response.
- **Return to Flat:** Each question starts on Bypass. Pressing the space bar toggles between Bypass and Your Response.
- **Absolute Identification:** Pressing the space bar toggles between Question and Bypass.
- **Return to Flat Memory:** Each question starts on Bypass. Pressing the space bar toggles between Bypass and Your Response.

Table 2.3 General keyboard shortcuts available with the Parametric Equalization ear training software module.

Key	*Function*
W	Selects Bypass (flat equalizer selection).
E	Selects Question equalizer.
R	Selects Your Response equalizer.
Space bar	Toggles among Bypass, Question, and Your Response depending on the practice type selected.
Q	Selects the frequency slider.
Up Arrow or Down Arrow	Changes the frequency of the selected band.
Left Arrow or Right Arrow	Changes the gain of the selected band.
Enter or Return	Checks your answer and moves to the next question. A frequency, gain, or Q slider needs to be focused for this key to work.
ESC	Toggles all audio on/off.

Table 2.4 Frequency selection keyboard shortcuts available with the Parametric Equalization ear training software module.

Number Key	*Frequency*
0	20 Hz, gain 0 dB (flat)
1	63 Hz
2	125 Hz
3	250 Hz
4	500 Hz
5	1,000 Hz
6	2,000 Hz
7	4,000 Hz
8	8,000 Hz
9	16,000 Hz

For exercises with only one gain option—for example, +12 dB—the gain is automatically set to the correct value when you move the frequency slider from 20 Hz to any other frequency. When you return the frequency slider to 20 Hz, it resets the gain to 0 dB. For exercises with more than one gain option—for example, both +12 and −12 dB—the gain drops to the lowest value for the Gain Combination—for example, −12 dB—when you move the frequency slider for the first time on a given question.

2.6 Working with the Parametric Equalizer Training Module

When you first open the Parametric Equalizer training module found at https://webtet.net, select Pink Noise in the Monitor Selection, and confirm that the quiz parameters at the top are set to their defaults:

- Practice Type: Matching.
- Frequency Resolution: 1 Octave.
- Number of Bands: One Band.
- Gain Combination: +12 dB.
- Q: 2.
- Frequency Range: 63 to 16,000 Hz.

These are the default settings when you open the module. Adjust the output level fader down about halfway to minimize clipping when your audio is boosted in the equalizer. Set a comfortable listening level with you monitor controller or headphone amp and make sure you are listening through decent headphones or loudspeakers. The audio you will hear is flat because the equalizer is bypassed. Press the space bar to hear the Question audio which has been boosted by 12 dB at one of the nine octave frequencies. Press the space bar again to hear Your Response, which should be flat if you have not moved any of the equalizer sliders. Adjust the frequency slider to hear the equalizer in Your Response. Scroll through each octave frequency to get a feel for the sound of each frequency. Once you change the frequency to anything other than 20 Hz, the gain will automatically jump to 12 dB, which is the default gain combination setting the module starts. Press the W and R keys to toggle between Bypass (flat) and Your Response, and compare the different timbres that result from a boost at each of the nine octave frequencies. Spend some time initially just listening to various frequencies, alternating between Bypass and Your Response equalizer. After you familiarize yourself with octave frequencies in pink noise, load in a sound file and do the same thing again with a music or speech recording.

When you audition a sound file, make note of what instruments or components of instrument sounds are affected by each octave frequency. For instance, a boost at 125 Hz may bring out the low harmonics in a snare drum or bass. On the upper end of the spectrum, 8 kHz may bring out crisp, metallic cymbal harmonics. If you audition a Baroque ensemble recording, you may find that a boost at 16 kHz can make a harpsichord more prominent. Boosts at specific frequencies can sometimes bring out individual instruments, voices, or sounds in a mix, and I provide examples in Section 2.9 below. Likewise, cuts at carefully chosen frequencies can help reduce masking and allow elements that would otherwise be obscured in a mix. Skilled mastering engineers take advantage of the power of equalization to subtly shift the balance or focus of a mix. Different recordings will be affected slightly differently by a given frequency, even if the recordings have similar instrumentation. For example, 63 Hz may bring out the bass in one recording, but in another recording a different frequency, say 125 Hz, may affect the bass more.

Depending on the frequency content and spectral balance of each individual instrument in a recording, a given equalizer setting may affect each mix differently. This is one reason why we must be attentive to what is required in each individual recording as opposed to simply relying on what may have worked in previous recordings. For example, just because a cut at 250 Hz may be effective on one snare drum in one recording does not mean that it will work on all snare drum recordings since it depends on the specific snare drum recording and the context of the mix and the other instruments.

2.6.1 *Practice Module Quiz Parameters*

The quiz parameters are the options that you can specify using the drop-down menus at the top of the screen. They include Practice Type, Frequency Resolution, Number of Bands, Gain Combination, Q, and Frequency Range, and they determine the focus of your training and the level of difficulty. We looked at these options in Section 2.5, and now I want to discuss how they relate to levels of difficulty and ways that you can use them in your training.

Since the accompanying software allows for maximum flexibility, it is possible to set the quiz parameters to produce questions across a range of levels of difficulty, from relatively easy to extremely difficult to impossible. Anyone would struggle to answer questions when all quiz parameters are set to difficult levels at the same time. For example, quiz parameters set to Absolute ID mode, one-third-octave resolution, three bands, Q of 8, all combinations of gain, and a time limit of 5 seconds using a sound file with dynamically changing timbre would be impossibly difficult.

The flexibility in quiz parameters allows us to adjust the difficulty of a single quiz parameter while the other quiz parameters are set to easier variations. For example, we could add a second or third frequency band while working with Matching mode, octave frequencies, a gain of +12 dB, a Q of 2, unlimited time, and pink noise as the sound source. Once we are comfortable with two or three bands, we may want to work with different sound files, with cuts (e.g., –12 dB), with smaller amounts of boost (e.g., +9, +6, or +3 dB), or with Matching Memory or Absolute ID.

The benefit of this flexibility is that you are fully in charge of your training quiz parameters, and you can set the difficulty level for yourself. One challenge with this flexibility is to figure out how to scaffold your training so that you can increase the difficulty level to make it appropriately challenging but not overwhelming and frustrating. In Section 2.8 we look at ways to vary difficulty level, and I offer some strategies you can use with your training.

2.6.2 *Practice Types*

The five practice types offer different ways of training with variations in level of difficulty. Below I describe the practice types and some considerations for each one.

Matching mode is the best Practice Type to start with. Even for more experienced listeners, Matching can be helpful when working with small gains (such as 3 dB) or with one-third-octave frequencies and multiple bands. One-third-octave frequencies can sometimes be difficult to discriminate from one another in the low-frequency range, depending on the sound file. Matching is a great place to start also because it allows you to listen to boosts at each frequency and explore timbral effects across the spectrum.

Matching Memory is more difficult than Matching mode. In this mode, once you listen to the Question and switch away to either Bypass or Your Response, you cannot go back to the Question. The goal with this mode is to try to increase your timbral memory for different equalizer settings by matching timbres by memory. It can be moderately to very difficult depending on the other quiz parameters that you choose. With Matching Memory mode, it is essential to listen closely to the Question before you switch to Your Response or Bypass. While still listening to the Question, set the equalizer parameters to your best guess. When you think you have identified the equalizer setting or are close, switch to Your Response to hear how close your guess is. If Your Response does not match the Question, this mode challenges you to remember the timbre of the Question, since it is no longer available to audition for that question, and adjust the equalizer settings to match your memory of the timbre.

In Return to Flat mode, the task is to cancel out the Question equalization by applying the opposite gain at the correct frequency. We can think of it as a "matching bypass" or "matching flat" mode. The task is to match Your Response to Bypass by switching between them and adjusting Your Response until they sound the same. This mode works differently than the other modes so it might be confusing at first. Here are the ways in which this mode is different from the other modes:

- Your Response will sound as if it has been equalized even before you move a slider. Unlike the other modes, the audio in Your Response equalizer is the question audio, which means it is not flat even if your equalizer gain is 0 dB. Since the goal is to cancel out a boost or a cut with the opposite gain, the equalized audio runs through your equalizer so that you can return it to flat.
- Because the task is to undo the Question equalization, the gain you apply in Your Response should be the opposite of the Question. For example, if your Gain Combination is +12 dB, ensure that the gain on Your Response is set to –12 dB. Conversely, if your Gain Combination is –12 dB, the gain on Your Response should be +12 dB. For single boosts or cuts, when you change the frequency, the gain slider on Your Response automatically jumps to the correct gain, that being the opposite gain of the Gain Combination.
- The goal is to match the timbre of Your Response with the timbre of Bypass. In other words, if you have Your Response equalizer set correctly, its timbre should match the timbre of Bypass. The timbre of the Question audio will match the timbre of Your Response audio only if the gain sliders are all at 0 dB.

Return to Flat Memory combines the Matching Memory and Return to Flat modes and offers an increased challenge over Return to Flat.

Absolute ID is the most challenging since it does not allow you to listen to Your Response. Because of the difficulty of this mode, starting with octave frequencies, +12 dB gain, one band,

and perhaps a restricted range of frequencies can help ease the transition when coming from an easier mode. After gaining competency, it makes sense to expand the frequency range, add another frequency band, and work with smaller amounts of gain and combinations of boosts and cuts.

In general, for any Practice Type, it is usually advisable to make changes to only one quiz parameter at a time so that you can become proficient before moving on to more difficult questions.

2.6.3 Frequency Resolution

2.6.3.1 Octave Frequencies

In terms of frequency resolution, there is no question you should start training octave frequencies. You'll want to work exclusively with octave frequencies for some time until you can confidently and consistently identify all nine octave frequencies using not only pink noise but also a range of different sound files. The goal is to develop a memory for the way each octave frequency affects timbre. Expect to spend significant practice time to develop this frequency memory. Eventually, after you become confident with octave frequencies, feel free to increase the difficulty level by selecting one-third-octave frequencies, but remember that octave frequency training remains relevant at any level of expertise. Even those who have attained proficiency with one-third-octave frequencies return to octave frequency resolution for review or when selecting higher difficulty levels on other quiz parameters, such as multiple bands, smaller amounts of gain, or new gain combinations.

One technique to help remember the timbre of octave frequencies is to associate them with vowel sounds. Associating specific vowel sounds with octave frequencies can help listeners identify frequencies because of the formant frequencies present in each vowel sound, as several researchers have noted (Letowski 1985; Miskiewicz 1992; Opolko and Woszczyk 1982; Quesnel 2001; Quesnel and Woszczyk 1994; Slawson 1968). While frequencies at the upper and lower extremes of the spectrum are not associated with any vowel due to the range of the human voice, five midrange octave frequencies roughly correspond to vowel sounds as listed in Table 2.5.

It is best to start with pink noise when first learning these associations. As you listen to a 12 dB boost at one of these five frequencies, make the vowel sound that corresponds to the boosted frequency and compare the timbre of the boosted frequencies with that of the associated vowel. Listen for the similarities in timbre between the two. The aim is to associate the timbre of each vowel sound and octave frequency unambiguously. Try it for each of the five frequencies and vowel sounds. After learning these associations, some people find that they can simply mouth the vowels and imagine the sounds rather than voice them. Either way, the strategy to associate vowels and frequencies helps increase the speed and accuracy with which we can identify these octave frequencies.

Table 2.5 The five mid-range frequencies and their associated vowel sounds. Try to either vocalize or silently hear in your head the sounds of the vowels below when each respective frequency is boosted.

Frequency	*Associated Vowel Sounds*
250 Hz	[u] as in b**oo**t /bOOt/
500 Hz	[o] as in b**oa**t /bOHt/
1,000 Hz	[ə] as in **a**bout /UHbOUt/
2,000 Hz	[ɛ] as in b**e**t /bEt/
4,000 Hz	[i] as in b**ea**t /bEEt/

2.6.3.2 *One-Third-Octave Frequencies*

Absolute identification exercises with one-third-octave frequency resolution are more difficult than those with octave frequencies because there are more one-third-octave frequencies—that is, there are 25 one-third-octave frequencies versus nine octave frequencies—and the one-third-octave frequencies are closer together. If you can confidently identify all nine octave frequencies with a range of different sound files and you want a new challenge, try one-third-octave frequency exercises.

Since the jump from nine octave to 25 one-third-octave frequencies is a big one, consider the following strategy to make the transition. Assuming a solid base of training with octave frequency timbres, think of these frequencies like anchors in the frequency spectrum to help learn the one-third-octave frequencies. With the octaves as anchors, each one-third-octave frequency is simply a neighbor slightly above or slightly below an octave frequency, as shown in Table 2.6.

Adjacent one-third-octave frequencies may not have identical timbres, but they are similar. Their similarity is one reason why absolute identification is so much more difficult with one-third octaves. We can use this similarity and our familiarity with the nine octave frequency timbres to expand our training to incorporate all 25 one-third-octave frequencies with the following strategy. When listening to a one-third-octave question, ask yourself this: Which octave frequency does it sound like? With further listening, try to determine if the frequency is exactly

Table 2.6 All octave and one-third-octave frequencies between 63 and 16,000 Hz. Octave frequency anchors are shown in bold and their respective one-third-octave frequency neighbors are shown above and below each octave frequency. Grouping the one-third-octave frequencies in this way can help in learning to identify them.

Frequency	*Octave Frequency Anchors and Their One-Third-Octave Frequency Neighbors*
16,000 Hz	**Octave frequency anchor**
12,500 Hz	One-third octave just below 16,000 Hz
10,000 Hz	One-third octave just above 8,000 Hz
8,000 Hz	**Octave frequency anchor**
6,300 Hz	One-third octave just below 8,000 Hz
5,000 Hz	One-third octave just above 4,000 Hz
4,000 Hz	**Octave frequency anchor**
3,150 Hz	One-third octave just below 4,000 Hz
2,500 Hz	One-third octave just above 2,000 Hz
2,000 Hz	**Octave frequency anchor**
1,600 Hz	One-third octave just below 2,000 Hz
1,250 Hz	One-third octave just above 1,000 Hz
1,000 Hz	**Octave frequency anchor**
800 Hz	One-third octave just below 1,000 Hz
630 Hz	One-third octave just above 500 Hz
500 Hz	**Octave frequency anchor**
400 Hz	One-third octave just below 500 Hz
315 Hz	One-third octave just above 250 Hz
250 Hz	**Octave frequency anchor**
200 Hz	One-third octave just below 250 Hz
160 Hz	One-third octave just above 125 Hz
125 Hz	**Octave frequency anchor**
100 Hz	One-third octave just below 125 Hz
80 Hz	One-third octave just above 63 Hz
63 Hz	**Octave frequency anchor**

one of the nine octave frequencies, or if it is slightly above or slightly below. For example, if a question sounds like a boost at or near 1,000 Hz, listen further to try to determine if the boost is at *exactly* 1,000 Hz or if it is at a nearby one-third-octave frequency. If it doesn't sound like it is at exactly 1,000 Hz, does it sound like it is slightly above—that is, at 1,250 Hz—or does it sound like it is slightly below, at 800 Hz?

Thinking of one-third-octave frequencies as neighbors just above and below the octave frequencies reduces the complexity since we can think of them as nine groups instead of 25 individual frequencies. The grouping also helps us relate the unfamiliar one-third-octave frequencies to the familiar octaves, but this assumes one has completed significant training with octaves before starting any one-third-octave training. Thus, it is crucial to become proficient and confident with all octave frequencies and to have worked with pink noise and a range of sound files before moving on to one-third-octave frequencies. If you feel generally proficient with octave frequencies but one-third-octave questions seem to give you difficulty, switch back to octaves and reacquaint yourself with them. Then go back to one-third-octave frequencies. Regularly going back to octave frequencies in the process of training with one-third-octave frequencies can help solidify the octave anchors and further differentiate between them and the one-third-octave frequencies between them. If you find that one-third-octave frequencies seem utterly impossible, even when you think of them as octave neighbors, it probably means you need to spend more time training with octaves before training with third octaves.

2.6.3.3 One-Third-Octave Matching

Technical ear training helps us learn to correctly identify the frequencies of boosts and cuts, with a goal to become proficient in Absolute ID mode. But it is also valuable to train our ability to hear very small differences in timbre—that is, our ability to discriminate or differentiate between very similar timbres. The Matching and Return to Flat modes can help in this area since the task with these two modes is to compare the timbres of two equalizer settings and match them. These modes are not overly challenging when working with octave frequencies, which is why they are great for beginners, but they can be considerably more difficult with third octaves. The choice of sound source also affects the difficulty. One-third-octave Matching and Return to Flat exercises are easiest with pink noise since it is a relatively continuous signal. Sound files of music recordings or speech, on the other hand, can be substantially more difficult than pink noise because of their dynamically changing spectra and level.

2.6.4 Number of Bands

If you are just beginning technical ear training, start with only one frequency band at a time. Without a solid base of experience answering questions with one frequency band, trying to answer questions with two or three bands can be frustrating and counterproductive. Take the time to develop confidence with one band at a time before moving to two bands.

When you do decide to work with two bands, begin each question by identifying the most obvious frequency that you hear. One frequency stands out more than the other, or you may hear them both equally well. If one frequency does stand out to you, try to match it and then compare Your Response to the Question. After you match one frequency, you should notice that frequency less when switching between the Question and Your Response, since the part of the timbre affected by that frequency is the same for both. The human auditory system is attuned to changes which means after you match one frequency, you should find it easier to identify the remaining frequency. It doesn't matter which slider you use to identify the two frequencies.

Typically, graphic equalizers are organized with low frequencies on the left and high frequencies on the right, but for this practice mode, the order does not matter.

2.6.5 Frequency Range

When you start training, it can be helpful to limit the frequency range to only three frequencies in the midrange, such as from 500 to 2000 Hz. Once you master these frequencies, feel free to expand the range one octave at a time. Take your time and get to know the timbre of each frequency. There is no rush. It is better to take your time and really get to know the sound of each octave frequency before moving to more difficult parameter combinations.

After you work from a small set of frequencies to the full list, you may find that some frequencies still give you trouble. Some people find that they mix up or have trouble differentiating 1,000 and 2,000 Hz or 2,000 and 4,000 Hz. When this happens, set the range to 1,000 to 4,000 Hz and spend some time reviewing just those frequencies. Low frequencies in the 63 Hz to 250 Hz range are often more difficult to identify correctly when practicing with music recordings, especially with third-octave frequencies. The low-frequency range can pose problems for a few reasons. First, music recordings do not always contain consistent levels for each frequency across the low-end range. Second, monitoring systems, whether loudspeakers or headphones, may not be capable of producing very low frequencies. Third, if our sound system does reproduce low frequencies accurately, room modes or standing waves—that is, the resonant low frequencies within a room—may interfere with what we hear by amplifying some frequencies and attenuating others. Using headphones can eliminate room mode problems, but headphones may not have a flat-frequency response or may be weak in their low-frequency response.

2.6.6 Gain Combination

As you might expect, larger changes in gain, such as 12 dB, are easier to hear than smaller changes in gain, such as 3 dB. Boosts are typically easier to identify than cuts. Work with 12 dB boosts until you become proficient with them.

An equalizer cut is more difficult to identify than a boost, since it can be harder to identify that which has been removed or reduced. It is generally easier to identify that which has been added, such as a boost. One strategy to identify cuts is to switch from the Question (the cut frequency) to Bypass (flat) and listen for the frequency that reappears in Bypass mode. The cut frequency that returns in Bypass mode may sound as if it has been boosted above normal when we switch from the Question to Bypass.

When working with a boost/cut Gain Combination, such as +12 and –12 dB, be aware that the human auditory system can be sensitive to *relative* changes in frequency response. What this means in practice is that you may hear a low-frequency *cut* incorrectly as a high-frequency *boost*, or you may hear a high-frequency *cut* as a low-frequency *boost*. This may seem unlikely, but this kind of error does happen on occasion, especially in the beginning stages of training.

2.7 Recordings for Practice

What makes a recording suitable for technical ear training? The main characteristics that make a track easier to use are:

- consistent instrumentation from beginning to end which creates a relatively even timbre over time,

- the presence of repetitive grooves, ostinato patterns, or other repeated musical figures,
- wide bandwidth that extends to at least 63 Hz in the low end and 16,000 Hz in the high end, and
- the presence of multiple instruments that cover the frequency spectrum; solo instrumental recordings are still useful but may not extend to the lowest or highest frequencies.

It can also be helpful to loop an excerpt of a track, such as a verse, chorus, or intro, to help maintain consistent instrumentation and timbres.

Recordings with extreme timbre variations such as those in the experimental electronic music genre can be exceptionally difficult sources to train with. With this type of recording, it is often difficult to know if a change in timbre is due to an equalizer setting or if it is inherent in the recording.

Uncompressed, lossless audio or Compact Disc-quality versions should be used for all exercises, rather than data-reduced versions. Also known as linear pulse-code modulation (PCM), uncompressed digital audio has a sampling rate of at least 44.1 kHz, a bit depth of 16 or 24 bits, and is stored in the WAV file format. Perceptually encoded versions, such as MP3, Advanced Audio Coding (AAC), or Ogg Vorbis, should never be used for equalizer exercises, even if the encoded version has been converted back to PCM or lossless audio. Once an audio file has been perceptually encoded, its quality has been degraded since audio data has been thrown away and cannot be recovered by converting back to linear PCM.

Below is a list of some commercially available recordings of various genres that are suitable for use as sound sources in the equalizer software practice module. They represent examples of high-quality recordings that have good spectral balance across a wide frequency range. There are many other recordings that are appropriate, of course, so I encourage you to explore other recordings and see what works for you.

- "Authors of Forever" by Alicia Keys from *ALICIA* (2020).
- "One Last Time" by Ariana Grande from *My Everything* (2014).
- "Straight" by Arild Anderson, Vassili Tsabropoulos, and John Marshall from *The Triangle* (2004).
- "Odyssey" by Beck and Phoenix from *Odyssey—Single* (2023).
- "Texas Hold 'Em" by Beyonce from *Cowboy Carter* (2024).
- "1–2-3" by Café Tacvba from *Jei Beibi* (2017).
- "Welcome to My Island" by Caroline Polacheck from *Desire, I Want To Turn Into You: Everasking Edition* (2023).
- "Marche pour la cérémonie des Turcs (Jean-Baptiste Lully)" by Le Concert des Nations from *Tous les matins du monde: Bande originale du film* [original soundtrack] (2002).
- "My Universe" by Coldplay and BTS from *Music of the Spheres* (2021).
- "Give Life Back to Music" by Daft Punk from *Random Access Memories* (2013).
- "Rapunzel" by Dave Matthews Band from *Before These Crowded Streets* (1998).
- "September" by Earth, Wind & Fire from *Greatest Hits* (1998).
- "Disco Ears" by Joshua Redman, Brad Mehldau, Christian McBride, and Brian Blade from *LongGone* (2022).
- "Swimming Pools (Drank)" by Kendrick Lamar from *goodkid, m.A.A.d city* (2012).
- "Stay" by The Kid LAROI & Justin Bieber –from *Stay—Single* (2021).
- "Teardrop" by Massive Attack from *Mezzanine* (1998).
- "IV. Vif" from Telemann: Concerto in G for 2 Viols, Strings and B.C. (52:G3) performed by Musica Antiqua Köln and Reinhard Goebel from *Telemann: String Concertos* (2000).
- "Modul 42" by Nik Bärtsch's Ronin from *Holon* (2008).
- "Modul 34" by Nik Bärtsch's Ronin from *Awase* (2018).

- "Copy of A" by Nine Inch Nails from *Hesitation Marks* (2013).
- "Panopticom (Bright-Side Mix)" or "Panopticom (Dark-Side Mix)" by Peter Gabriel from *I/O* (2023).
- "Freedom" by Rage Against the Machine from *Rage Against the Machine* (1992).
- "Over (feat. Yebba)" by Robert Glasper from *Black Radio III* (2022).
- "Time" by Sarah McLachlan from *Afterglow* (2003).
- "Better Broken" by Sarah McLachlan from *Better Broken* (2025).
- "Peg" by Steely Dan from *Aja* [Original Recording Remastered] (1977/1999).
- "Gaslighting Abbie" by Steely Dan from *Two Against Nature* (2000).
- "The Hounds of Winter" by Sting from *Mercury Falling* (1996).
- "Digital Witness" by St. Vincent from *St. Vincent* (2014).
- "On the Sunny Side of the Street (feat. Cassandra Wilson)" by Terence Blanchard from *Let's Get Lost: The Songs of Jimmy McHugh* (2001).
- "Fast Car" by Tracy Chapman from *Fast Car* (1988).
- "Blinding Lights" by The Weeknd from *After Hours* (2020).

Lossless audio recordings can be purchased and downloaded from Qobuz www.qobuz.com, HDtracks www.hdtracks.com, and ProStudioMasters www.prostudiomasters.com.

Author and recording engineer, Mike Senior, hosts a site with a growing number of free multi-track downloads: www.cambridge-mt.com/ms-mtk.htm.

Many DAWs come with recordings of solo instruments that can be useful with the EQ practice and test software.

2.8 Progression Through Equalizer Training Exercises

The equalizer module allows you to select any exercise parameter combination for practice and testing, and what follows is a possible progression from easy to difficult. I have divided the progression into two groups: octave frequencies and third-octave frequencies. For each step following Step 1, I show only the quiz parameters that have changed from the prior step. Start with Step 1 and take the time to become proficient and comfortable with a step before moving to the next one. Do not skip steps since the later steps are much more difficult and without enough practice they will be impossible to complete.

2.8.1 *Octave Frequencies, One Band, Matching and Absolute ID*

Step 1:

- Begin your training with the following quiz parameters:
  - Monitor Selection: Pink Noise.
  - Practice Type: Matching.
  - Frequency Resolution: Octave.
  - Number of Bands: 1.
  - Gain Combination: +12 dB.
  - Q: 2.

- Select small ranges of frequencies in different parts of the spectrum and then gradually expand your working range as in the suggested sub-steps below.
  - Frequency Range sub steps:
    - 500 to 2,000 Hz.
    - 63 to 250 Hz.
    - 4,000 to 16,000 Hz.
    - 250 to 4,000 Hz.
    - 125 to 8,000 Hz.
    - 63 to 16,000 Hz.

Step 2:

- Monitor Selection: Sound File.
- Begin working with sound recordings of your choice. Work with a variety of different recordings of different genres. Feel free to work with reduced frequency ranges until you get to know each sound file.

Step 3:

- Practice Type: Absolute ID.
- By now you should be familiar with all nine octave frequencies, so it is time to work in Absolute ID mode. Go back to smaller frequency ranges as needed. As you gain proficiency, expand the frequency range.

Step 4:

- Gain Combination: –12 dB.

Step 5:

- Gain Combination: +12/–12 dB.

Step 6:

- Gain Combination: +9/–9 dB.

Step 7:

- Gain Combination: +6/–6 dB.

2.8.2 *Octave Frequencies, One Band, Return to Flat*

Step 8:

- Practice Type: Return to Flat.
- Gain Combination: +12 dB.

Step 9:

- Gain Combination: –12 dB.

Step 10:

- Gain Combination: +12/–12 dB.

Step 11:

- Gain Combination: +9/–9 dB.

Step 12:

- Gain Combination: +6/–6 dB.

2.8.3 Octave Frequencies, Two Bands, Matching and Absolute ID

Step 13:

- Practice Type: Matching.
- Number of Bands: Two Bands.
- Gain Combination: +12 dB.

Step 14:

- Practice Type: Absolute ID.

Step 15:

- Gain Combination: +12/–12 dB.

Step 16:

- Gain Combination: +9/–9 dB.

Step 17:

- Gain Combination: +6/–6 dB.

2.8.4 Octave Frequencies, Three Bands, Matching and Absolute ID

Step 18:

- Practice Type: Matching.
- Number of Bands: Three Bands.
- Gain Combination: +12 dB.

Step 19:

- Practice Type: Absolute ID.

Step 20:

- Gain Combination: +12/–12 dB.

Step 21:

- Gain Combination: +9/–9 dB.

Step 22:

- Gain Combination: +6/–6 dB.

2.8.5 One-Third-Octave Frequencies, One Band, Matching and Absolute ID

Step 23:

- Practice Type: Matching.
- Frequency Resolution: 1/3 Octave.
- Number of Bands: One Band.
- Gain Combination: +12 dB.

Step 24:

- Gain Combination: +12/–12 dB.

Step 25:

- Gain Combination: +9/–9 dB.

Step 26:

- Gain Combination: +6/–6 dB.

Step 27:

- Practice Type: Absolute ID.
- Gain Combination: +12/–12 dB.

Step 28:

- Practice Type: Matching.
- Number of Bands: Two Bands.
- Gain Combination: +12 dB.

Step 29:

- Practice Type: Absolute ID.
- Number of Bands: Two Bands.
- Gain Combination: +12 dB.

Step 30:

- Practice Type: Matching.
- Number of Bands: Three Bands.
- Gain Combination: +12 dB.

Step 31:

- Practice Type: Absolute ID.
- Number of Bands: Three Bands.
- Gain Combination: +12 dB.

2.8.6 *Time Limit*

In a recording studio or live sound venue, time is of the essence. As audio engineers, we must often make quick yet accurate decisions about sound quality and audio signal processing. To help prepare for these real-world situations, you can apply a time limit in the practice module so that you can practice equalization parameter identification with speed as well as accuracy.

Sometimes a time limit is useful because it forces us to respond with our first impression rather than spend too much time thinking and second-guessing ourselves. Students who have used the practice module often report to me that overthinking a question results in mistakes and that their first impressions are often the most accurate.

To increase difficulty further, use the timer to focus on the speed with which you respond to questions.

2.9 Octave Frequency Boosts and Their Effect on Recordings

To help memorize the timbral effect of each octave frequency, note the instruments and voices that are affected for each frequency for each recording that you train with. To do this, boost an octave frequency and identify which instrument or voice is most affected by that frequency. Then do the same thing with each of the remaining eight octave frequencies. For each recording, keep written notes to describe the way each frequency affects the instruments and voices in a recording, and refer to your notes as you train. The goal of this exercise is to help memorize the timbre of each frequency by associating it with a specific element in a recording. If we know, for example, that a 2,000 Hz boost brings out the vocals and snare on a particular recording, it can help us remember that frequency. When I give listening quizzes in my class, I encourage students to refer to their notes about the timbral effect of each frequency with each recording.

The frequency content of most sound sources spans more than one octave so it can be helpful to describe the effect more specifically. For example, although 250 Hz may bring out the chest sound of a vocal, 4,000 Hz may bring out the vowel clarity.

Below I describe some ways that the nine octave frequencies affect different recordings for me. All are with boosts of 12 dB and Q = 2.

2.9.1 63 Hz

A boost at 63 Hz can produce a pleasant effect with most recordings, especially when listening at quiet volumes, since it helps compensate for the ear's reduced sensitivity to low frequencies at low sound levels. But in recording and mixing projects, the low end can get out of control if we're not paying close attention to it. Low-frequency rumble, thumps, or other excessive low-frequency content can detract from a mix, especially when played back over a system with good low-frequency extension.

"Teardrop" by Massive Attack from Mezzanine *(1998)*
The already prominent low frequencies of the kick drum get boosted even more with a boost at 63 Hz.

"1-2-3" by Café Tacvba from Jei Beibi *(2017)*
The bass, kick drum, and floor tom are all affected by a boost at 63 Hz in this track. The effect seems to be most prominent on the low end of the bass and low end of the floor tom than it does on the kick drum.

2.9.2 125 Hz

Although 63 and 125 Hz often both affect the bass and kick drum, they affect these elements in a mix differently. To my ears, a boost at 125 Hz can sound "thuddy" or "boomy."

"Peg" by Steely Dan from Aja *(1977)*
A boost at 125 Hz on this track again brings out the bass more than the kick drum. Still, the kick is affected, and this boost gives it a thuddy or punchy quality, making this portion of the kick drum frequency spectrum overly loud in the mix.

"Fast Car" by Tracy Chapman from Tracy Chapman *(1988)*
A boost at 125 Hz on this track again brings out the bass a bit more than the kick drum, but the kick drum and bass sound more balanced on this track. I don't hear the same thuddy sound on the kick drum.

2.9.3 250 Hz

VOWEL SOUND: [U] AS IN B**OO**T /BOOT/

In general, the 250 to 500 Hz range can be a tricky one. On the one hand, this octave provides important midrange frequencies to a mix so that it doesn't sound thin. But because so many instruments and voices have harmonics or overtones in this range, these can add up to easily overpower other frequency ranges and make a mix sound cluttered or muddy.

"My Universe" by Coldplay and BTS from Music of the Spheres *(2021)*
A boost at 250 Hz brings out the low end of the vocals as well as some articulation in the bass.

"September" by Earth, Wind & Fire from The Best of Earth, Wind & Fire Vol. 1 *(1978)*
A boost at 250 Hz has a similar effect on this track, bringing out the bass articulation and low end of the vocals.

2.9.4 500 Hz

VOWEL SOUND: [O] AS IN B**OA**T /BOHT/

The first quality I hear with a 500 Hz boost is hollowness. Many instruments and voices overlap in this important midrange frequency.

"Gaslighting Abbie" by Steely Dan from Two Against Nature *(2000)*
Almost all instruments in this track have harmonics or overtones in the 500 Hz range. The lower end of the snare, mid-range articulation of the bass, low end of the vocals, guitars, keyboards, woodwinds, and bit of the mid-range on the kick drum are all brought out with a boost at 500 Hz.

"Hate to Lose Your Lovin'" by Little Feat from Let It Roll *(1988)*
Several elements on this track including piano, guitars, vocals, and snare are affected by a boost at 500 Hz.

2.9.5 1,000 Hz

VOWEL SOUND: [ə] AS IN **A**BOUT /UHBOUT/

A boost at 1,000 Hz can make a mix sound closer. Often piano and keyboards are accentuated with 1,000 Hz, but many instruments and voices overlap in this range.

"Blinding Lights" by The Weeknd from After Hours *(2020)*
The vocals, melodic synths and pads, and syncopated percussion parts are all affected. The snare is also partially affected. I find it interesting that a 1,000 Hz boost also brings out the reverberation on this track.

"Disco Ears" by Joshua Redman, Brad Mehldau, Christian McBride, and Brian Blade from LongGone *(2022)*
A boost at 1,000 Hz brings out the piano and soprano saxophone, but it also seems to affect the drums and midrange harmonics on the acoustic bass.

2.9.6 2,000 Hz

VOWEL SOUND: [ɛ] AS IN B**E**T /BET/

A boost in the 2,000 to 3,000 Hz range can sometimes be used in the mastering stage to bring out a vocal. Also, snare drums usually have most of their characteristic "crack" sound near 2,000 Hz.

"breathin" by Ariana Grande from Sweetener *(2018)*
A boost at 2,000 Hz really brings out the vocals as well as the snare. Percussion parts and the guitar solo also become more prominent with a boost at this octave frequency.

"HERicane" by Lucky Daye from Algorithm *(2024)*
Like the Ariana Grande track, a boost at 2,000 Hz brings out the vocals, snare, and guitar. I am also hearing more reverberation with this boost engaged.

2.9.7 4,000 Hz

VOWEL SOUND: [I] AS IN **BEA**T /BEET/

In general, a boost at 4,000 Hz can make a well-balanced recording sound harsh. Since the ear canal has a natural resonance in the 3,000 to 4,000 range, we are most sensitive to these frequencies.

"Better Broken" by Sarah McLachlan from Better Broken *(2025)*
A boost at 4,000 Hz brings out the vocals, snare, hi-hat, and the vinyl crackle effect. It also makes the delay effect on the vocal more audible.

"Over (feat. Yebba)" by Robert Glasper from Black Radio III *(2022)*
A boost at 4,000 Hz brings out the vocals, mainly by affecting the sibilance of the vocals (i.e., the "ess" sound and breathiness). It also brings out the snare and hi-hat.

2.9.8 8,000 Hz

"Copy of A" by Nine Inch Nails from Hesitation Marks *(2013)*
A boost at 8,000 Hz brings out the edgy, high-frequency programmed percussion parts, making them sound piercing. It also affects the sibilance on the vocal and the high frequencies of the snare drum.

"Give Life Back to Music" by Daft Punk from Random Access Memories *(2013)*
A boost at 8,000 Hz, brings out the cymbals most but it also affects the guitar, synth parts, vocoder, and to a lesser extent the kick.

2.9.9 16,000 Hz

If a recording does not have much content in the 16,000 Hz range, it can be difficult to hear a boost or cut there. Mixing and mastering engineers sometimes add a small boost with a high shelf in this range to give an airiness and openness to a recording.

"Marche pour la cérémonie des Turcs (Jean-Baptiste Lully)" by Le Concert des Nations from Tous les matins du monde—Bande originale du film *(2001)*
In this Baroque music recording, a boost at 16,000 Hz brings out details and clarity in the harpsichord articulation. High frequencies in the strings are also highlighted.

"The Hounds of Winter" by Sting from Mercury Falling *(1996)*
A boost at 16,000 Hz brings out details in the high overtones of the cymbals, sibilance and airiness on the vocal, and the snare drum.

These are just a few examples. I encourage you to discover for yourself how each octave frequency affects different recordings. If you keep notes about your discoveries, this can help you learn and memorize the distinctive timbres produced by each of the octave frequencies.

2.10 Summary

Fully sweepable parametric equalization is perhaps our most important tool as audio engineers. It is possible to learn how to identify boosts and cuts by ear through practice. The available software practice module can serve as an effective tool for progress in technical ear training and critical listening when used for regular and consistent practice.

Notes

1 Sonic Visualiser is a free, open-source application for Windows, Linux, and Mac: https://sonicvisualiser.org.
2 Reaper, ReaPlugs VST FX Suite: www.reaper.fm/reaplugs.
3 For example, MC Squared System Design Group's Room Mode Calculator: www.mcsquared.com/classic/modecalc.htm.

Bibliography

Bristow-Johnson, Robert. 1994. "The Equivalence of Various Methods of Computing Biquad Coefficients for Audio Parametric Equalizers." *97th Convention of the Audio Engineering Society*, Preprint 3906.

Fletcher, Harvey, and W.A. Munson. 1933. "Loudness, Its Definition, Measurement and Calculation." *Journal of the Acoustical Society of America* 5: 82–108. https://doi.org/10.1121/1.1915637.

ITU-R. 2015. *Methods for the Subjective Assessment of Small Impairments in Audio Systems*. BS.1116-3. International Telecommunication Union.

Letowski, Tomasz. 1985. "Development of Technical Listening Skills: Timbre Solfeggio." *Journal of the Audio Engineering Society* 33: 240–44.

Massenburg, George. 1972. "Parametric Equalization." Paper presented at 42nd Convention of the Audio Engineering Society. *42nd Convention of the Audio Engineering Society*.

Meyer, Jürgen. 2009. *Acoustics and the Performance of Music: Manual for Acousticians, Audio Engineers, Musicians, Architects and Musical Instrument Makers*. Fifth Edition. Translated by Uwe Hansen. Springer.

Miskiewicz, Andrzej. 1992. "Timbre Solfege: A Course in Technical Listening for Sound Engineers." *Journal of the Audio Engineering Society* 40: 621–25.

Opolko, Frank J., and Wieslaw R. Woszczyk. 1982. "A Combinative Microphone Technique Using Contact and Air Microphones." *72nd Convention of the Audio Engineering Society*, Paper number 1948.

Quesnel, René. 2001. "A Computer-Assisted Method for Training and Researching Timbre Memory and Evaluation Skills." PhD Thesis, McGill University. https://escholarship.mcgill.ca/concern/theses/h702q794m.

Quesnel, René, and Wieslaw R. Woszczyk. 1994. "A Computer-Aided System for Timbral Ear Training." *96th Convention of the Audio Engineering Society*, Paper number 3856.

Slawson, A. W. 1968. "Vowel Quality and Musical Timbre as Functions of Spectrum Envelope and Fundamental Frequency." *Journal of the Acoustical Society of America* 43 (1): 87–101.

Toole, Floyd E. 2026. *Sound Reproduction: The Acoustics and Psychoacoustics of Loudspeakers, Rooms and Headphones*. 4th ed. With Sean Olive and Todd Welti. Routledge. https://doi.org/10.4324/9781003477495.

Chapter 3

Spatial Attributes and Reverberation

3.1 Spatial Attributes

Reverberation can create distance, depth, and spaciousness in recordings, whether it is captured with microphones during the recording process or added later during mixing. The use of reverberation in recordings has evolved into distinct conventions for various music genres and eras. Although specific reverberation techniques are not necessarily consistent from one musical genre to another, the general principles of reverberation are the same. Film and game sound also make extensive use of reverberation to reinforce visual scenes with a parallel auditory perspective or to provide information about off-camera actions or scenes.

In classical music recording, microphones are often positioned to blend direct sound from instruments and voices with indirect sound—that is, reflected sound and reverberation. The idea is to represent the natural sound of musicians performing in a reverberant space. As such, it is important to listen closely to the balance of the dry and reverberant sound and adjust microphone positions to optimize the balance of direct sound and reverberation. Microphone distance from the instruments can change the direct sound to reverberation ratio. Closer microphone placement focuses on the direct sound, making the recording drier. Moving microphones farther away from the instruments increases reverberation and decreases direct sound level.

Pop, rock, electronic, and other styles of music that use predominantly electric instruments and computer-generated sounds are not usually recorded in reverberant acoustic spaces, although there are some spectacular exceptions. Rather, engineers often create a sense of space with artificial reverberation and delays after recording music with close microphones in a studio space with dry acoustics. Artificial reverberation and delay can be used to mimic real acoustic spaces or to create completely unnatural-sounding spaces. It may not suit a project to position every instrument or voice to sound as though they are at the front edge of the virtual sound stage, and reverberation can help create a sense of depth and distance.

We can think of recorded sound images as a photograph or a painting. Photographers create a sense of depth by focusing on a few elements in the foreground while blurring other elements in the mid-ground and background. Delay and reverberation are the key tools that help us create a sense of depth and distance in a recording. More reverberation on a source makes it sound farther away while dryer elements remain to the front of our sound image. Not only can we make sounds seem farther away and create the impression of an acoustic space, but we can also influence the character and mood of a recording with careful use of artificial reverberation, delays, and simulated early reflections. In addition to depth and distance control, we can control the angular location—that is, the left–right position, also known as azimuth—of sound sources through standard amplitude panning. Spatial attributes of audio signals include characteristics of the sound sources and the

DOI: 10.4324/9781003461173-3

acoustic spaces in which sources are located. In other words, direct sound and indirect sound are both included when we talk about spatial attributes.

With stereo speakers, we have two dimensions within which to control sound source location: distance (near to far) and angular location (azimuth). With elevated loudspeaker arrays such as those found in Dolby Atmos and IMAX movies, theme parks, and audio research environments, or with 3D virtual reality audio over headphones, we can reproduce audio that includes the dimension of height.

Acoustic spaces, whether simulated or real, include reverberation, early reflection patterns, maybe also prominent and/or delayed echoes, and for smaller rooms, standing waves or room modes. These characteristics help listeners perceive the size of the space. Artificial reverberation parameters such as decay time, predelay time, early reflection patterns, size, and diffusion provide the ability to vary the qualities of a simulated room.

3.1.1 Correlation and Spatial Attributes

Spatial attributes also include correlation which can affect the spatial continuity of a sound image. Correlation characterizes the similarity of two signals, and it is measured with a correlation or phase meter on a scale that typically ranges from –1 to +1. When a correlation meter registers +1, it means that the left and right channels are identical, although they can be different amplitudes. When it registers –1, it means that the left and right channels are identical except that one of them is opposite polarity, or phase inverted.[1] Most stereo recordings have a correlation that dynamically fluctuates between 0 and +1 and with occasional jumps toward –1.

Left and right channel correlation affects the perceived width in a recording. Two perfectly correlated channels will result in a mono image. Stereo signals with a correlation of –1 create an artificially wide-sounding stereo image because the channels are opposite polarity. It is easier to hear the effect of negatively correlated channels over loudspeakers than over headphones. Sitting in the ideal listening position, we tend to localize them to the sides of our head or outside of the loudspeaker locations. If we move ever so slightly to the left or right of the ideal listening location, the sound image will shift quickly to one side or the other. Negatively correlated sound images seem unstable and difficult to localize. It is important to listen for artificially wide stereo mixes or elements within a mix, which would indicate that there may be a polarity problem somewhere in the signal path that needs to be corrected. Some amount of opposite polarity in a music recording is not a problem per se, but it is important to keep in mind that opposite-polarity signals cancel when summed to mono. Playing a recording over a single, mono Bluetooth speaker, for example, sums the left and right channels and cancels any opposite-polarity portions of a signal. We discuss more on listening to opposite-polarity channels in the discussion of mid-side processing in Section 3.9.

Decorrelated channels—that is, when the correlation or phase meter reads 0—tend to create a split or wide stereo image where two sound images are located only to the sides at the left and right speakers, with little in the center. If you listen to decorrelated pink noise over stereo speakers, you may notice very little in the center of the stereo image. The image is not typically wider than the speakers, as it is with a negatively correlated image. If you do localize a stereo image in the center, it will likely consist of mainly low frequencies. High frequencies are clearly located at the speakers in a decorrelated image.

In addition to a correlation meter, a goniometer, sometimes called a vectorscope, is also useful for monitoring the stereo image width. It measures the relationship between the phase of two channels and displays it as a two-dimensional graph called a Lissajous figure. The shape of a Lissajous

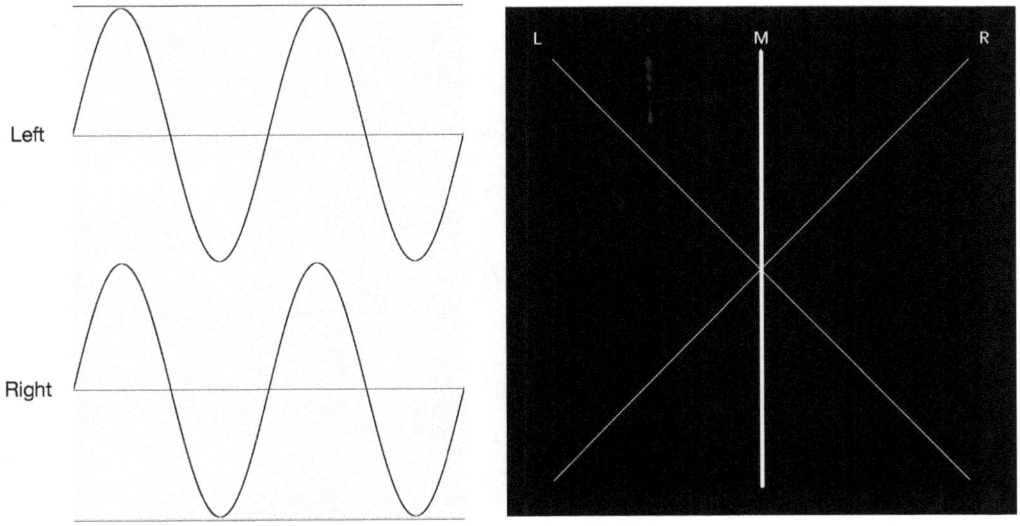

Figure 3.1 Two identical sine tone signals are shown on the left side of the figure. A goniometer, shown on the right, visualizes the stereo image width as a vertical line in the center of the meter. The correlation of these two channels is +1. In other words, the channels are correlated. (The goniometer image is a screenshot of the Apple Logic Pro MultiMeter plugin.)

figure indicates phase coherence and provides a visual indication of stereo image width and left–right channel balance. Originally, this measurement was only available using an oscilloscope in X–Y mode, but goniometers or vectorscopes are commonly found in DAW multimeter plugins, usually in combination with a phase or correlation meter.

To get an idea how a goniometer represents a stereo image, we will look at a sine tone test signal. Starting with a sine tone at 1,000 Hz panned center, a goniometer displays a vertical line in the center of the meter, as shown in Figure 3.1. The meter represents where we would localize this center-panned sound—that is, directly in the center of the stereo image. Since the left and right channel signals are identical, the correlation of the left and right channels is +1.

If the sine tone is panned to the left, the goniometer shows a line tilting to the left at a 45° angle from the center, as shown in Figure 3.2. If we listen to a signal panned to the left, we localize it in the left speaker or headphone. Since the signal is present in the left channel only, the correlation of the left and right channels is 0.

If the sine tone is panned back to center but the polarity of one of the output channels is inverted, the goniometer shows a horizontal line in the center, as shown in Figure 3.3. The horizontal line represents negatively correlated left and right channels. Since the left and right channel signals are identical but opposite polarity, the correlation is –1.

While sine tones are useful for illustrating some basic goniometer conditions, in practice we are more likely to encounter more complex signals such as music, speech, and sound effects. Figure 3.4 shows goniometer screenshots of a moment in time of two different pop stereo recordings. You can see in the top of the figure that the Lissajous figure is primarily located in the center of the goniometer display, which means the correlation between the left and right channels is high and the stereo image is narrow. Because many prominent elements of most pop recordings, such as vocals, bass, snare drum, and kick drum, are often panned center, the meter reflects the stereo

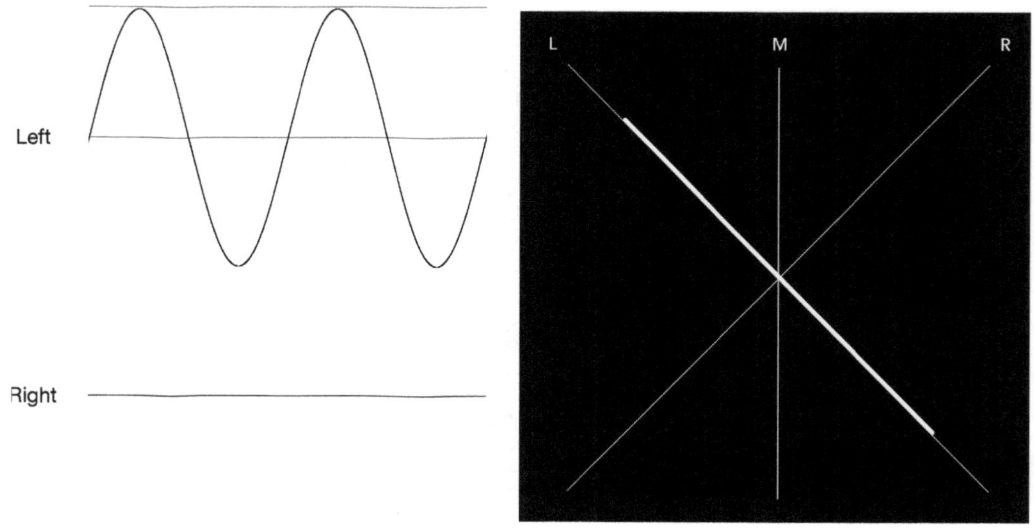

Figure 3.2 A sine tone signal in the left channel only is shown on the left side of the figure. A goniometer, shown on the right, visualizes the stereo image width as a straight line at a 45° angle in the center of the meter. The correlation of these two channels is 0. In other words, the channels are decorrelated. (The goniometer image is a screenshot of the Apple Logic Pro MultiMeter plugin.)

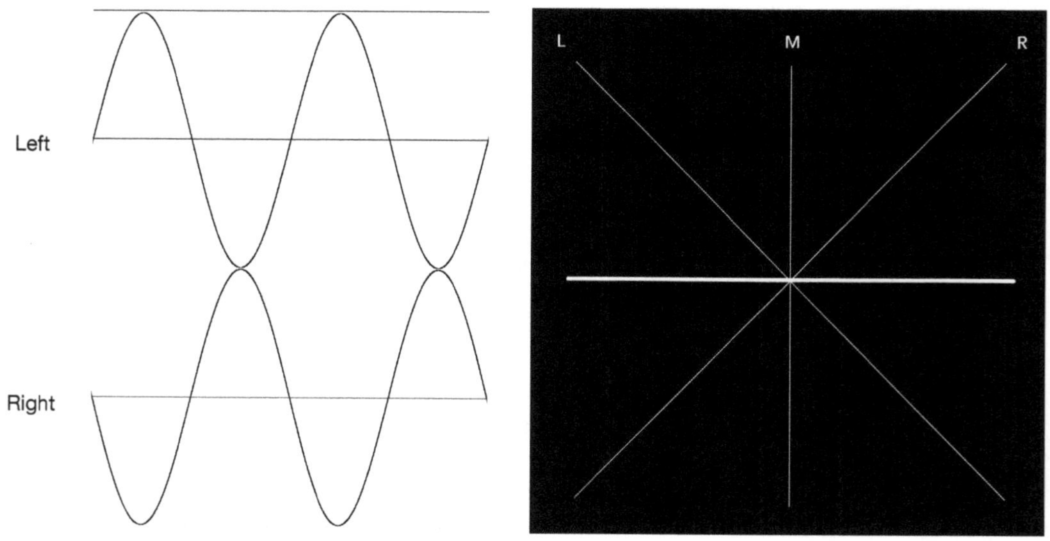

Figure 3.3 Two opposite-polarity sine tone signals are shown on the left side of the figure. A goniometer, shown on the right, visualizes the stereo image width as a straight horizontal line in the center of the meter. The correlation of these two channels is –1. In other words, the channels are negatively correlated. (The goniometer image is a screenshot of the Apple Logic Pro MultiMeter plugin.)

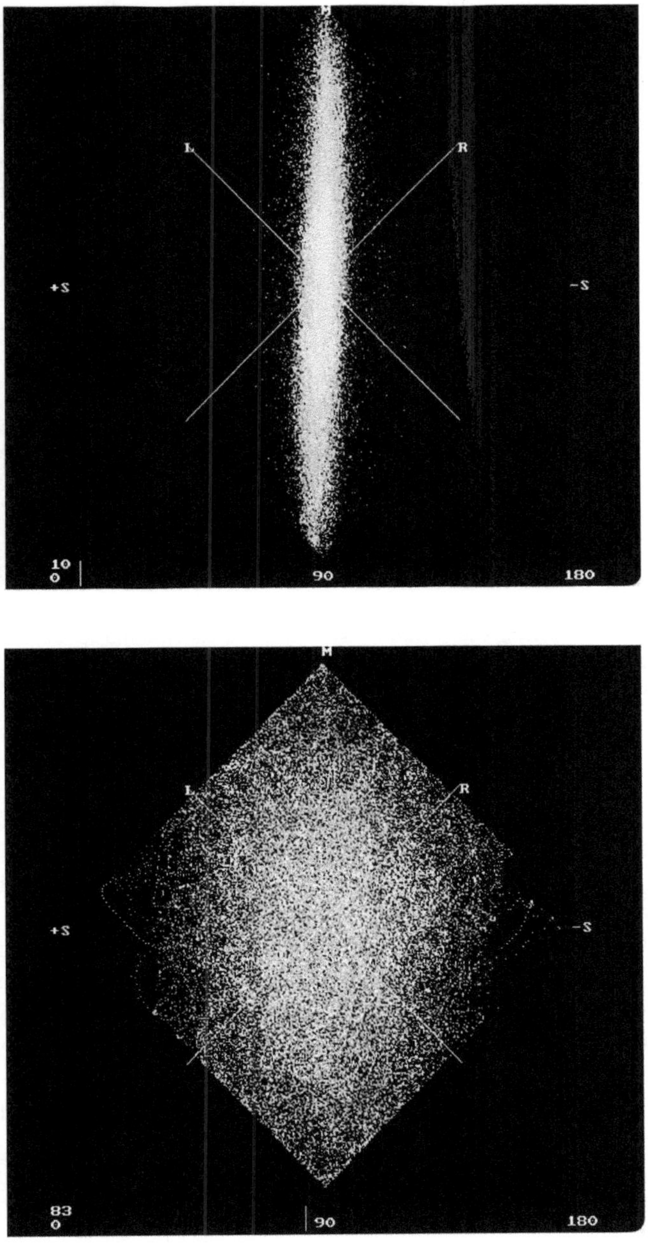

Figure 3.4 Two goniometers showing stereo image widths of two different pop music recordings. The goniometer readings show a narrow stereo image on top and a wide one on the bottom. The correlation meter at the bottom of each goniometer ranges from 0° to 180°. Correlation of most stereo recordings fluctuates over time, but at the moments these screenshots were taken, the correlation meter of the top meter was reading 10, which means the left and right channels are highly correlated, and the correlation meter of the bottom was reading 83, which means the left and right channels are nearly decorrelated. A correlation of 90 on this meter represents perfect decorrelation. (The images are screenshots of the Cockos REAPER Goniometer plugin.)

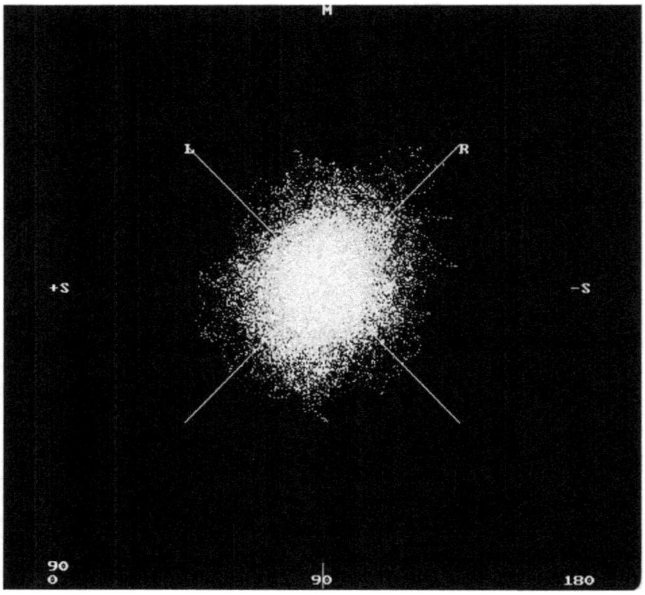

Figure 3.5 A goniometer showing the stereo image width of a classical music recording. The stereo signal is widely spread across the meter and is reflected in the wide stereo image of the recording. The correlation meter reads 90, which means the channels are decorrelated. (The image is a screenshot of the Cockos REAPER Goniometer plugin.)

image that we hear: a center-heavy mix. The bottom of Figure 3.4 shows a goniometer meter with a different pop recording. This recording has a much wider stereo image, and the width is reflected in the wide Lissajous figure in that goniometer display.

With different mixes we get different Lissajous patterns in the meter. Figure 3.5 shows another stereo mix, this time of a Baroque music recording. Note the wide representation on the meter, which is reflected in the wide stereo image. Classical music recordings will typically produce this type of image on a goniometer because recordings in this genre generally rely primarily on stereo microphone techniques rather than panning sources to the center.

Note that the correlation meter shown at the bottom of each goniometer image in Figures 3.4, 3.5, and 3.7 ranges from 0 to 180 degrees. This corresponds to a range of 1 to –1 in the normalized correlation scale.

In Section 7.2 we look at goniometer displays of specific recordings.

3.1.1.1 Listening Exercise: Negatively Correlated (Opposite-Polarity) Channels

For this exercise, open a DAW, create a new stereo track, and import any stereo recording. Creating this effect on most DAWs is a three-stage process. The idea is to mix a stereo track to mono, so that the left and right channels are identical, reverse the polarity on one of the channels, and then pan the now identical but reverse polarity channels to left and right. Figure 3.6 shows a block diagram of the signal flow for this exercise.

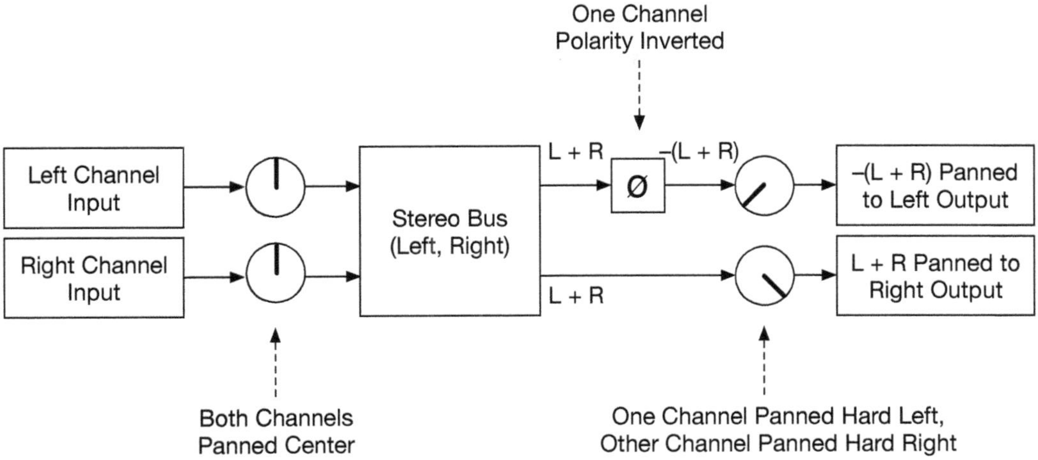

Figure 3.6 A block diagram showing the signal path to convert a left/right stereo recording to two negatively correlated signals. The left and right channels are summed (both are panned to center), making them identical. Then the polarity of one stereo bus channel is inverted and panned left. The other channel is panned right without polarity inversion.

Depending on the workstation that you are working with, there are different ways to accomplish this:

- **Apple Logic Pro:** Insert the Gain plugin on the stereo track and turn the Mono button ON in the plugin. Insert a second Gain plugin after the first one and turn the Phase Invert ON for either the left or right channel but not both. If you convert to mono and phase invert within just a single Gain plugin, you get a different result–that is, you get the side component, which we discuss in Section 3.9 below. The way to get negatively correlated channels is to use two separate Gain plugins.
- **Avid Pro Tools:** Pan the left and right channels of a stereo track to center. Insert the Trim plugin on the Master Fader track, not on the input track, and within that plugin turn the Master Link OFF and reverse the polarity of either the left or right channel by clicking on the polarity button, which is labeled as ø.
- **Cockos REAPER:** Right-click on the stereo track panner, select Dual pan as the pan mode, and pan both channels to center. Insert the Channel Polarity Control plugin in the Mixer Master, not on the track channel, and under Polarity Mode, select either Invert Left or Invert Right.

For recordings with strong center images, you may need to reduce the track channel fader to avoid clipping, since the level can increase by 6 dB for any center-panned parts of the stereo image.

Listen to the wide stereo effect that the opposite polarity or out-of-phase condition creates. It creates a highly unusual effect, so once you hear it, you will likely remember it and recognize it immediately when you hear it again.

3.1.1.2 Listening Exercise: Decorrelated Channels

In a DAW, create a stereo track and add a pink noise generator plugin to it. Depending on the workstation, the plugin will have different names. For example, it is called Test Oscillator in Logic Pro, Signal Generator in Pro Tools, and Pink Noise Generator in REAPER. It might be found in the Utilities or Other section of the plugins. After you add a pink noise generator to the stereo bus, it should be sending out two decorrelated channels of pink noise. The Logic Pro Test Oscillator may not automatically generate separate decorrelated pink noise signals for left and right, since it seems to default to a single mono, correlated signal for left and right outputs. One way to force it to generate independent left and right signals is to switch to a sine wave and back to pink noise for one of the channels. After you do that, it should generate decorrelated signals.

How do we know if the left and right signals are decorrelated? One way is to listen. If you hear a narrow, mono sound in the center of the stereo image, the channels are identical and are not decorrelated. Another way to know is to use a goniometer, such as the Logic Pro MultiMeter or REAPER Goniometer plugins. Pro Tools does not include a goniometer, but third-party plugins are available.

With decorrelated pink noise being generated, invert the polarity of one channel of the stereo bus, using a plugin such as the Logic Pro Gain plugin, the Pro Tools Trim plugin, or REAPER Channel Polarity Control plugin as described above. There should be no audible effect or measurable change in the correlation when you invert the polarity on either channel because the channels are decorrelated. Inverting the polarity has no audible effect on a stereo image when the left and right channels are decorrelated.

When measuring decorrelated pink noise, a goniometer will show a random distribution evenly spread across the stereo image, as shown in the Lissajous pattern in Figure 3.7. In this case at least, the visual image does not completely match what I hear. When I listen to decorrelated pink noise,

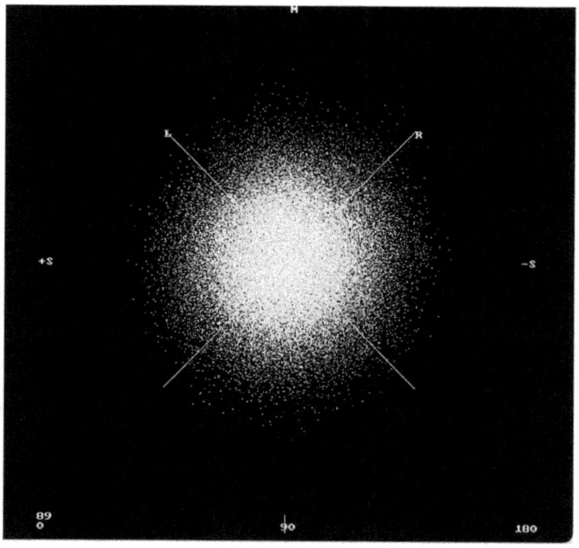

Figure 3.7 A goniometer showing the stereo image width of decorrelated pink noise. The signal is evenly distributed and widely spread around the center of the meter. The correlation meter reads 89, which means the two channels of pink noise are decorrelated. (The image is a screenshot of the Cockos REAPER Goniometer plugin.)

it seems as though the sound is originating mainly at the speakers with a hole in the middle of the stereo image. Although a goniometer gives us some clues about what we hear in a stereo image, it may not always represent precisely our perception of stereo width. This is another reason why we cannot rely solely on meters to make final decisions about stereo image, tonal balance, and sound quality; we must use our ears.

We can use recording techniques to create decorrelated signals which can be panned to make wide stereo images. For example, we can create decorrelated signals by recording one musical part—for example, a vocal, guitar, or keyboard part—and then re-recording the same musical part in unison a second time on a different track. The resulting recordings of the two unison musical parts are decorrelated even though they are the same notes played in time together. When panned hard left and hard right, they will produce a wide stereo image.

3.2 Analysis of Perceived Spatial Attributes

Spatial attributes refer to the azimuth, elevation, and distance of sound sources, as well as information about the acoustic space—that is, the environment or enclosure, in which sounds are produced. Because we have two ears, we rely on interaural time differences, interaural intensity differences, and filtering by the pinnae or outer ears to determine the location of a sound source (Moore 1997). Localizing sound reproduced over loudspeakers is somewhat different from localization in natural listening.

As audio engineers, we want to be attuned to spatial processing that may be present, however subtle. Panning, delay, and reverb affect the balance and blend of elements in a mix, which in turn influence the way in which listeners or audience members perceive a musical recording, a movie, or a live event and react to it emotionally. For example, long reverb times can create drama and excitement, as though the music is anthemic and emanating from a big space. Alternatively, short reverberation times can create a warm and intimate feeling or, conversely, a stark and cold sound image.

Reverb is important not only in classical music recordings but also in recordings of other music genres. Phil Spector's Wall of Sound recordings from the 1960s made use of reverberant spaces to create emotional impact in his productions. The quintessential song in this production style is "Be My Baby" by the Ronettes from 1963. A couple of decades later, with the help of producers Brian Eno and Daniel Lanois, U2's landmark albums *The Joshua Tree* and *The Unforgettable Fire* employed extensive use of delays and reverb to create the sense of big open spaces. Eno and Lanois had become well known for their ambient music recordings only a few years prior, and they brought some of these spatial processing and experimental electronic music treatments to subsequent mainstream pop music recordings. German record producer Manfred Eicher and his label ECM use reverberation more prominently on their jazz recordings than many other jazz labels do, which gives their recordings a distinctive sound. Indie pop/folk bands such as Fleet Foxes, The Lumineers, and Sigur Rós have produced albums with clearly noticeable washes of reverb. Kendrick Lamar and SZA's "All The Stars" displays generous use of reverb especially on the vocals, which creates a large, lush space for the song. The long, smooth reverberation decay on Billie Eilish's "What Was I Made For?" effectively supports her breathy vocals. The selective use of long reverb times on vocals and piano in Taylor Swift's song "The Fate of Ophelia" helps create depth, creates a juxtaposition with the comparatively dry drum parts, and helps the backing vocals blend with the lead vocals.

Although some listeners may perceive prominent reverb in music recordings as gimmicky, I often find that reverb makes a recording more interesting and engaging. But simply adding more reverberation does not fix a bad recording. Reverberation that does not blend well with the music, makes the music muddy, or does not have any apparent musical role can detract from the listening experience by distracting us from the music. Production choices should be guided by the needs of the music. Regardless of the way we may feel about reverb use in general, we can still listen to the way reverberation, echo, and delay are used in commercial recordings and try to emulate some of the myriad ways reverb is used.

3.2.1 *Spatial Layout and Masking*

Masking is the phenomenon by which a louder sound can make a quieter sound less audible. One of the ways to reduce masking is to spatially separate sounds. Sound sources in a mix panned to the same location may partially or completely mask other sound sources panned to the same location. Pan the sounds apart and a quieter sound that was previously masked may become audible even without changing its level. Mono recordings are more difficult to mix because everything is in the same place, which means masking is more of an issue for this format. Also, mono recordings require more compromises than stereo or multichannel recordings because there is no spatial separation in mono recordings and some elements will not be as audible as they can be in stereo or multichannel formats. But masking is not something that we need to always minimize or avoid. We can use masking to help blend elements in a mix. The spatial layout of sources can influence the clarity and cohesion of a sound image.

The amplitude envelope of a sound can also affect how much it masks another sound. Reverberation in a recording can seem inaudible or at least difficult to hear when it blends with and is partially masked by sustained sounds. Transient elements, on the other hand, such as percussion and drums, allow reverberation to be heard more clearly because reverberation usually lingers much longer than transient elements like percussion. The short decay of a transient sound makes it possible to hear reverberation tails and early reflections that are present.

3.2.1.1 Listening Exercise: Hearing Reverb in Your Work

When you add reverb to a mix, mute and unmute the reverberation to hear its contribution to a mix. Try listening to one track at a time with the reverb. This is usually easier to do if the reverb is on an auxiliary track and you are using aux sends to route audio to it. Does the reverb seem more audible when mixed with some sound sources, such as those that are more transient in nature, compared to other sources? It can be difficult to hear a reverberation tail when the decay time is short or if the mix contains mostly sustained sounds. If you cannot hear the reverberation decay specifically, do you hear any changes to overall spaciousness or depth? Is it helping elements in the mix blend? If you do hear the reverberation clearly, is it mixed too high or is the decay time too long, making the mix sound cluttered and muddy?

3.2.2 *Spatial Attributes of Sound Sources*

The spatial attributes of sound sources consist of three main categories:

- angular location,
- distance, and
- spatial extent.

3.2.2.1 *Angular Location and Panning*

A sound source's angular location or azimuth is its perceived location in a stereo image, generally between the left and right loudspeakers. If sources are spread out across a stereo image, masking is reduced and the clarity of each sound source is optimized. Masking is also more likely to occur when sources occupy not only the same spatial location but also the same frequency range.

Microphone signals and tracks can be panned to specific locations between loudspeakers using conventional constant-power panning found on most mixers, or they can be positioned by delaying a signal's output to one loudspeaker channel relative to the other loudspeaker output. But delay-based panning is rarely found on mixers since it is not as effective as amplitude-based panning and because it creates additional issues. One problem with delay-based panning is that its effectiveness depends highly on a listener's location relative to the loudspeakers. When a listener moves to the side of the sweet spot in a loudspeaker setup, the time of arrival differences change. When time of arrival differences change, the perceived locations of delay-panned sources also change. Another problem with delay-based panning is that summing to mono can introduce comb filtering, which affects the timbre of sources that are delay-panned. We discuss comb filtering in more detail below in Sections 3.3 and 3.4.

Although delay-based panners are uncommon, spaced stereo microphone techniques, such as ORTF, NOS, and A-B, inherently include interchannel time differences. Stereo microphone techniques specify that microphone signals should be panned hard left and right, and microphone spacing causes time of arrival differences between them. Although the time differences are small—for example, a maximum of 0.5 milliseconds for the 17 cm (6.7 in.) spacing of ORTF—the interchannel time difference combines with interchannel amplitude differences to create a more natural source placement in the stereo image. This is similar to natural hearing which combines interaural time and level differences due to the fact that human ears are spaced apart on opposite sides of the head. The time difference reinforces our perception of source location that the amplitude difference provides. Sound source locations in a stereo image will depend on the stereo microphone technique used and the respective locations of each source. Spaced stereo microphone techniques such as ORTF will produce a wider sound image than a coincident technique such as X-Y. This difference in stereo image width is partly because the X-Y technique does not produce an interchannel time difference. Stereo microphone techniques such as ORTF and A-B have stood the test of time and can offer a natural width and spatial continuity that is not possible, or at least very difficult to achieve, with a single panned microphone signal and reverberation. The late Bruce Swedien, legendary recording engineer best known for his collaboration with Michael Jackson and the producer Quincy Jones, reported that he used stereo microphone techniques almost exclusively in his recordings. Perhaps that is one reason why the recordings he engineered sound so good.

3.2.2.2 *Distance*

Although perception of absolute distance of sound is often inaccurate, relative distance within a recording can give it depth. Music recorded in an acoustically live space may exhibit a natural sense of depth, like the depth an audience member would hear in the same space. The effect of depth and relative distance in classical music recordings can happen quite naturally with a stereo pair of microphones in front of an ensemble on a stage. Musicians at the front of the stage sound closer, as the microphones are closer to them, than musicians near the back of the stage who are further away from the microphones. Sound sources close to microphones are louder and drier—that is, less reverberant—than those further away.

When we make recordings in acoustically dry spaces such as recording studios, we can create depth by controlling the relative levels of tracks, applying low-pass filters, adding delays, and mixing in artificial reverberation. Sound source distance can be controlled by adjusting physical parameters such as the following.

In natural listening:

- **Direct sound level:** Listeners judge quieter sounds to be farther away because the sound intensity of direct sound naturally drops by 6 dB for each doubling of distance from a source. This cue can be ambiguous for the listener because a change in loudness *could* be the result of a sound source's change in distance, or it could be due to the sound source getting louder or quieter. Or a sound level could remain the same as a sound source moves away and simultaneously gets louder. A listener may not be able to judge distance accurately just by listening.
- **Reverberation level:** When a listener moves away from a sound source in a room or hall, the direct sound level decreases while the reverberant sound remains roughly constant. A change in the direct-to-reverberant level affects the perception of distance. More reverberation and less direct sound typically mean that a source is further away.

In recordings:

- **Microphone distance:** Moving microphones farther away decreases the direct-to-reverberant ratio and therefore creates a greater sense of distance.
- **Room microphone placement and level:** Microphones placed at the opposite end of a room relative to musicians will pick up reverberant sound primarily. Room microphone signals can be used like a reverberation effect to create depth in a mix.
- **Low-pass filtering:** When sound travels through air, high frequencies are naturally attenuated more than low frequencies because of air absorption. We can mimic this effect with low-pass filters on direct sounds and on reverberation. Furthermore, the acoustic properties of surfaces in a room affect the spectrum of reflected sound reaching a listener's ears. Some materials, such as fabric upholstery, drapes, and carpet, absorb more high frequencies than other materials, such as wood, tiles, or drywall.

3.2.2.3 Spatial Extent

The spatial extent of a source is its perceived width. Some sound sources can be precisely localized within a mix, and we can point directly to their virtual locations within a stereo image because they are narrow and focused. Other sound sources may be wider, have fuzzier edges, and have more ambiguous locations. A related concept in concert hall acoustics research is called apparent source width or ASW, which is related to the strength, timing, and direction of side reflections. Stronger reflections from the side of a concert hall can cause listeners to perceive a sound source to be wider than if the side reflections are not present (Barron 2010).

As with concert hall acoustics, the perceived width of sources reproduced over loudspeakers can be influenced by adding early reflections, whether recorded with microphones or generated artificially. Early reflections generally begin to arrive within a few milliseconds of the direct sound, and because their arrival time follows so soon after the direct sound, they tend to fuse perceptually with the direct sound. In other words, we do not hear early reflections as separate echoes of the direct sound. As a result, early reflections, depending on their angle of arrival, can produce a stereo image that is wider than just the dry sound on its own. With loudspeaker reproduction, the perceived width of a sound image will vary with the microphone technique used, the sound source, the acoustic environment in which it is recorded, and any artificial early reflections and reverberation that may be added.

As we discussed above, a stereo image correlation of 0—that is, decorrelated left and right channels—creates a wide image that seems to originate mainly in the left and right loudspeakers with no phantom image in the center. We can affect left–right correlation, and therefore the width of a stereo image, with the spacing of a pair of microphones. In most cases, two microphones placed close together will produce highly correlated signals, except for certain cases with the Blumlein technique that I describe in the next paragraph. Because pairs of coincident microphones occupy nearly the same physical location, the acoustic energy reaching both will be almost identical. As we move them apart, correlation will decrease. A small spacing of an inch or two (a few centimeters) will decorrelate high frequencies but not low frequencies. As microphones are spaced further apart, decorrelation at low frequencies increases. Microphone spacing and the lowest frequency of correlation are inversely proportional because as we go lower in frequency, wavelengths increase, thus requiring greater spacing for low frequencies to be decorrelated. In other words, as we widen a pair of microphones, the resulting stereo image also widens, as correlation decreases, assuming one mic is panned hard left and the other is panned hard right.

The Blumlein stereo microphone technique, which uses coincident figure-8 microphones angled 90° apart, creates a slightly more complicated stereo image. Sound arriving at the front of the microphones is in phase for the two microphones, which means the two channels are correlated. The same is true for sound arriving at the back of the microphones; it is also in phase and correlated. Sound arriving at the sides of the microphones is picked up by each microphone at the same time, which means it is in phase, but the polarities of the microphone outputs are opposite. For

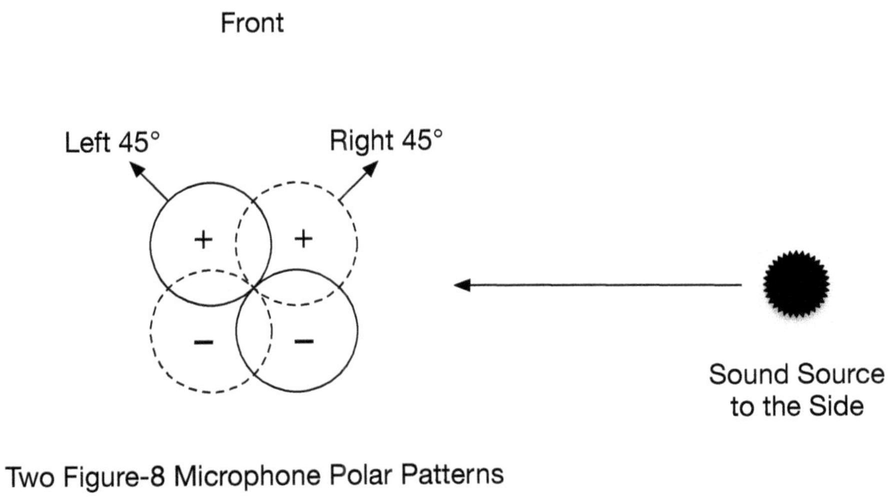

Figure 3.8 A Blumlein stereo microphone technique uses two coincident figure-8 microphones angled 90° apart, as shown by the polar patterns on the left of this diagram. The microphone aimed 45° to the left is shown with a solid line polar pattern, and the microphone aimed 45° to the right is shown with a dashed line polar pattern. Sound arriving from 90° to the right, shown by a black spiky circle on the right side of the illustration, is picked up equally by both microphones, but because the lobes of the figure-8 polar patterns are opposite polarity at the sides of this stereo microphone technique, the resulting stereo image is negatively correlated.

example, sound arriving directly from the right side of a Blumlein pair is picked up equally by the front, positive lobe of the right-facing microphone and by the rear, negative lobe of the left-facing lobe. As a result, sounds arriving from the side are negatively correlated in the stereo image. See Figure 3.8, which shows the polar patterns of the figure-8 microphones and a sound source arriving from the side.

3.2.3 *Spatial Attributes of Acoustic Spaces and Sound Stages*

The spatial attributes of acoustic spaces and sound stages consist of three main categories:

- reverberation decay character,
- spatial extent (width and depth) of the sound, and
- spaciousness.

3.2.3.1 Reverberation Decay Character

The acoustic environment or sound stage may consist of a real acoustic space captured with room microphones, or it may consist of a virtual sound stage created by mixing artificial reverberation with dry recordings. We might choose to use a single reverberation for all sounds or tracks in a project, or we might choose a variety of different reverberation plugins to accentuate differences among the elements in a mix. For instance, in non-classical music it is common to treat vocals and solo instruments with a different reverberation from the rest of the instruments in the band or ensemble.

Decay time is perhaps the most common parameter in artificial reverberation algorithms. Although reverberation decay time is often not adjustable in a real acoustic space, some halls and studios have three-sided panels on the walls and ceiling that can rotate to expose one of three different sound absorbing or reflecting materials. The orientation of the panels determines the amount of sound absorption which in turn controls the reverberation decay time. Longer reverberation times are typically more audible than shorter reverberation times for a given reverberation level. And as we discussed above, transient sounds such as drums or percussion expose decay time more than sustained sounds, allowing us to hear the rate of decay more clearly, because there is less masking with transient sounds.

Some artificial reverberation algorithms incorporate small amounts of time delay modulation to make reverberation decay sound more realistic. Depending on its design and user settings, artificial reverberation risks sounding metallic and ringy, and a small amount of modulation can help minimize these negative effects. The idea is that in a real acoustic space, moving air currents and slight variations in air temperature from one location to another within the room can affect the way sound propagates. Even though temperature differences may be small within a room, the speed of sound in air is determined by temperature, so small amounts of delay time modulation in artificial reverberation is one way to mimic this effect of small variations in the speed of sound. Artificial reverberation can sound unnaturally smooth, and subtle modulation of delay times within the building blocks of a reverberation algorithm can help create the illusion that the reverb is real, or at least less artificial.

3.2.3.2 Spatial Extent (Width and Depth) of the Sound Stage

A sound stage is the acoustic environment within which we hear a sound source, and in analytic listening it is helpful to differentiate the sound stage from the sound source. The environment may be a recording of a real space, or it may be something that has been created artificially using delay and artificial reverberation. Sound stage and sound source may be more difficult to differentiate in audio from a stereo microphone technique compared to that of a dry track mixed with artificial reverberation in which we can add or subtract the sound stage with a fader on a DAW.

3.2.3.3 Spaciousness

Spaciousness represents the perception of physical and acoustical characteristics of a recording space, and in concert hall acoustics, it is related to listener envelopment. We can use the term "spaciousness" to describe the feeling of space within a recording and the sense that we perceive sound to be arriving from every direction around us. Surround sound and 3D audio recordings are likely to have more listener envelopment and spaciousness than stereo recordings because there are loudspeakers or virtual sources, in the case of 3D audio over headphones, located behind the listener.

3.2.4 *Spatial Attributes of Sound Images Over Loudspeakers and Headphones*

Also included with the category of spatial attributes are the overall impressions and characteristics of a stereo image reproduced by loudspeakers or headphones. A stereo image is the illusion of sound source localization from loudspeakers, known as phantom images. Although there are only two loudspeakers for stereo, the human binaural auditory system allows us to hear phantom images at locations between the loudspeakers. They are called phantom images because they seem to originate from locations in which there is no speaker. Likewise with headphones, it is possible to generate virtual sound images in locations other than at the sides of the head where the headphone drivers are located. In this section, we consider the overall qualities of stereo and spatial audio images rather than those specific to source and sound stage.

3.2.4.1 Coherence and Relative Polarity Between Channels

Despite widespread use of stereo and multichannel playback systems among consumers, mono compatibility continues to remain critically important, mainly because people listen to music through a variety of mono sources such as computers, wireless Bluetooth speakers, and mobile devices. Checking mixes for mono compatibility means listening for changes in timbre or mix balances. Typically, timbral changes result when destructive interference between the left and right channels produces comb filtering or when opposite-polarity content is attenuated when summed to mono. If both left and right channels are identical but opposite polarity, typically a worst-case scenario, they will cancel completely when summed to mono. If both channels are identical, or completely correlated, then the mix is monophonic and not truly stereo. Most stereo mixes include some combination of mono or correlated and stereo or decorrelated components, like the examples we saw in Section 3.1. To review, we can describe the correlation between signal components in the left and right channels along a scale between –1 and +1:

- **Correlation = +1:** In this condition, the left and right channels are identical, and the stereo image is monophonic.
- **Correlation = 0:** In this condition, the left and right channels are different. They could be completely unrelated musically or they could be unison—that is, the same musical notes—recorded a second time and panned apart. Two recordings of the same musical part will be decorrelated

because the audio signals are not identical. Although they may have musical correlation, their waveforms are not correlated, which results in a wide stereo image.

- **Correlation = –1:** In this condition, the left and right channels are identical but are opposite in polarity. This is also known as negative correlation.

Sometimes a stereo track within a multitrack mix has opposite-polarity channels panned hard right and left. This could be due to opposite-polarity stereo line outputs from electric instruments or due to a polarity-flip cable that was used by mistake during a recording. Often stereo line outputs from electronic instruments are not truly stereo but mono. When one output is opposite polarity relative to the other, the two channels will cancel when summed to mono. If such a signal is present in the full mix, a phase or correlation meter may not register it strongly enough to give an unambiguous visual indication, which means we must always use our ears.

When we check mono compatibility of a mix, we can simulate what a listener will hear when they listen through a mono playback system. If we discover that summing to mono cancels significant components of a mix or changes the timbre in a negative way, we can address the issues before the recording gets released.

3.2.4.2 *Spatial Continuity of a Sound Image from One Loudspeaker to Another*

An overall attribute of a mix is the continuity and balance of a sound image from one side to the other. An ideal stereo image will be balanced between left and right and will not be focused in any one location. As we saw in the Lissajous patterns in Figure 3.4, pop and rock music mixes may have a strong center component because of the number and strength of instruments that are typically panned center, such as kick drum, snare drum, bass, and vocals. Or they may have a wide stereo image with sources panned more equally across the stereo image. Classical and acoustic music recordings usually have a wide, full stereo image, as we saw in the Lissajous pattern in Figure 3.5. It is possible for the center image to be deficient, and sometimes this is referred to as having a "hole in the middle" of the stereo image.

3.3 Basic Building Blocks of Digital Reverberation

When using artificial reverberation, one challenge is to translate between subjective impressions of the virtual space and controllable parameters such as decay time, predelay time, and early reflections. Although we may want to control sound source distance using reverberation, it is rare to find a parameter labeled "distance" in a conventional digital reverb processor. To make a sound source seem more distant, we need to control distance indirectly by adjusting reverberation parameters in a coordinated way until we have the desired sense of distance. The challenge is to translate between objective parameters of reverberation and the sense of space, envelopment, and distance that we want to create.

Choice of reverberation parameter settings depends on several things such as the transient nature and width of our dry sound sources, as well as the decay and early reflection characteristics of a reverberation algorithm. Professional engineers often adjust parameters until the reverb sounds right for a mix, rather than simply pulling up a preset they used on a previous mix and assume that it will work. A particular combination of parameter settings for one source and reverberation usually cannot simply be duplicated for an identical distance and spaciousness effect with a different source or reverberation algorithm.

3.3.1 Delay: Comb and All-Pass Filters

Although simple in concept, delays serve as fundamental building blocks for a wide variety of complex effects including reverberation, flangers, chorus, and phaser effects. We can use delays to create a sense of space by mixing a straight delay—that is, without feedback—with a dry signal, and this effect can supplement or substitute reverberation. With shorter delay times—for example, less than around 10 to 35 milliseconds depending on the audio signal—the auditory system tends to fuse direct and delayed sounds, and we localize the combined sound based on the location of the first-arriving direct sound. Helmut Haas discovered that a single reflection added to a speech signal fused perceptually with the dry sound unless the reflection arrived more than approximately 25 to 35 milliseconds after the dry sound, at which point we perceive the delayed sound as an echo or separate sound. The phenomenon is known as the Haas effect, the precedence effect, or the law of the first wavefront.

When a signal is mixed with a delayed version of itself, and the delay time is greater than about 25 to 35 milliseconds, depending on the audio signal, it is possible to hear the delayed signal as an echo of the direct sound. The actual amount of delay time required to create a distinct echo depends on the nature of the audio signal being delayed. Transient, percussive signals like xylophone reveal distinct echoes with shorter delay times—that is, less than 30 milliseconds—whereas sustained, steady-state sound requires much longer delay times, typically more than 50 milliseconds, to create an audible echo.

Figure 3.9 shows a block diagram of a signal being mixed with a delayed version of itself, and the associated impulse response. This arrangement of a delay line is called a feedforward comb filter.

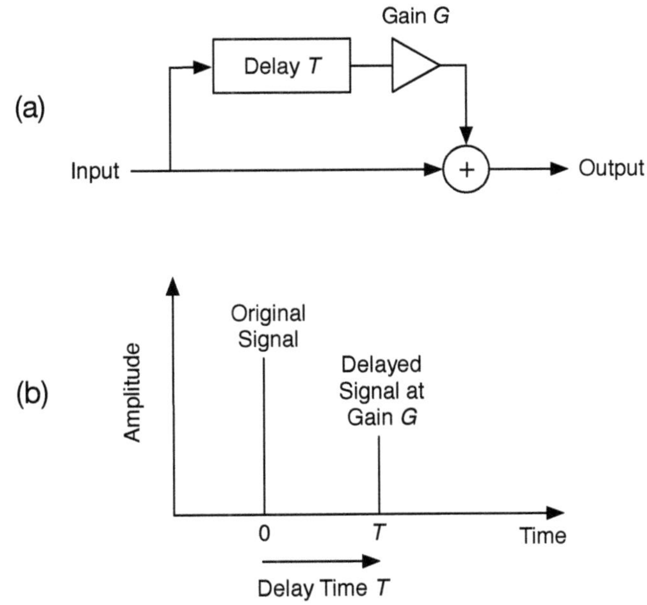

Figure 3.9 The top part (a) shows a block diagram of a signal combined with a delayed version of itself, also known as a feedforward comb filter. The delay time amount is represented by the variable T, and gain amount by G. The bottom part (b) shows the impulse response of the block diagram with a gain of 0.5: the original signal followed by a delayed version of itself at half the amplitude.

Figure 3.10 shows the frequency spectrum of a comb filter with a delay time of 0.5 milliseconds plotted on a linear scale (top) and a logarithmic scale (bottom). The notches are due to destructive interference between the original signal and the delayed version of the signal, and the notch frequencies depend on the delay time. As the delay time changes, the notch frequencies shift.

By simply delaying an audio signal and mixing it with the original, non-delayed signal, the result is either comb filtering when using short delay times, or echo when using longer delay times. Artificial reverberation plugins mimic early reflection patterns of sound in a real room by combining hundreds of delayed versions of a signal in complex and sophisticated ways. Chorus and flange are two other effects that use comb filters and modulated delay times.

Figure 3.11 shows a block diagram of a delay with feedback and its associated impulse response. We can see that the shape of this feedback comb filter's decay looks a little bit like the decay of sound in a room since the level gets lower over time. A single feedback comb filter will not sound like real reverb. To make it sound like actual reverb, designers combine many feedback comb filters in parallel all set to slightly different delay times and gain amounts.

When combined, a feedforward and feedback comb filter create what is known as an all-pass filter, as shown in Figure 3.12. All-pass filters have a flat frequency response, thus the name, but can be set to produce a decaying time response. As we will see below, they are an essential building block of digital reverbs.

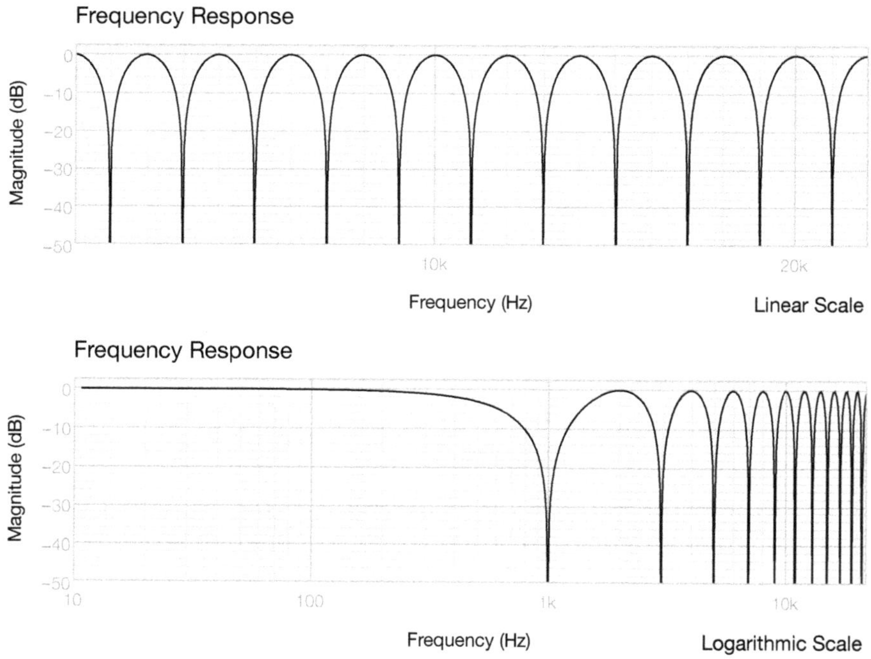

Figure 3.10 The frequency spectrum of a feedforward comb filter with a delay time of 0.5 milliseconds plotted on a linear scale (top) and a logarithmic scale (bottom). Notches appear every 2,000 Hz, starting at 1,000 Hz due to destructive interference. The two plots in this figure represent exactly the same frequency response, just two different x-axis scales. We normally see frequency spectrum in a logarithmic scale in equalizers, but I have included the linear scale to more clearly show the equal spacing of the notches.

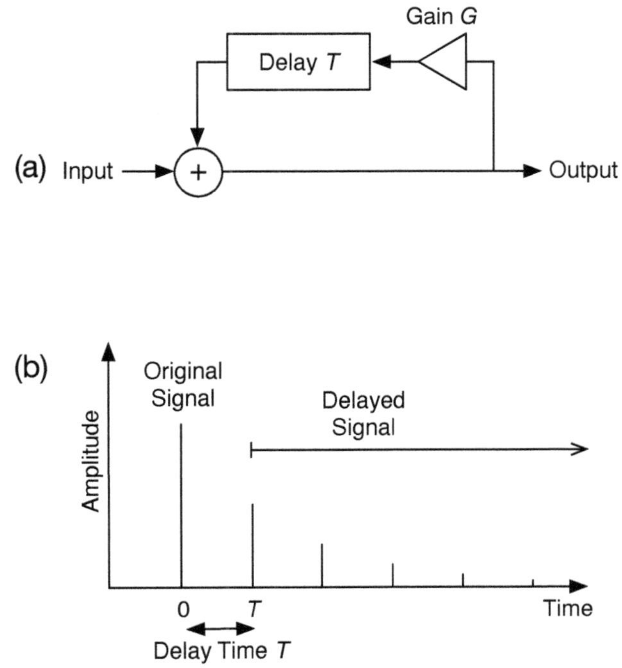

Figure 3.11 The top part (a) shows a block diagram of a signal combined with a delayed version of itself with the output connected back into the delay, also known as a feedback comb filter. The delay time amount is represented by the variable T, and the gain amount by G. The bottom part (b) shows the impulse response of the block diagram with gain G = 0.5: the original signal followed by a repeating delayed version of itself where each subsequent delayed output is half the amplitude of the previous one.

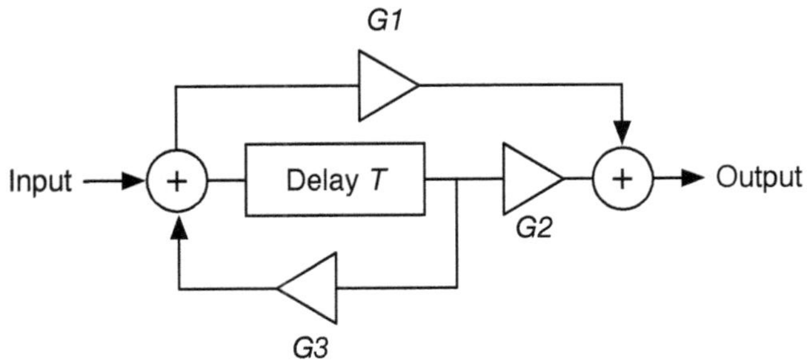

Figure 3.12 A block diagram of an all-pass filter, which is a combination of a feedforward comb filter and a feedback comb filter. All-pass filters have a flat frequency response, but they can be set to produce a decaying time response. There is one delay time, T, and three gain variables: G1 = non-delayed signal gain, G2 = feedforward delay signal gain, G3 = feedback delay signal gain.

3.3.2 Reverberation

Whether originating from a real acoustic space or a plugin, reverberation is a powerful effect that can help sounds blend and provide a sense of spaciousness, depth, cohesion, and distance in recordings. It can have an almost magical property to transport a listener into the world of a music recording, movie, or game more effectively than almost any other effect since it can create the illusion of being immersed in an environment different from our physical surroundings. With artificial reverberation, we can create the illusion of sources performing in a shared space together or in different acoustic spaces. Reverberation along with delay can be layered to accentuate and highlight a specific musical part such as a lead singer, drummer, or keyboardist, or it can be varied over the course of a song to contrast song sections such as verse and chorus.

Reverberation can be used to make recorded sound seem more natural. For example, the sound of an instrument or singer recorded with a close microphone creates an intimate or perhaps even uncomfortable feeling when listening over loudspeakers or headphones. Listeners do not normally stand a few inches in front of a singer while they sing, but a closely positioned microphone gives us a close-sounding perspective we do not normally hear from an acoustic performance. Listening to a close-miked singer over headphones without any added reverberation can cause us to localize the singer's voice inside our head. This is not something listeners are accustomed to hearing acoustically from a live music performance, and it can make them feel uncomfortable. Concert goers hear live music performances at least several feet if not tens of feet away from the performers. At normal distances between a performer and an audience member, reflected sound from walls, floor, and ceiling of a room along with reverberation fuse perceptually with direct sound. When recording a performer with a close microphone, reverberation can be added to create the perception of a more comfortable distance between the listener and sound source.

Like most other types of audio processing, how well reverberation supports a project depends on the way it is used. Reverberation can also negatively impact a recording when mixed too high or with a decay time that is excessively long for the context. It can reduce the clarity of a mix, washing it out, or, in the case of speech, affect intelligibility. The quality and amount of reverberation need to be optimized to suit and best support the artistic goals for each project.

There are two main types of digital artificial reverberation: algorithmic and convolution. Probably the most common type of digital artificial reverberation is algorithmic reverberation, which can create complex and realistic-sounding reverberation decay using delay as its basic building block. Digital reverberation plugin designers combine elaborate and complex delay networks, all-pass filters, and comb filters, along with features like equalization, low- and high-pass filters, and delay time modulation.

Manfred Schroeder (1962) was the first to propose the use of comb and all-pass filter to generate artificial reverberation. Figure 3.13 shows a block diagram of Schroeder's digital reverberation which specifies four parallel comb filters connected to two all-pass filters in series.

Although simple in concept, the original Schroeder reverb continues to form the basis for contemporary algorithmic reverberation algorithms today. To create artificial reverberation algorithms, plugin designers create large networks of interconnected comb and all-pass filters. Typically, they also manually tune delay times and gain levels for each stage of processing to create reverberation effects to their own specifications. The main advantages of algorithmic reverberation are computational efficiency, the ability to create highly realistic-sounding reverberation decays with them, and the range of controllable parameters most models provide that allow users to customize decay shapes, early reflection patterns, and frequency-dependent controls.

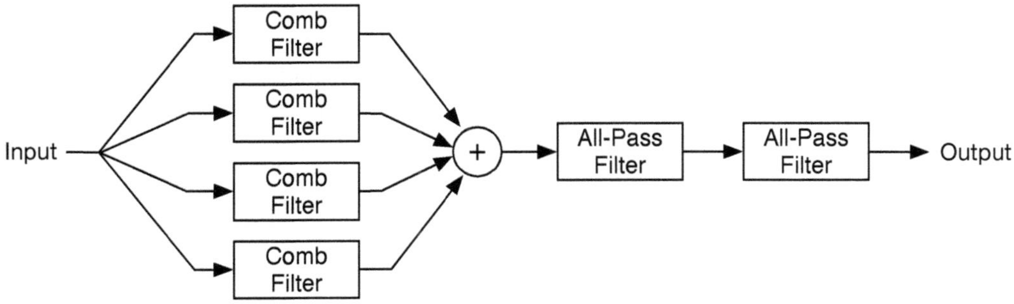

Figure 3.13 A block diagram of Manfred Schroeder's original digital reverberation algorithm, showing four comb filters in parallel that feed two all-pass filters in series. This design continues to form the basis for much more complex contemporary conventional reverb algorithms.

Algorithmic digital reverberation units usually allow several parameters to be controlled. Although these parameters vary from one manufacturer to another, a few of the most common include the following:

- Reverberation time (RT60).
- Delay time.
- Predelay time.
- Room shape or size: This can allow some control over early reflection patterns.
- Low-pass filter cut-off frequency.
- High-pass filter cut-off frequency.
- Reverberation time multipliers for different frequency bands.
- Gate: Useful for creating a gated reverb sound, typical parameters include threshold, attack time, hold time, release or decay time, and depth.

Although most digital reverberation algorithms represent simplified models of the acoustics of a real space, they are widely used in audio production to help augment a recorded acoustic space or to create a sense of spaciousness that did not exist in the original recording due to close-miking techniques.

The other main type of digital reverberation uses acoustic fingerprints of real rooms as its basis. Called convolution reverberation, it uses a mathematical process that combines two signals, a dry signal and a reverberation decay signal, by applying the features of one signal to another. In theory, any two audio signals can be convolved, although in practice convolution is typically only used for reverberation (with rare exceptions in experimental electronic music composition). The reverberation decay signal of a room is usually obtained by taking what's called an impulse response of the room. There are different ways to capture a room's impulse response, but one way is to generate an impulse—that is, a loud but very short sound such as a balloon bursting or a starter pistol firing—in an acoustic space and record the sound decaying in the room. The resulting reverberation is the response of the acoustic space to the impulse, and that space's reverberation signature can be combined or convolved with a dry signal. When we convolve a voice recording—say, with an impulse response from a concert hall—we create a new signal that sounds like a recording of the vocalist singing in the actual concert hall from which the impulse response was taken. As a result, it sounds highly realistic.

Hardware units capable of convolution-based reverberation have been commercially available since the 1990s, and software plugins are commonly included with DAWs. Convolution reverberation is sometimes called "sampling" or "IR" reverb because a sample or impulse response of a real acoustic space is convolved with a dry audio signal. The resulting reverb from a convolution reverberator is arguably more realistic-sounding than that from a conventional algorithmic digital reverberation which uses comb and all-pass filters. The main drawback of convolution reverberation plugins is that they usually do not provide as much flexibility to control parameters as is possible with algorithmic reverberation.

In the following subsections, we will look more closely at reverberation time, predelay time, and reverberation plugin presets.

3.3.2.1 Reverberation Time (RT60)

Reverberation time is defined as the amount of time it takes the sound level to drop by 60 dB after a source stops sounding; thus it is referred to as RT60. Wallace Clement Sabine, who created the field of architectural acoustics, proposed the first equation for calculating reverberation time (in seconds) of a real acoustic space (Howard and Angus 2017):

$$RT60\ (seconds) = 0.161 * \frac{V}{S\alpha} \tag{3.1}$$

V = volume in m^3, S = surface area in m^2, and α = average absorption coefficient of surfaces.

One difficulty of using this formula is that the RT60 will always be some value greater than zero even if all the surfaces in the room are perfectly absorptive—that is, if the absorption coefficient (α) is 1.0. In other words, using the Sabine equation we would calculate a reverberation time greater than 0 seconds for an anechoic chamber, even though anechoic chambers have no reverberation. The Sabine reverberation formula is easy to use but is typically only valid for absorption coefficient (α) values less than about 0.3.

Another way to calculate reverberation time is to use the Norris–Eyring reverberation formula, a slightly more complicated variation of the Sabine equation but that is applicable to a wider range of values (Howard and Angus 2017):

$$RT60 = \frac{-0.161 * V}{S * \ln(1-\alpha)} \tag{3.2}$$

V = volume in m^3, S = surface area in m^2, and α = absorption coefficient surfaces.

It is helpful to have an intuitive sense of the sound of various decay times. A decay time of 2 seconds, for example, will have a much different sonic effect than a decay time of less than 1 second.

3.3.2.2 Predelay Time

Predelay time in artificial reverberation is typically defined as the time delay between the direct sound and the onset of reverberation. Predelay can be used to give the impression of a larger space even with a short decay time. In a real acoustic space with no physical obstructions between a sound source and a listener, there will always be a short delay between the arrival of direct and reflected sounds. The longer this initial delay is, the larger we perceive the space to be, but artificial reverberation allows us to extend the time between direct sound and reverberation to a much longer time than is possible in an acoustic space.

3.3.2.3 *Digital Reverberation Presets*

Most digital reverberation units, whether in plugin or hardware form, offer hundreds if not thousands of reverberation presets. What may not be immediately obvious to the novice engineer is that there may only be a handful of unique algorithms for a given reverberation plugin or unit. The presets simply give variations in parameter settings for each of the algorithms. Presets are individually named to indicate an application like vocals or drums, or an acoustic space such as a large hall, bright room, studio, or theater. All the presets for one algorithm use identical processes, which means we can match the sound of one preset to another if we match their parameters.

Because engineers may adjust many reverberation parameters to create the most suitable reverberation for each application, it makes sense to start with any preset and then take the time to tune the parameters to a project's needs, rather than spend time searching for the perfect preset. The main drawback of trying to find the right preset for each instrument and voice during a mix is that the reverberation sound that you need might not exist in a preset. Or if something close does exist, it will likely require parameter adjustments anyway, so why not just start by adjusting parameters. As we edit parameters, we learn the capabilities of an artificial reverberation, the effect that each parameter can have on the sound, and how to get the most of it. In the parameter editing phase for an unfamiliar reverb, I find it helpful to turn parameters to the extremes of their ranges to make sure I can hear their contributions and then dial them to the settings I want.

This is not to say that presets are useless. They aren't. We can learn more about the capabilities of a reverb algorithm by going through factory presets to understand the capabilities of the plugin and to get ideas from the manufacturer about ways to get the most out of it. Searching through endless lists of presets may not be the best use of a mixing session when under time pressure, but it can be useful to listen carefully to presets during downtime.

3.3.2.4 *Variations in Digital Reverberation*

Despite the fact that conventional digital reverberation algorithms use the same building blocks—that is, combinations of delays, comb filters, and all-pass filters—any number of different sounding reverberation algorithms can be created. Unlike equalization, reverberation plugins vary widely in their implementations and therefore in the way they sound. The number of delay lines, comb filters, and all-pass filters used in a plugin and the ways in which these building blocks are interconnected within an algorithm vary from one model to another. Plugin designers have a great deal of latitude to create plugins that meet their specifications since there is no standard room shape or size to model mathematically. In practice, this usually means that each artificial reverb plugin has its own distinct sonic characteristics and qualities, even when set to the same reverb time as another reverb plugin model. Algorithmic reverb is not created from precise models of the behavior of sound in a room, so it is impossible to know exactly how the reverberation decay time, the RT60, of a given artificial reverberation algorithm relates to the decay time of sound in a real room. Some reverb plugins are more realistic than others, but realism is not always the goal.

Because of the variety of reverb algorithms, reverb tail decays vary from model to model in many ways, but the differences are perhaps clearest in their temporal envelopes and amount of stereo spread. To illustrate differences in temporal envelopes, Figure 3.14 shows the impulse responses of five different commercially available reverberation plugins, each with decay time set to 2.0 seconds and predelay time set to 0.0 milliseconds. Reverb plugins rarely share all the same controllable parameters, but the parameters reverb time and predelay are the most common ones. Even with identical reverb time and predelay time settings, the reverberation decay plots in Figure 3.14

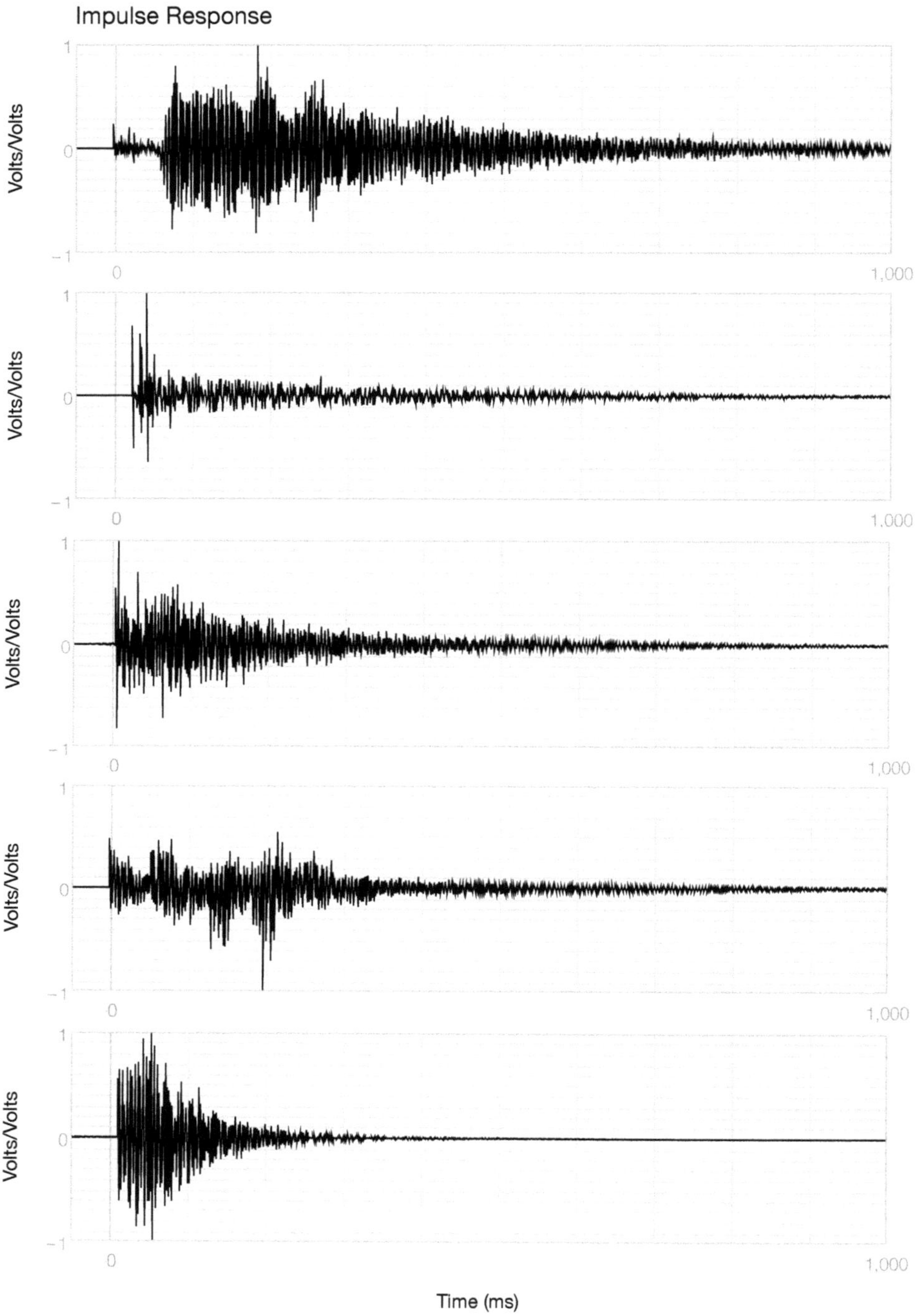

Figure 3.14 Impulse responses of five different commercially available reverb plugins all set to a reverberation time of 2.0 seconds and a predelay time of 0.0 milliseconds. Plots are scaled to the same x- and y-axis extents so they can be compared. Not only do the reverb decays look different, but they also sound distinctly different. (Impulse response measurements were taken with FuzzMeasure.)

show distinctly different amplitude envelopes. As you might guess, these reverb plugins also *sound* different from one another.

Digital reverb algorithms do not sound the same because there are many ways to design a reverb algorithm. Besides that, thousands of real acoustic spaces may be modeled, and within any given space, reverberation character depends on the sound source and listening or recording locations. But reverb algorithm models do not need to correspond to real acoustic spaces; thus there is a great deal of room for creativity in reverb plugin design. This is one reason why it is worth exploring different reverb models to find out what works best for a given project. Reverberation is a powerful sonic tool to create the aural illusion of space and depth, and the potential to provide listeners with the sense that they have been transported away from their immediate physical surroundings and into the world of a musical artist, film, or game.

3.4 Listening for Comb Filters in Unexpected Places

Although comb filters are an integral part of algorithmic reverberation, they can also arise unexpectedly outside of reverberation and can negatively affect a recording. As such, it is important to be listening vigilantly for comb filtering. Once a signal is recorded with comb filtering, there is no way to undo the effect on the frequency response of the signal. Ideally, it is best if we can notice any comb filtering in the moment while there is still a chance to correct or undo it. Let's look at four ways that comb filtering can occur.

3.4.1 Comb Filtering with Parallel Processing

One of the ways in which comb filtering can inadvertently occur is through signal routing in a DAW. Plugins take some finite amount of time to do their processing, and this time delay is called latency. Different plugins have different amounts of latency depending on their processing and the ways in which they have been programmed. One example where plugin latency can be a problem is in parallel compression of a drum track, as shown in Figure 3.15.

A compressor on a track—drums in this case—adds latency. The uncompressed drums do not have any latency. When we mix the two together without latency compensation, comb filtering results. Fortunately, most DAWs include automatic latency compensation to mitigate this effect. This means that, in our example, the DAW delays the uncompressed drum track by an amount equal to the compressor plugin latency on the parallel track. With latency compensation, all channels are synchronized precisely based on the plugins that are inserted. Without latency compensation, comb filtering occurs when the two tracks—compressed and uncompressed, in this example—are mixed.

While DAWs automatically apply latency compensation in most cases, it may still be possible to incur comb filtering with more elaborate signal routing such as when using auxiliary sends, for which a DAW may be unable to calculate the net latency. For parallel processing with an analog compressor, no latency compensation is needed since the processing time is practically instantaneous in the analog domain.

3.4.2 Comb Filtering When Monitoring Through an Audio Interface

Signal routing through an audio interface while recording is another situation that can produce comb filtering. While this may not affect the recorded signal, it does affect what we hear through

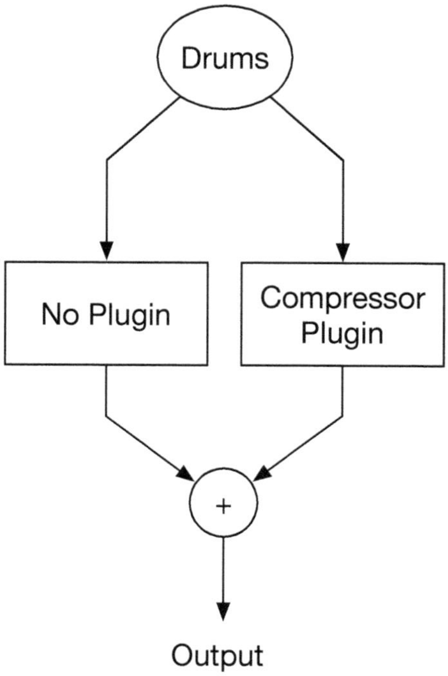

Figure 3.15 A drum track processed with parallel compression in a digital audio workstation may incur comb filtering.

the monitor output. Comb filtering can occur because audio interfaces usually allow monitoring of both a live microphone input and a live microphone input after it has been routed through a DAW and back to the audio interface, as shown in Figure 3.16. The path through the DAW adds some amount of latency which is greater than the latency of the direct-to-monitor signal path. Therefore, a comb filter is created when a live microphone signal is mixed with itself after it has gone through the DAW and back out, even without any plugins inserted. The solution is to monitor the microphone signal either directly from the input or after it has gone through the DAW but not to monitor both paths at the same time. Although the figure shows a USB interface, this effect can occur regardless of the interface connection type, be it USB, Dante, Ethernet, or Thunderbolt.

A similar comb filtering effect can also occur during Zoom calls when using an external microphone with an audio interface that has loopback inputs. If an external USB audio interface with loopback functionality is selected as the microphone in Zoom, the software selects all audio inputs from the external audio interface, including loopback inputs (if they are available on an interface). When this happens, it causes comb filtering. If you use an external microphone with a USB audio interface for Zoom calls and your audio interface has loopback inputs, you could be inadvertently sending comb filtered audio to others on the call. Comb filtering can be eliminated by unselecting all audio inputs in Zoom except the channel in which the microphone is connected. Comb filtering occurs because the live microphone is sent to Zoom along with a slightly delayed version through the loopback output. It's

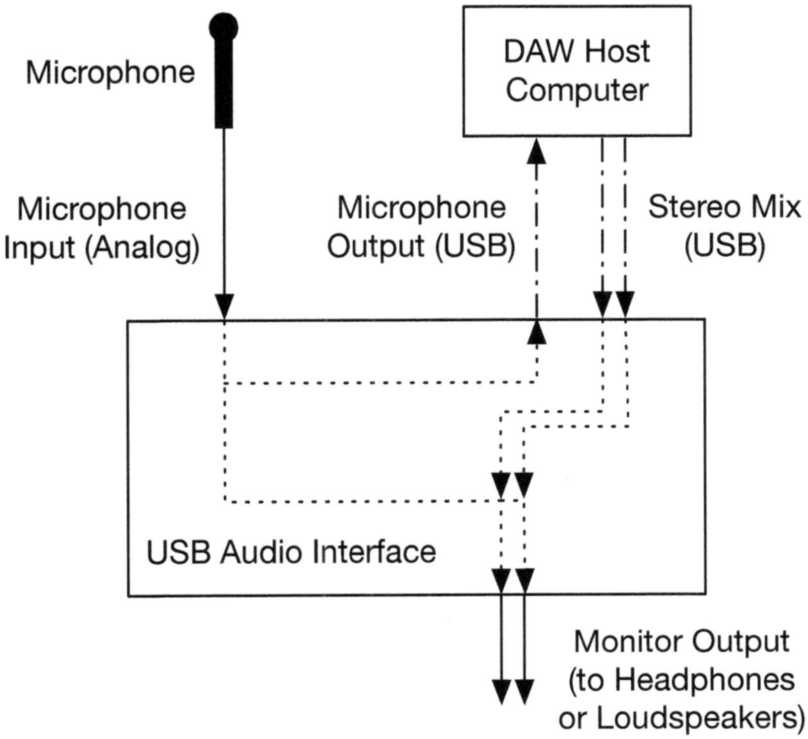

Figure 3.16 A diagram of a microphone signal routed through a USB audio interface that causes comb filtering at the monitor output. The microphone input signal can be routed directly to the monitor output, and it can also be routed to the monitor output after it has gone through the DAW and incurred some amount of latency. This routing results in comb filtering since the two versions of the microphone signal are mixed in the interface. The solution is to monitor the microphone signal either directly from the input or after it has gone through the DAW, but not to monitor both paths at the same time. Although the figure shows a USB interface, this effect can occur regardless of the interface connection type (USB, Dante, Ethernet, Thunderbolt, etc.).

always helpful to test your microphone in Zoom by listening to it before using it in a call to make sure the audio is clean.

3.4.3 *Comb Filters with Microphones Near Walls*

Some recording setups can create acoustical comb filters depending on microphone and sound source proximity to a reflective surface such as a wall or window, microphone polar pattern, and sound source radiation pattern. For example, Figure 3.17 shows a sound source and microphone positioned close to a wall. The reflection arrives slightly after the direct sound, and the direct and reflected sound sum at the microphone capsule. Even with the natural drop in sound level over the reflected path and the attenuation due to microphone polar pattern off-axis response, comb filtering can still occur. The notches of the comb filter may not be as deep as a comb filter

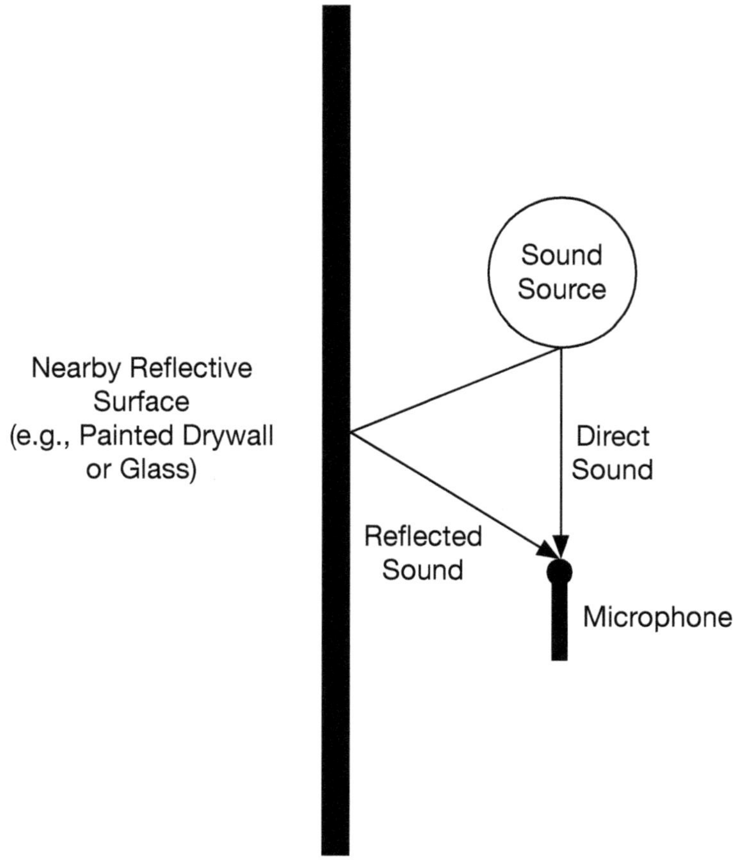

Figure 3.17 A diagram of a sound source and microphone positioned close to a hard reflective surface such as a wall or window. In this situation, the reflected sound level is still relatively close to the direct sound level. When the direct and reflected sound combine at the microphone capsule, comb filtering can occur.

created in a DAW or audio interface since the reflected sound will be attenuated, but the effect can still be audible.

3.4.4 Comb Filters with Multiple Sources and Microphones Close Together

Another situation that can result in comb filtering is when recording or doing live sound reinforcement for multiple singers or instrumentalists positioned close together with multiple spaced microphones on stands. Figure 3.18 shows placement of singers and microphones and the paths from each singer to the microphones. Each microphone can pick up the sound of not only the singer to which it is aimed but also adjacent singers. The problem is the time of arrival differences for each microphone. For example, sound from singer 2 reaches microphone 1 slightly later than it reaches microphone 2. Because of the difference in time of arrival, comb filtering occurs when the two microphones are mixed. To avoid comb filtering, either place microphones closer to the singers or space the singers and their microphones further apart. Sometimes it may be preferable to maintain

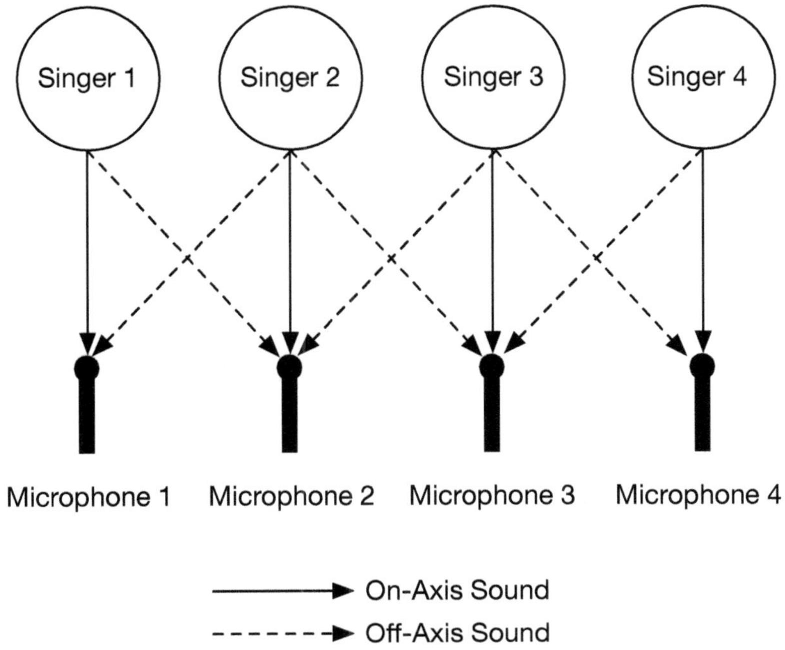

Figure 3.18 A diagram of four singers standing close together with a microphone for each singer. In this setup, each microphone picks up sound from adjacent singers. Because of the differences in time of arrival to each microphone, comb filtering can occur when the microphone signals are mixed.

more distant microphone placement such as for certain live sound or classical music recording, while at the same time the singers may want to be close to each other to hear each other better and there may not be room to space them apart. Another option is to position the musicians so that they face one another to take advantage of microphone polar patterns and/or move the microphones closer to the respective musicians.

Another recording setup that can create comb filtering is that of a singer who plays acoustic guitar while singing. In this situation, the voice can bleed into the guitar microphones and vice versa. Again, we can reduce comb filtering by positioning microphones closer, by orienting microphones to best take advantage of their polar patterns, or by selecting microphones with a different polar pattern—for example, figure-8 instead of cardioid.

Listen closely for comb filtering during sound check. If it is present, you want to be aware of it while you still have a chance to mitigate its effects.

3.5 Reverberation in Spatial Audio

My informal research and listening experience seem to indicate that, in general, spatial audio formats can accommodate higher levels of reverberation than two-channel stereo can, while maintaining an acceptable level of clarity. More formal tests need to be conducted to verify this point, but it may make sense from what we know about spatial masking. As we discussed earlier, spatial separation of two sound sources reduces the masking that occurs when they are in the same place (Kidd et al. 1998; Saberi et al. 1991). The effect seems to be consistent for real sound sources as well as for virtual sound sources panned across a multichannel loudspeaker array. It appears that because of the larger spatial distribution of sound in multichannel audio, relative to two-channel stereo, reverberation is less likely to obscure or mask dry sound, and therefore reverberation can be more prominent in multichannel audio. It could be that reverberation is more crucial in multi-channel audio mixes because the format offers a much greater possibility to re-create a sense of immersion in a virtual acoustic space than two-channel stereo.

Regardless of the format in which we are working, we can benefit from a systematic training method to learn to match parameter settings of artificial reverberation by ear and to further develop the ability to consistently identify subtle details of sound reproduced over loudspeakers.

3.6 Reverberation in Recordings

Reverberation is found on almost every recording, but the specific ways that it is used varies widely. On many recordings, reverberation decay time and level are held constant, but on other recordings reverberation level and decay time change with a song's sections or to emphasize certain lyrics. Reverberation is not always the same for all instruments and voices in a recording. An instrument or voice may have more reverb than others, or its reverberation decay time may be different. Here are a few recordings that feature reverberation in unique or contrasting ways.

"Be My Baby" by The Ronettes from Be My Baby: The Very Best of the Ronettes *(1963/2011)*
A classic track from 1963 produced by Phil Spector, known for his "wall of sound" production style. Mixed in mono, the cavernous-sounding space is most apparent on the drums, tambourine, and castanets.

"Owner of a Lonely Heart" by Yes from 90125 *(1983)*
The intro starts with a reverberant distorted guitar and soon follows with a much drier guitar sound. The snare drum has a short, tight reverb which contrasts with the long reverb tail on the vocal.

"Kiss" by Prince and The Revolution from Parade (Music from the Motion Picture Under the Cherry Moon) *(1986)*
Most of the elements in this track are very dry and contrast with the highly reverberant kick drum, which has a low-pass filtered reverberant decay that expands wide. The room reverb on the vocal sounds like that of a real space with its short decay.

"Where Will I Be" by Emmylou Harris from Wrecking Ball *(1995)*
Electric guitars with spacious reverberation and delays start the track, inviting the listener and setting the stage for the performance. The vocal has a more subtle reverberation with a long tail. In contrast, the drum set is much drier.

"Al Birwa" by Anouar Brahem from The Astounding Eyes of Rita *(2009)*
Consistent with most recordings on the ECM label, there is a noticeable but not excessive reverberation tail on this track. It gives the music a wonderful space and coherence without making it muddy or cluttered.

"If You Need to, Keep Time on Me" by Fleet Foxes from Crack-Up *(2017)*
The solo strummed acoustic guitar intro showcases an echoey reverb which sounds as though it has a substantial predelay. The reverberation in this track is unusually prominent for the genre, and it gives the sense that the song was performed in a large space.

"Mariella" by Khruangbin and Leon Bridges from Texas Moon – EP *(2022)*
The drums are mixed with what sounds like a spring reverb, and the snare drum has a splash of reverb that decays in the left channel only. The vocal has a tight, short reverb with an audible predelay, which gives it a slap back echo effect.

"doomsday" by Lizzy McAlpine from five seconds flat *(2022)*
This track starts with a dry vocal and strumming acoustic guitar. During the first verse, the reverb level is brought up significantly in an obvious and dramatic way, as though it is one of the musical parts. The reverb on the vocal is removed and added again at different points in the song. There are also contrasts in the use of reverb between the vocals and other elements in the mix such as the drums, guitar, backing vocals, and strings.

"About You" by The 1975 from Being Funny in a Foreign Language *(2022)*
The track starts off with a moderate amount of reverberation and then is suddenly drenched in reverberation, especially on the vocals. The reverberation level and long decay time makes the vocals sit back in the mix. High frequencies in the reverb decay allow us to hear the reverb splashes from the vocal sibilants.

"What Was I Made For?" (Single From the Motion Picture "Barbie," 2023) by Billie Eilish
An expansive, spacious reverberation envelopes the breathy vocal sound with a long decay. This contrasts with the relatively dry piano.

3.7 Technical Ear Training Software Module: Reverberation

The "Technical Ear Trainer—Reverb" software module, found at https://webtet.net/apcl/#/reverberation, is designed to focus on hearing subtle details and parameters of artificial reverberation. Although not focused on real room acoustics, its goal is to help you improve your critical listening skills for real room acoustics as you become more sensitive to reverberation decay times, echoes, reflections, and source locations. The computer randomizes the exercises and gives a choice of difficulty and parameters for an exercise. It works in much the same way as the equalization module works.

Just as it is critical to learn to recognize spectral resonances using equalization, it is equally important to improve our perception of artificial reverberation. Research has demonstrated that listeners can learn reverberation for a given room. It was found that listeners can improve their ability to localize sound sources in a reverberant environment with experience in a given room (Shinn-Cunningham 2000). Other work in training listeners to identify spatial attributes of sound has been conducted as well. Neher et al. (2003) have documented a method of training listeners to identify spatial attributes using verbal descriptors for the purpose of spatial audio quality evaluation. Other researchers have used graphical assessment tools to describe the spatial attributes of reproduced sound, such as Ford et al. (2003) and Usher and Woszczyk (2003).

The included training software has an advantage in that the goal is to compare one spatial scene with another by ear; there is no requirement to translate an auditory sensation to another sensory modality or means of expression, such as drawing an image or choosing a word. The software asks users to compare and match two sound scenes, with a given set of artificial reverberation parameters. Additionally, this method has ecological validity, as it mimics the process of a sound engineer sculpting sonic details of a sound recording by ear rather than through graphs and words.

3.7.1 Sound Sources

I encourage you to begin the reverb training with simple, transient, or impulsive sounds such as solo percussion—a single snare drum hit is a great sound source, for example. After you become comfortable with a single drum hit as a sound source, move to more complex sounds such as speech and music recordings. In the same way that we use pink noise in equalization ear training because it exposes the spectral changes better than many music samples, percussive or transient sounds are best for training with time-based effects processing. Reverberation decay time is easier to hear with transient signals than with continuous or steady-state sources, which tend to mask or blend with reverberation, making the decay more difficult to hear.

3.7.2 User Interface

The graphical user interface (GUI) shown in Figure 3.19 provides a control surface for you to interact with the system.

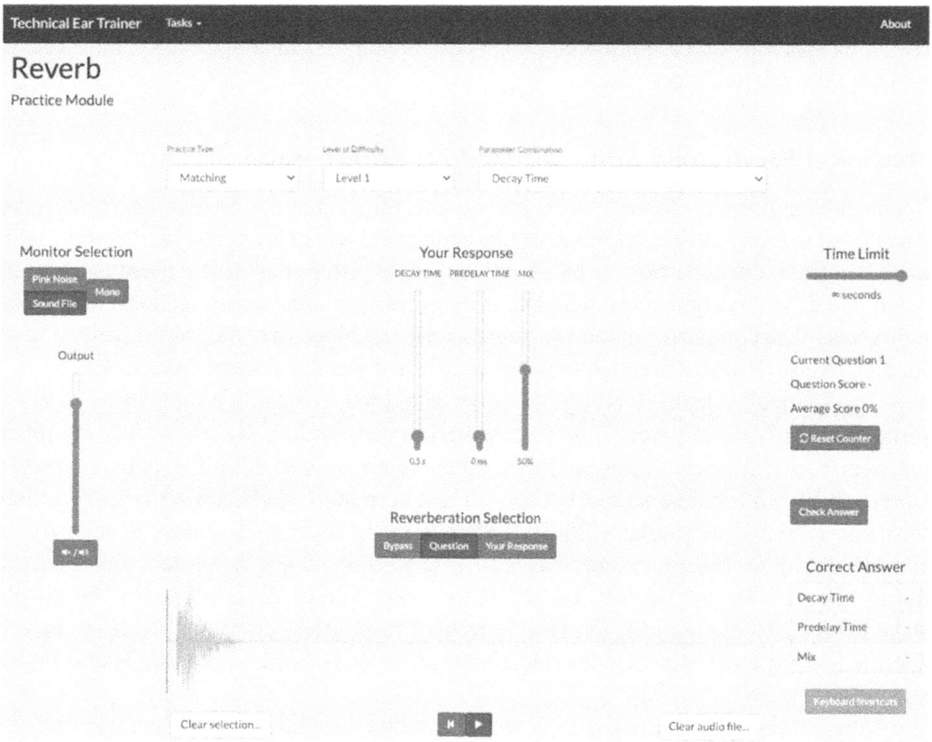

Figure 3.19 A screenshot of the user interface for the Reverb technical ear trainer.

With the GUI you can:

- choose the level of difficulty,
- select the parameter(s) with which to work,
- choose and load a sound file,
- adjust parameters of the reverberation,
- toggle between the reference and your answer,
- control the overall level of the sound output, and
- submit a response to each question and move to the next example.

The graphical interface also keeps track of the current question and the average score up to that point, and it provides the score and correct answer for the current question.

3.8 Getting Started with Practice

The training curriculum covers a few of the most common parameters in digital reverberation units, including the following:

- reverb decay time,
- predelay time,
- mix (reverberation level), and
- delay time.

As with the equalizer module, your task with the exercises and tests is to duplicate a reference sound scene by listening and comparing your answer to the reference and making the appropriate changes to the parameters until they sound the same. The software randomly chooses parameter values based on the level of difficulty and test parameters you choose, and it asks you to identify the reverberation parameters of the reference by adjusting the appropriate parameter to the value that most closely matches the sound of the reference. You can toggle between the reference question and your answer either by clicking on the switches labeled Question and Your Response or by pressing the space bar. Once the two sound scenes are matched, you can click on the Check Answer button or hit the Enter key to submit your answer and see the correct answer. Clicking on the Next button moves to the next question.

3.8.1 Reverb Decay Time

Decay times range from 0.5 seconds to 2.5 seconds with an initial resolution of 1 second and increasing in difficulty to a resolution of 0.25 seconds.

3.8.2 Predelay Time

Predelay time is the amount of time delay between the direct or dry sound and the beginning of early reflections and reverberation. Predelay times vary between 0 and 200 milliseconds, with an initial resolution of 40 milliseconds and decreasing to a resolution of 10 milliseconds.

3.8.3 Mix Level

There are two main ways of adding reverberation to a track in a DAW. One way is to insert a reverb plugin on an auxiliary track and use aux sends to route audio from each mix channel to the aux channel. Then the level of the reverberation in the mix can be adjusted using the aux channel fader. The other way is to insert the reverb plugin directly on a track's fader channel and adjust the mix level within the plugin to determine the amount of reverb going to the output bus. Using an aux fader for reverb is most useful when you want to add the same reverb to several tracks since it allows you to adjust the reverb parameters for all the tracks using only one plugin.

The reverb training system allows you to practice learning various mix levels of reverberation. A mix level of 100% means that there is no direct, unprocessed, sound at the output of the algorithm, whereas a mix level of 50% represents an output with equal levels of processed and unprocessed sound. The mix value resolution at the easiest level of difficulty is 50% and progresses up to a resolution of 12.5%, covering the range from 0 to 100% mix.

3.8.4 Delay Time

Delay times range from 0 milliseconds 200 milliseconds with an initial resolution of 40 milliseconds and increasing in difficulty to a resolution of 10 milliseconds.

3.8.5 Levels of Difficulty

Below is a listing of the possible decay times, predelay times, mix levels, and delay times for each of the three levels of difficulty.

Level 1 (easiest):

- **Decay Times:** 0.5, 1.5, or 2.5 seconds.
- **Predelay Time:** 0, 40, 80, 120, 160, or 200 milliseconds.
- **Mix:** 0, 50, or 100%.
- **Delay Time:** 0, 40, 80, 120, 160, or 200 milliseconds.

Level 2:

- **Decay Times:** 0.5, 1.0, 1.5, 2.0, or 2.5 seconds.
- **Predelay Time:** 0, 20, 40, 60, 80, 10, 120, 140, 160, 180, or 200 milliseconds.
- **Mix:** 0, 25, 50, 75, or 100%.
- **Delay Time:** 0, 20, 40, 60, 80, 100, 120, 140, 160, 180, or 200 milliseconds.

Level 3 (most difficult):

- **Decay Times:** 0.5, 0.7, 1.0, 1.2, 1.5, 1.7, 2.0, 2.2, or 2.5 seconds.
- **Predelay Time:** 0, 10, 20, 30, 40, 50, 60, 70, 80, 90, 100, 110, 120, 130, 140, 150, 160, 170, 180, 190, or 200 milliseconds.
- **Mix:** 0, 12, 25, 37, 50, 62, 75, 87, or 100%.
- **Delay Time:** 0, 10, 20, 30, 40, 50, 60, 70, 80, 90, 100, 110, 120, 130, 140, 150, 160, 170, 180, 190, or 200 milliseconds.

3.9 Mid-Side Matr.ixing

A mid-side matrix may not be the first thing that comes to mind as a tool for technical ear training, but it is useful to analyze and deconstruct recordings, since it can help highlight components of a mix that might be difficult to pick out otherwise. Mathematician and audio engineer Michael Gerzon (1986, 1994), probably best known for his invention of the 360° surround sound recording and reproduction format known as Ambisonics, is credited with developing stereo matrixing and Blumlein shuffling techniques that can be used to enhance and rebalance correlated and decorrelated components in a mix.

Although mid-side processing takes its name from the mid-side (M/S) stereo microphone technique, the principles can be used to convert any stereo recording to its mid and side components,

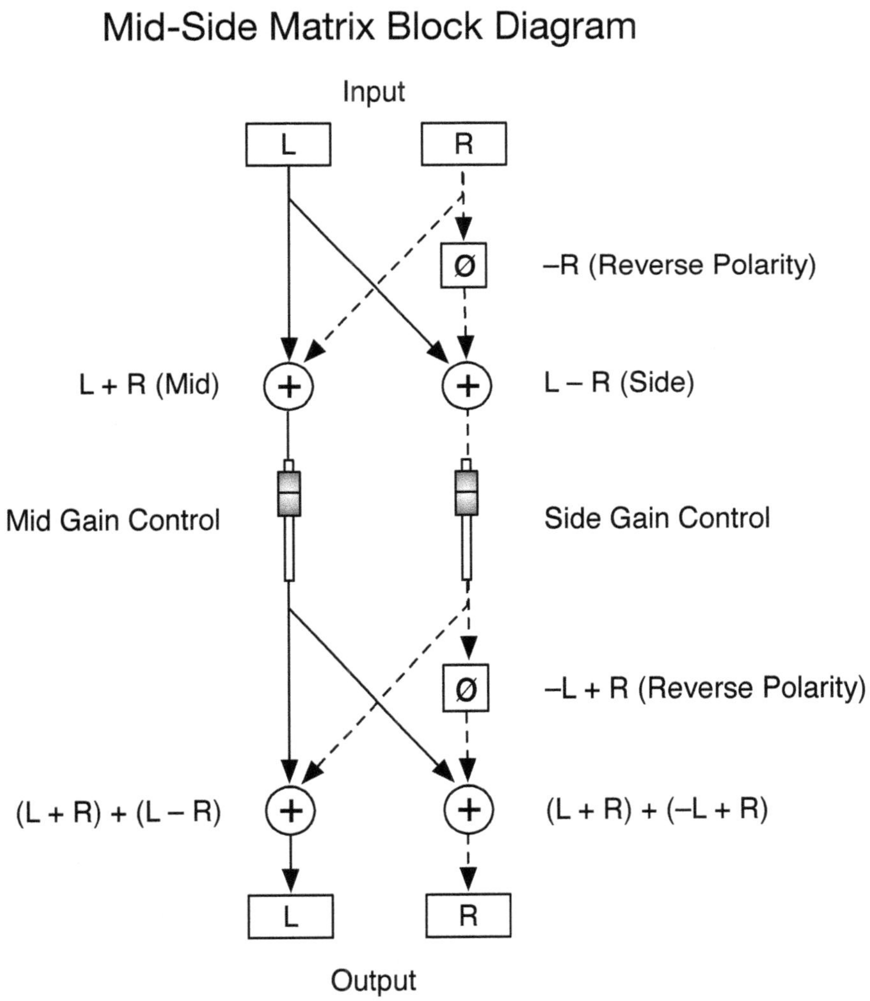

Figure 3.20 A block diagram to convert left and right stereo signals into mid (left + right) and side (left – right) signals, and subsequent mixing back into left and right channels. Left and right signals enter at the top, are converted to mid and sides with independent fader control in the middle of the diagram, and then are converted back to left and right prior to output at the bottom.

separate from the microphone technique. Figure 3.20 shows a block diagram of a matrix to convert a stereo (left/right) mix to mid and side components and then back to stereo again. The mid component is derived by mixing the left and right channels together and the side component is derived by subtracting the left and right channels.

Some plugins have built-in M/S processing, but it is possible if a bit cumbersome to derive the mid and side components manually on a mixer. One way to implement a mid-side matrix on a mixer uses six mixer channels, as shown in Figure 3.21.

To create a mid-side matrix on the mixer, split the left and right channels of a stereo mix and route them to three channels each, alternating left and right. Pan channel 1, the left channel input, and channel 2, the right channel input, to center. Because the faders on channels 1 and 2 control the mid component level, they must be level matched and grouped together. Pan channel 3, the left channel input, and channel 4, the right channel input, hard left and reverse the polarity (or phase) on channel 4. Pan channel 5, the left channel input, and channel 6, the right channel input, both hard right and reverse the polarity on channel 5. Because the faders on channels 3, 4, 5, and 6 work together to control the side component level, they must be level matched and grouped together.

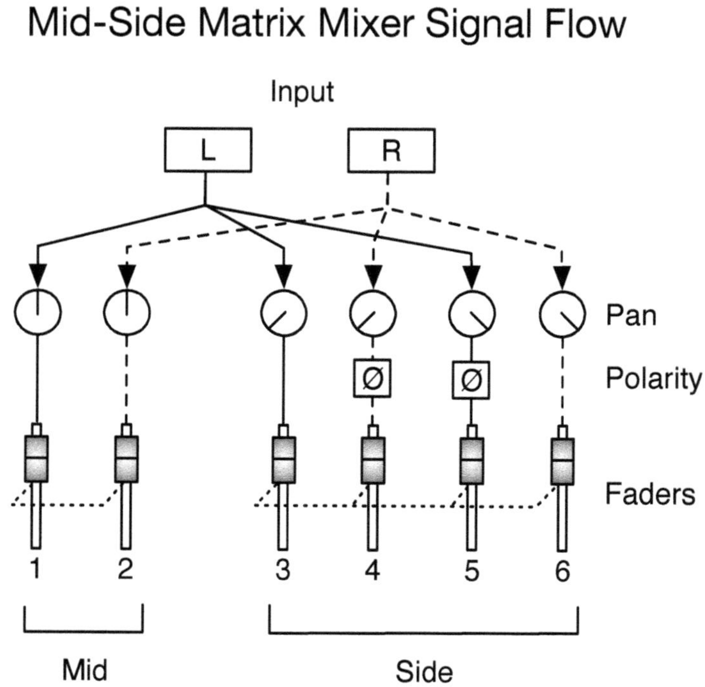

Figure 3.21 This diagram shows one way to route signals on a mixer to convert left and right stereo signals into mid (left + right) and side (left – right) signals. Channels 4 and 5 have their polarity reversed but none of the other channels do. Note that the pair of faders for the mid signal should always be level matched, as should the four faders for the side signals, as indicated by the dotted lines connecting these two groups of faders.

With a stereo mix split into mid and side components, it becomes easy to rebalance mid and side, and apply processing to them independently. As long as the two mid faders are grouped and level matched with each other, and the four side faders are grouped and level matched with each other, the original stereo mix remains intact when all six faders are at the same gain.

3.9.1 The Mid or Sum Component

The mid component of a stereo mix is a sum of the left and right channels, which means it includes any signals panned center or to the sides but does not include signals that are opposite polarity. Relating the concept back to our discussion at the beginning of the chapter, the mid component includes correlated and decorrelated signals from the stereo but not negatively correlated signals. Any components of the stereo mix which are opposite polarity will cancel through destructive interference and will be attenuated completely in the mid component.

For most recordings, the mid component contains most if not all the elements of the stereo mix, which means the difference between the original stereo mix and the mid component is usually subtle since the mid component is simply the stereo mix summed to mono.

3.9.2 The Side or Difference Component

The side component, in contrast to the mid component, may sound distinctly different from the stereo mix, depending on the recording. The side component of a stereo mix is the difference between the left and right channels, which means it includes opposite-polarity signals and any panned signals, but does not include signals panned to the center. In other words, the side component includes only decorrelated and negatively correlated signals but not correlated signals.

In pop music recordings, usually lead vocals, bass, kick drum, and snare drum are panned center, which means they are partially or entirely cancelled in the side component. Listening to the side component on its own usually helps reveal elements in music recordings which may be difficult to hear clearly in the stereo mix, such as delays, reverberation, and quieter panned elements.

I also want to clarify one point about the side signal and the way we use it in this context. There is only one side channel, but because the mixer version of the mid-side matrix converts stereo to mid-side and then back to stereo, the side component consists of the side signal panned to the left and an opposite-polarity version of that same side signal panned to the right. The side signals on their own are negatively correlated and will sound unnaturally wide and have unstable locations when listening on loudspeakers. If you are familiar with the mid-side microphone technique, you will know that this is the way we monitor the side microphone when recording so that it gets matrixed with the mid microphone to properly create a stereo image.

3.9.2.1 Exercise: Listening to Mid-Side Processing

The practice module "Technical Ear Trainer—Mid-Side," available on the companion website https://webtet.net/apcl/#/midside, offers an easy way to audition the mid and side components of any stereo recording. This ear trainer module provides a slider labeled Width Control to control the relative levels of mid and side. In this module, the side signal output is sent to both left and right loudspeaker or headphone channels but with opposite polarity, consistent with the mid-side microphone technique matrix. With the slider position to the right at 100% side (wide), the left and right outputs are negatively correlated (correlation = –1) because the left channel is producing the original side component, and the right channel is producing an opposite-polarity version of the same signal.

Try loading a sound file, move the width control slider to the right (wide, 100% side), and listen for elements that may be masked when listening to the regular stereo (left/right) format. Listening to just the side component, you may be able to hear elements of a mix such as stereo reverberation (mono reverberation is cancelled in the side component), punch-ins/edits, distortion, dynamic range compression, vocal layers, delay effects, instrumental parts, and fader level changes that are not audible or that may be difficult to hear in the stereo mix. What kinds of sounds and elements in a mix can you uncover by listening to just the side component?

3.10 Summary

This chapter covers spatial attributes of sound, focusing primarily on reverberation and mid-side processing. The goal of the reverberation practice module is to systematically familiarize you with aspects of artificial reverberation, predelay, and delay. By comparing two audio scenes by ear, we can match one or more parameters of artificial reverberation to a reference randomly chosen by the software. We can progress from comparisons using percussive sound sources and coarse resolution between parameter values to more steady-state musical recordings and finer resolution between parameter values. Often very minute changes in reverberation parameters can have a significant influence on the depth, blend, spaciousness, and clarity of the final mix of a sound recording. After spending time training with the reverberation module, you may notice more qualities of the reverberation present in recordings and acoustically.

Note

1 The term "opposite polarity" is sometimes referred to as "180° out-of-phase," but technically these two terms are not equivalent since shifting a signal 180° out-of-phase requires a time shift or delay. Inverting the polarity simply means the negative part of a signal becomes positive and the positive part becomes negative.

Bibliography

Barron, Michael. 2010. *Auditorium Acoustics and Architectural Design*. 2nd ed. Spon Press.

Ford, Natanya, Francis J. Rumsey, and Tim Nind. 2003. "Evaluating Spatial Attributes of Reproduced Audio Events Using a Graphical Assessment Language." Paper presented at AES 24th International Conference on Multichannel Audio, Banff, Alberta, Canada. *AES 24th International Conference on Multichannel Audio*.

Gerzon, Michael A. 1986. "Stereo Shuffling: New Approach—Old Technique." *Studio Sound*.

Gerzon, Michael A. 1994. "Applications of Blumlein Shuffling to Stereo Microphone Techniques." *Journal of the Audio Engineering Society* 42 (6): 435–53.

Howard, David M., and Jamie Angus. 2017. *Acoustics and Psychoacoustics*. 5th ed. Routledge.

Kidd, Jr., Gerald, Christine R. Mason, Tanya L. Rohtla, and Phalguni S. Deliwala. 1998. "Release from Masking Due to Spatial Separation of Sources in the Identification of Nonspeech Auditory Patterns." *Journal of the Acoustical Society of America* 104 (1): 422–31.

Moore, Brian C.J. 1997. *An Introduction to the Psychology of Hearing*. 4th ed. Academic Press.

Neher, T., Brookes, T., and Rumsey, F. 2003. Unidimensional simulation of the spatial attribute "ensemble depth" for training purposes. Part 1: Pilot study into early reflection pattern characteristics. Proceedings of the Audio Engineering Society 24th International Conference, Banff, Canada.

Saberi, Kourosh, Lynda Dostal, Toktam Sadralodabai, Vicki Bull, and David R. Perrott. 1991. "Free-field Release from Masking." *Journal of the Acoustical Society of America* 90 (3): 1355–70.

Schroeder, Manfred R. 1962. "Natural Sounding Reverberation." *Journal of the Audio Engineering Society* 10 (3): 219–23.

Shinn-Cunningham, Barbara. 2000. "Learning Reverberation: Considerations for Spatial Auditory Displays." Paper presented at International Conference on Auditory Display, Atlanta, GA. *Proceedings of the 2000 International Conference on Auditory Display*.

Usher, John, and Wieslaw R. Woszczyk. 2003. "Design and Testing of a Graphical Mapping Tool for Analyzing Spatial Audio Scenes." Paper presented at AES 24th International Conference on Multichannel Audio, Banff, Alberta, Canada. *AES 24th International Conference on Multichannel Audio*.

Chapter 4

Dynamic Range Control

4.1 Mix Balances, Levels, and Dynamic Range

The balance of a mix has a direct effect on the way listeners or audiences perceive a musician's performance, a movie director's artistic vision, or a game designer's intentions. Mix balance is about the relative levels of the elements in a project. If one or more elements in a mix are too loud or too quiet, listeners may not be able to hear important parts well enough or they may think the emphasis is on a different part than the artist, director, or designer intended. To balance tracks in a mix, an engineer adjusts levels and controls the dynamic ranges of many different tracks using faders, compressors, limiters, expanders, and gates. In addition to the relative balance of elements in a mix, the overall dynamic range and fluctuations in loudness matter too. The term "dynamic range" describes the difference between the loudest and quietest levels of an audio signal. The amount of dynamic range control needed for a music recording depends significantly on genre. What may be expected in one genre may be completely unsuitable for another, depending on the conventions that have developed over time and what may be current.

In the classical music world, conductors and composers work hard to find an optimal ensemble balance for each performance and piece of music. Achieving an appropriate balance of a musical ensemble is essential for expressing an artist's musical intention. If an instrumental part within an ensemble is not loud enough to be heard clearly, listeners do not receive the full impact of a piece of music. Overall balance depends on the control of individual vocal and instrumental amplitudes in an ensemble. Classical music recordings usually do not benefit from highly controlled dynamic range because listeners expect dynamic range variation in the genre and too much dynamic range control can make it sound too constrained and unnatural. Furthermore, artifacts from compression and limiting are usually not welcome in the genre. Although signal processing artifacts such as distortion, limiting, equalization, and delays are often an expected part of pop, rock, and electronic music, processing of these sorts is usually avoided in classical music recordings. In an ideal sense, classical music recordings should not sound like recordings at all but should mimic the ideal concert hall experience from the best seat in the house. Since classical music is performed acoustically, except for pieces that integrate electronic sounds with acoustic sounds, the experience of recorded classical music mimics the live concert experience with the full dynamic range left intact.

For most non-classical music genres, on the other hand, at least some amount of dynamic range control is desirable. For pop, rock, electronic, hip-hop, country, metal, and related genres, significant dynamic range compression is desirable, partly to make the recordings sound loud and partly to produce timbres consistent with a recording's genre. Without appropriate amounts of compression and limiting, a recording will not sound like other recordings in its genre. Dynamic range

DOI: 10.4324/9781003461173-4

control can help not only to smooth out audio signal levels but also to act like a glue that helps add cohesion to various musical parts in a mix.

When we mix a multitrack project, we can have direct control over the relative levels of each track and therefore also over the musical expression that listeners ultimately hear. Depending on the dynamic range of the tracks in a project and their relative balance, we may need to adjust the level of specific instruments or voices to maintain consistent balance from the beginning to the end of a track. For recorded tracks that have an excessively wide dynamic range, we can use faders to compensate for variations in level to produce a consistent perceived loudness. The process can involve manually (with a fader) boosting levels during quiet sections and attenuating levels during loud sections. In this way, fader level adjustments made moment-to-moment through the duration of a recording, whether through drawing automation curves or moving the faders, are essentially a form of manual dynamic range compression. Dynamic range controllers—that is, compressors/limiters and expanders/gates—adjust levels automatically based on an audio signal's level, and we can use them on individual audio tracks and on a full mix. We can also use a hybrid of fader control and dynamic range controllers.

Some signals have an inherently wide dynamic range, while others are relatively narrow. Distorted guitars, for example, generally have a smaller dynamic range, because distortion is created by aggressively limiting or clipping the guitar signal with instantaneous attack and release times. A close-miked lead vocal, on the other hand, can have an extremely wide dynamic range. In some cases, a singer's dynamic range may vary from a loud scream to just a whisper within a single song. If the fader of a vocal track with a wide dynamic range is set to one level and left for the duration of a piece with no compression, limiting, or other level change, there will be moments in the song when the vocal will be much too loud and other moments when it will be too quiet. When a vocal is too loud relative to other instruments in a mix, it can make listeners uncomfortable, and they may then want to turn the entire mix down. In the opposite situation, a quiet vocal is difficult to hear and understand, which creates an unsatisfying musical experience for listeners. It is likely to be impossible to find an optimal static fader level without using compression for a sound source as dynamic as pop vocals, unless the singer intentionally sings within a narrow dynamic range. To compensate for the wide dynamic range of a singer, some engineers manually adjust the fader level for each word or phrase that a singer sings in addition to using compression. Although some tracks do call for such detailed manual control of fader level, compression is still helpful as a first stage of control to achieve consistent, intelligible, and musically satisfying levels, especially for vocals that have a wide dynamic range.

Consistent levels for each instrument and vocal in a pop music recording help communicate the musical intentions of an artist more effectively than levels that have a wide dynamic range. Even though most pop music recordings have very limited dynamic range, dynamic contrasts in a performance are still essential to help convey musical emotion, even with heavy compression. It begs the question: if the level of a vocal track is adjusted so that the loudest (*fortissimo* or *fff*) passages are the same level or loudness as the quietest (*pianissimo* or *ppp*) passages, how do we hear any dynamic contrast? How do we hear differences in the intensity of a performance if the dynamic range is flattened? For most instruments and voices, the timbre varies with the dynamic level. As the level increases from soft to loud, high-frequency harmonics increase in number and strength. So even with a heavily compressed vocal performance, changes in dynamic level are still perceptible because of the way timbre changes with level. Fortunately, even with extreme compression and limiting, it is usually still possible to convey the original intensity of a performance. But this does not mean that engineers always completely flatten the dynamic range of a track or a mix. They certainly do not in most cases. Overuse of compression and limiting can leave a performance sounding lifeless and dull, even in genres such as pop music where compression and limiting are expected.

If you have studied the theory behind the operation of a compressor, you understand a compressor is basically just a fancy volume control. A compressor's main action is to adjust the amplitude of a signal. Technically, this is called amplitude modulation. In the case of a compressor or limiter, the amplitude modulation it applies is based on the amplitude envelope of either the audio signal itself or that of another audio signal. A compressor simply applies gain reduction that varies over time based on a signal's level and the settings of the parameters: threshold, ratio, attack time, and release time.

4.2 Compressors/Limiters and Expanders/Gates

Dynamics processing such as compression, limiting, expansion, and gating all offer means to sculpt and shape audio signals in unique and time-varying ways. The processes are time-varying and nonlinear because the gain reduction they apply varies over time, and it depends on the input (or sidechain) signal level which also changes over time. Dynamic range controllers belong to a group of sound processing effects that are adaptive, meaning that the amount or type of processing is determined by some component of the signal itself (Verfaille et al. 2006). The amount of gain reduction applied to a signal in a compressor or expander is dependent on the level of the signal itself or on the level of a secondary signal known as a sidechain or key input. Because dynamics processors are time-varying, compressing two tracks individually first and then mixing them together is not the same as mixing the two tracks together first and then compressing the mix (Smith 2007).

By contrast, equalization and reverberation are linear processes since the type, amount, or quality of processing remains the same regardless of the input signal characteristics. The nonlinear, time-varying characteristic of dynamics processing also means that once it is applied to a signal, it cannot be undone without going back to the original signal, assuming there is an uncompressed version. There is no way to completely undo the effect of compression on a track if it has been recorded that way. This is unlike reciprocal peak/dip equalization, for example, in which we can undo alterations with equal parameter but opposite gain settings. Compression and limiting do not offer such transparent flexibility. Here again, we are reminded that critical listening can help us make decisions about dynamics processing before committing to a setting that may not be able to be undone later on.

Dynamic range compression is effectively the modulation of a signal's amplitude which is synchronized with the signal's amplitude fluctuations themselves. Because gain reduction is synchronized with the amplitude envelope of the audio signal itself, the gain reduction or modulation can be difficult to hear. It can be difficult to discern if the modulation we hear was part of the original signal or not. Amplitude modulation can be almost inaudible because a compressor reduces signal amplitude as a signal's original amplitude rises. Without hearing the original signal, we do not know exactly how a signal varied dynamically before compression. Compression or limiting can be easier to hear when we set the parameters of a device to their maximum or minimum values, such as a high compression ratio, a short attack time, a long release time, and a low threshold.

Although the gain reduction of dynamics processing can be inaudible or transparent, other types of amplitude modulation can be much more obvious. For example, a tremolo effect, typically used on an electric guitar, modulates a guitar signal's amplitude with a sine, triangle, or square wave. Because the modulator is periodic and not synchronous with the guitar signal, it is audible as an intentional effect. With tremolo processing, the goal is usually to highlight the effect rather than make it transparent or inaudible. With compression, the goal is sometimes to make the gain reduction inaudible and sometimes to make it clearly obvious.

Through the action of gain reduction, compressors can create audible effects, such as timbre changes, which can contribute meaningfully to the sound of a recording. In other situations, dynamic range control can be applied without creating any artifacts or changing the timbre of sounds. Sometimes the goal is to turn down the loud parts in a way that controls the peaks but that does not distract the listener with artifacts or timbral changes. In either case, we need to know what the artifacts sound like to decide how much or little dynamic range control to apply to a recording. On many dynamic range controllers, the user adjustable parameters are interrelated to a certain extent and affect how we use and hear them.

Other forms of dynamic processing such as expansion and gating increase the dynamic range by attenuating lower amplitude sections of a recording. In contrast to a compressor or limiter, an expander attenuates a signal when its level falls below the threshold. Expanders or gates are commonly used when mixing drums for pop and rock music. Even if each component of an acoustic drum kit is close-miked, there can still be some leakage from adjacent drums into each microphone.

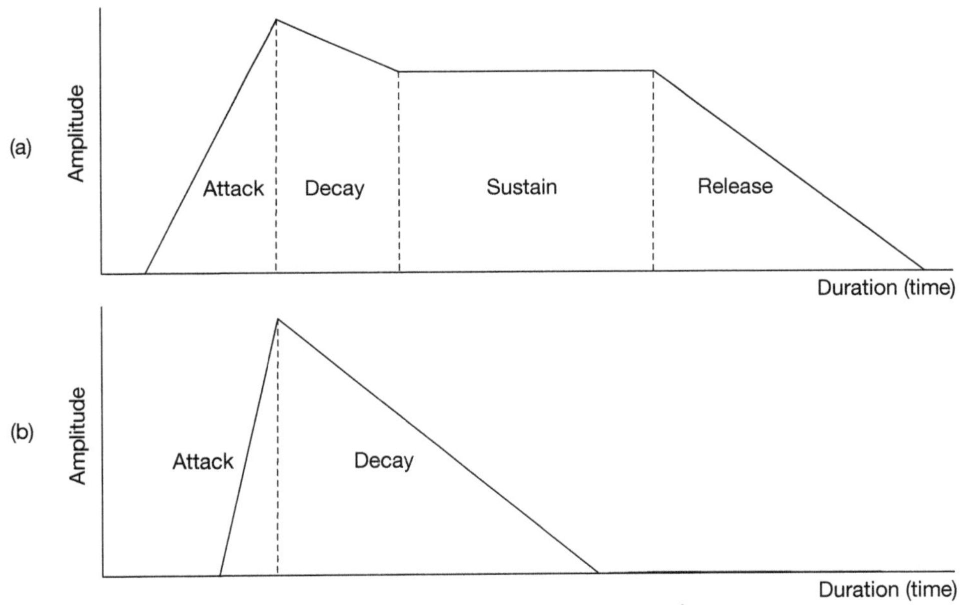

Figure 4.1 The top graph (a) shows the four components of an ADSR, attack, decay, sustain, and release, amplitude envelope that shape a synthesized sound. A note's amplitude begins with the attack and then it transitions to the decay portion. The note's sustain is determined by the duration that it is held. As soon as the note stops being held, the release portion of the envelope begins. The bottom graph (b) shows an amplitude envelope for an acoustic sound, such as from a plucked string, a piano note, or a drum hit, which can have a relatively fast attack but immediately starts to decay after being struck. Actual attack and decay times vary across instruments and even within the range of a single instrument. For example, a low piano note will have a much longer decay than a high piano note, assuming the piano key is held to allow the string to vibrate.

To reduce this effect, an expander or gate can be used to attenuate a microphone signal between hits on the microphone signals of each drum.

It may be worth making a clarification of terminology here since the words "attack" and "release time" can mean different things depending on the context. If you happen to be familiar with conventional sound synthesis theory, you may know that the amplitude envelope of a synthesized sound is usually described in terms of four main properties: attack, decay, sustain, and release, or simply ADSR. Figure 4.1(a) shows a generic ADSR amplitude envelope. The attack time in this context refers to the time it takes for a synthesized note's onset to rise from silence to its peak amplitude. Acoustic instruments, by contrast, each have their own inherent attack times dictated by the physical properties of a vibrating string, column of air, membrane, or plate. Some instruments, such as piano or percussion, have a short attack or rise time, while other instruments, such as a bowed cello, produce a slightly longer attack time.

While the term "attack time" with respect to an instrument or synthesized sound refers to the quick rise in amplitude at a note onset, the attack time of a compressor, by contrast, refers to a gain reduction or a drop in amplitude. The term "release time" also has opposite meanings. A synthesized note's decay and release refer to the note's reduction in level, whereas the release time on a compressor refers to a level increase, as a compressor returns a signal to unity gain following a gain reduction.

In the following sections, I use the word "attack" to refer to both a note's onset and a compressor's attack time, and I use the word "release" to refer to the decay after a note stops and a compressor's release time. I have worked to make sure the context makes it clear which one I am using.

Most compressors include several parameters including threshold, ratio, attack time, release time, and knee, which we look at below.

4.2.1 *Threshold*

Each make and model of compressor has its own unique "sound." This sonic signature is due to a few factors. One factor is the signal detection circuit or algorithm a compressor uses to determine an input audio signal's level, and whether it is above or below the threshold. If the measured level is above the threshold, the compressor attenuates the signal. The ratio determines the amount of attenuation to apply.

On most compressors, the threshold is variable, although on some models it is fixed. For those with fixed thresholds, the input gain is variable to allow the input signal to be boosted up to the threshold. With a fixed threshold design, there may be less make-up gain to apply at the end, which can reduce noise introduced by an analog compressor's circuit. Compressors apply gain reduction once an input or sidechain signal's level goes above the threshold. A sidechain or key input provides the option to use an alternate signal to determine the gain reduction applied to the main audio signal input. Compression to the input signal is triggered when the sidechain signal rises above the threshold, regardless of the input signal level.

4.2.2 *Ratio*

The compression ratio determines the amount of gain reduction applied once the signal rises above the threshold. It is the ratio of input level to output level in decibels above the threshold. For instance, with a 2:1 compression ratio, a compressed signal level's level is half the level (in decibels) of the signal before it is compressed, when we look at just the signal level above the threshold. So,

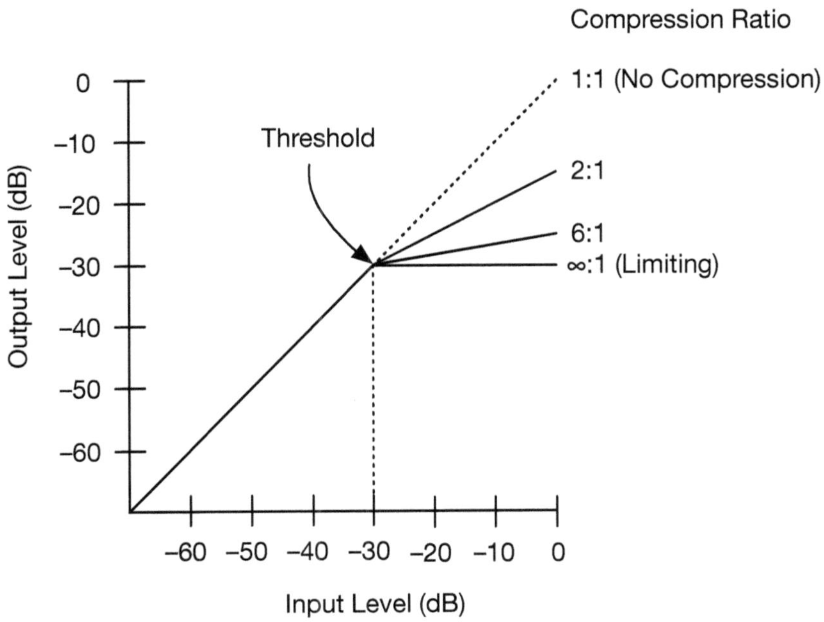

Figure 4.2 The transfer function of a compressor with various compression ratios. As input levels (x-axis) rise above the threshold, output levels (y-axis) rise by a smaller amount depending on the compression ratio.

if the uncompressed signal level is 12 dB above the threshold, the compressed signal level would be 6 dB above the threshold, for a 2:1 ratio.

A compressor set to a ratio of about 10:1 or higher is generally considered to be equivalent to a limiter. Compressors set to higher ratios apply more gain reduction when a signal goes above the threshold making the compression more apparent. Figure 4.2 shows the transfer function—that is, input to output level—with ratios of 1:1, 2:1, 6:1, and infinity:1 which is equivalent to limiting.

4.2.3 Attack Time

Other factors that contribute to a compressor's character are the attack and release curves. Although a compressor begins to reduce the gain of the audio signal as soon as its amplitude rises above the threshold, it usually takes some amount of time to achieve maximum gain reduction. The actual amount of gain reduction applied depends on the compression ratio, how far the signal is above the threshold, and the specified attack time. In practice, attack time can help either add definition to a transient—that is, make it more prominent—or reduce the definition of transients by rounding off note onsets.

With appropriate adjustment of attack time, a recording can sound punchier. Some classic analog compressors, such as the Universal Audio LA-2A, do not have an adjustable attack time. Others have limited options for attack time, such as the Solid State Logic E-Series Dynamics Module, which has a switch to select between fast and normal attack times, and the API 225L Compressor, which has a switch to select among fast, medium, and slow attack times.

4.2.4 *Release Time*

The release time is the time that it takes for a compressor to stop applying gain reduction after an audio signal has dropped below the threshold. As soon as the signal level falls below the threshold, the compressor begins to return it to unity gain, and the release time determines the amount of time it takes to reach unity gain.

4.2.5 *Knee*

The knee describes the transition from levels below the threshold, in which no gain reduction is applied, to levels above the threshold, in which gain is reduced. A smooth transition from one to the other is called a soft knee, whereas an abrupt change at the threshold is called a hard knee.

4.2.6 *Level Detection*

Dynamics processors use the decibel as the unit of measure of audio signal levels. One reason a decibel scale is used in audio is because it is a logarithmic scale. The human auditory system, like other senses, interprets changes in loudness in approximately a logarithmic way. Therefore, the decibel as a measurement scale seems to correlate roughly with the way we perceive sound levels. The second main reason for using decibels is because it reduces the huge range of audible sound levels into a more manageable range. Human perception of sound levels ranges from the threshold of hearing—that is, the quietest sound we can hear—at about 0.00002 pascals (20 micropascals) to the threshold of pain at around 20 pascals. Pascals are a unit of pressure that measure force per unit area. This range of pascals, 20 micropascals to 20 pascals, represents a factor of 1 million. When this range is converted to decibels, it scales from 0 to 120 dB sound pressure level (SPL), which is a much more meaningful and manageable range.

Compressors and limiters attenuate an input signal only if the input or sidechain signal level increases above the threshold. To determine if or when a signal crosses the threshold, the compressor needs to measure the signal's level. The two most common representations of audio signal level are peak and RMS, which is short for root-mean-square. The peak level is simply the instantaneous amplitude of the signal at any given time. Commonly represented by peak meters in a DAW, peak level is essential to know precisely how close a signal is to the 0 dBFS digital clip point. The disadvantage of peak levels is that they do not correspond to the way we perceive loudness since our auditory systems are slower to respond to sound than a peak meter. RMS, on the other hand, is analogous to the average, even though it is not simply the average. The reason why we cannot simply take the average of an audio signal is because audio signals vary between positive and negative values. One half of any sound wave is a positive pressure, and the other half is a negative pressure. Assuming a signal's duration is at least one cycle of the component frequencies measured, the average of any audio signal is zero. This is not much help if we want to know the level of a signal. The root-mean-square or RMS solves this problem by squaring the signal first, which makes all the negative signal levels positive. Next, the mean or average of the signal level is calculated over some amount of time, often in the range of about 50 to 300 milliseconds, depending on the model. The final step in the RMS process takes the square root of the mean. RMS levels have the advantage that they more closely match our perception of signal levels than peak meters do. The main disadvantage is that RMS meters cannot be used to accurately indicate when a transient audio signal is clipping.

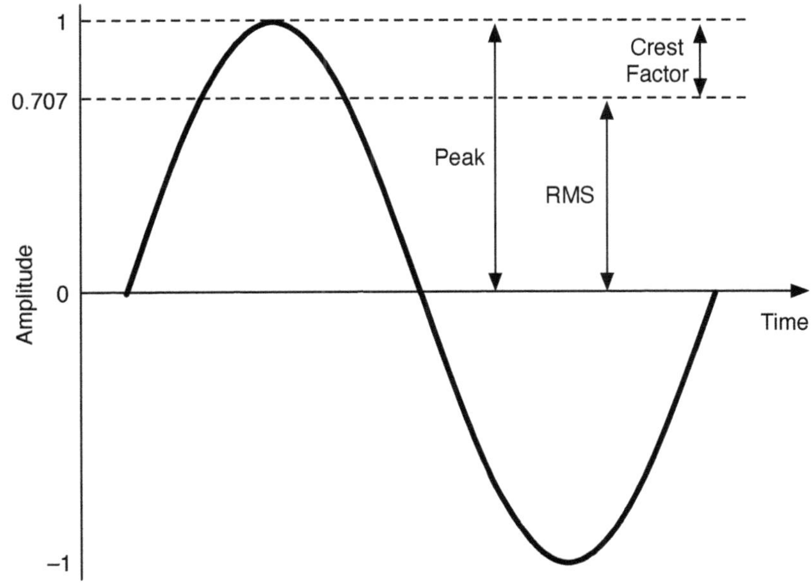

Figure 4.3 The RMS value of a sine wave is always 70.7% of the peak value, which is the same as saying that the RMS value is 3 dB below the peak level. This is only true for a sine wave. The crest factor is the difference between the peak and RMS levels, usually measured in decibels; thus a sine wave has a crest factor of 3 dB.

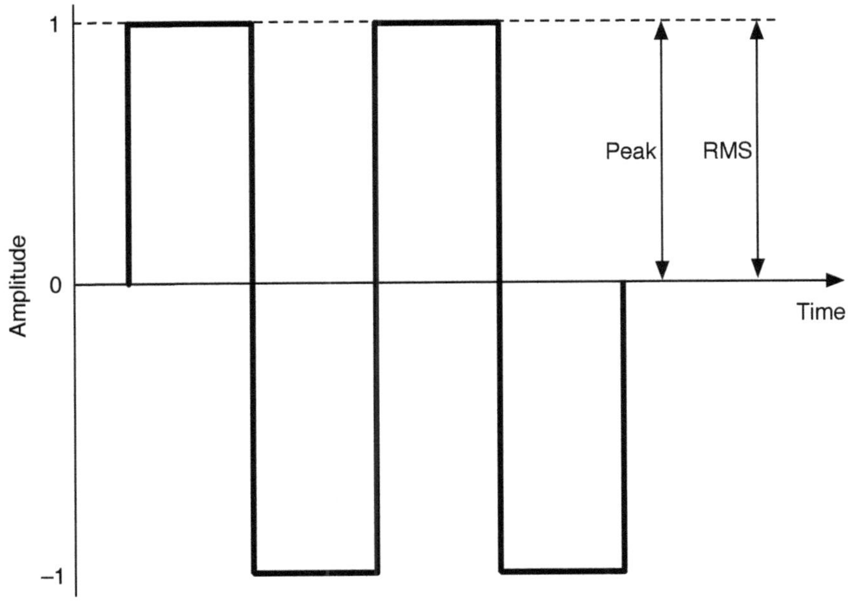

Figure 4.4 A square wave has equal peak and RMS levels, so the crest factor is 0 dB.

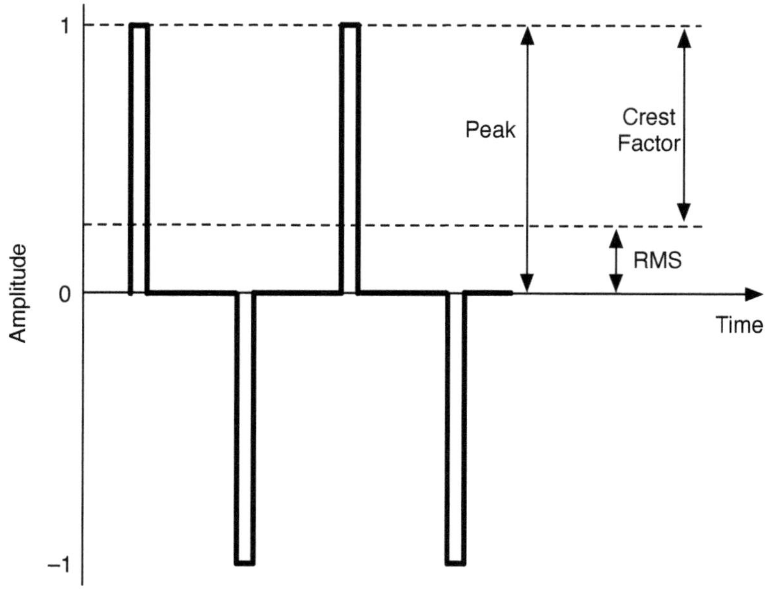

Figure 4.5 A pulse wave is like a square wave except that the amount of time the signal is at its peak level is shorted. The length of the pulse determines the RMS level, where a shorter pulse will give a lower RMS level and therefore a larger crest factor. The RMS level shown in the figure is approximate.

To get a better understanding of RMS, we will look at three different test signals. For a sine wave, the RMS is easily calculated. It is always 3 dB below the peak level. A 3 dB reduction of a sine wave is equivalent to multiplying the peak level by 0.707. For a square wave, the RMS and peak levels are the same, which makes it unique among waveshapes. For a pulse wave, the RMS is much lower than the peak level.

When we talk about RMS and peak levels, it brings up another relevant attribute of a signal: crest factor. The crest factor is simply the difference in decibels between the peak and the RMS. For a sine wave, the crest factor is 3 dB. For a square wave, the crest factor is 0 dB. Figures 4.3, 4.4, and 4.5 illustrate peak, RMS, and crest factor levels for three different signals: a sine wave, a square wave, and a pulse wave. A pulse wave is like a series of transient signals. The peak level of a pulse wave is much higher than its RMS. This means that the peak level of pulse wave signal may cross a compressor's threshold, but the RMS level may not, depending on the threshold and type of level detection used in the compressor.

A sine tone, by definition, is a signal that does not vary in frequency content or level. Most real-world audio signals, however, do vary in their frequency content and level over time. This means that the RMS levels of speech, music, and sound effects cannot be calculated by simply subtracting 3 dB from the peak level. They need to be measured continuously. VU, short for volume unit, meters give the RMS level for a sine tone and approximate the RMS for more complex real-world signals.

Compressors typically use either peak or RMS detection circuits but some models do use both. Compressors designed by renowned engineer and producer George Massenburg—namely, the GML Model 8900 Dynamic Range Controller (analog; now discontinued) and the Massenburg

Design Works MDWDRC2 and MDWDRC3 dynamic range controller plugins—use a combination of both RMS and peak level detection. The Cockos ReaComp compressor plugin included with REAPER provides the ability to control a parameter labeled "RMS size" which ranges from 0 to 1,000 milliseconds and determines the compressor's responsiveness. It seems likely that an RMS size of 0 milliseconds is equivalent to peak level. The Tokyo Dawn Labs Kotelnikov Mastering Compressor plugin offers parameters that control peak crest, release peak, and release RMS, to help reduce pumping and breathing effects.

Because peak level is higher than RMS for almost all signals, compressors using peak level detection will attenuate input signals more quickly and to a greater extent than those using RMS signal detection. The type of signal detection used also determines whether transient signals are compressed and to what extent. A compressor using peak detection can respond to very short transients, allowing them to be compressed, if wanted. On the other hand, a compressor that uses RMS level detection responds more slowly and is less likely to be able to respond quickly enough to compress transient signals as aggressively as one with peak detection.

4.2.7 *Visualizing the Output of a Compressor*

To fully understand the effect of dynamics processing on an audio signal, it is helpful to look beyond the input/output transfer function commonly seen with explanations of dynamics processors as we saw in Figure 4.2. Visualizing a compressor's output over time, when input levels change, can give a better understanding of the ever-critical parameters that do not appear in a compressor's transfer function: attack and release time.

To view and test the effect of a compressor's attack and release times on an audio signal, a specific test signal called a *step function* is ideal. A step function is a signal whose amplitude changes quickly from a low level to a high level or vice versa and then stays at the new amplitude for some period. By using a step function, it is possible to illustrate how a compressor responds to an immediate change in the amplitude of an input signal and eventually settles to its target gain. For the following visualizations, an amplitude-modulated sine wave acts as a step function, as shown in Figure 4.6. In this case, the modulator is a square wave with a period of 1 second, which means the peak amplitude of the sine wave is switched between 1, which corresponds to 0 dB, and 0.25, which corresponds to –12 dB, every half second.

Figure 4.7 shows the step response of a compressor for (a) slow, (b) medium, and (c) fast attack and release times. These kinds of responses are usually not published with compressors' specifications, but we can visualize them by recording their output when an amplitude-modulated sine tone is used as the input signal.

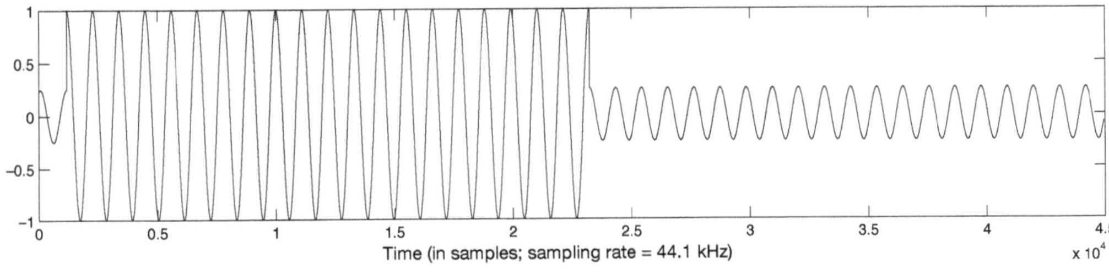

Figure 4.6 This figure shows a step function, an amplitude-modulated sine wave, which can be used to test the attack and release times of a compressor.

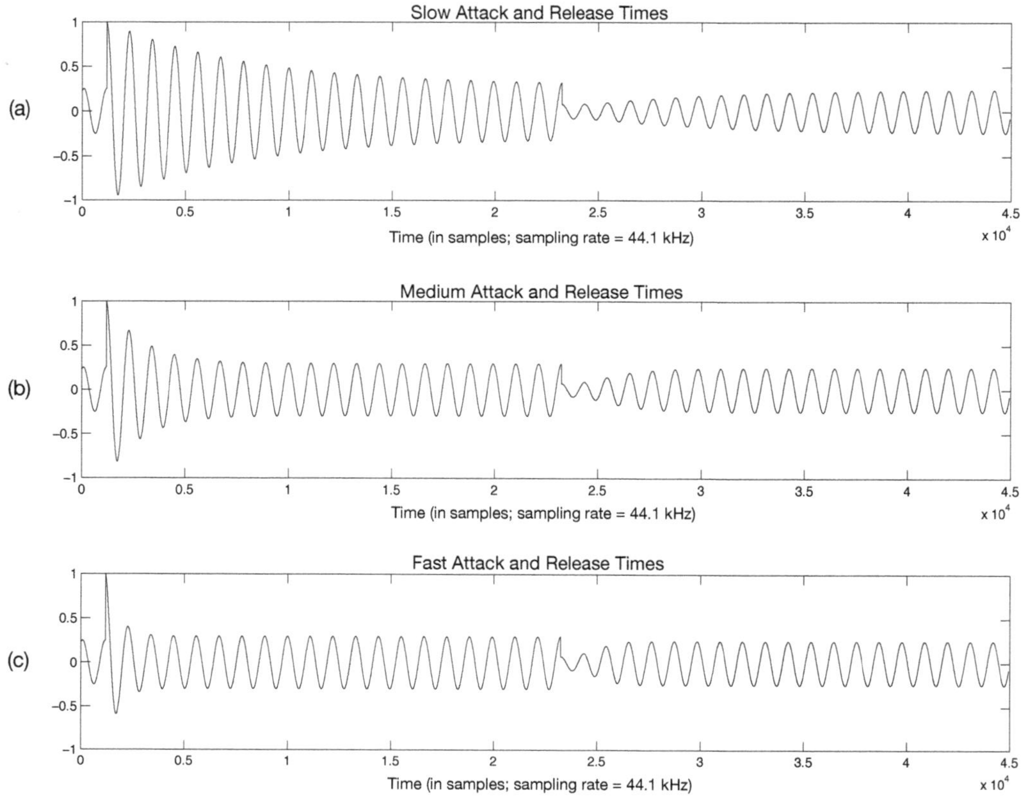

Figure 4.7 The step response of a compressor showing three different attack and release times: (a) slow, (b) medium, and (c) fast.

The step responses of most analog and digital compressors look like those in Figure 4.7, but there are exceptions. For example, Figure 4.8 shows the step function of an analog compressor in which there appears to be an overshoot in the amount of gain reduction in the attack before the signal settles into a constant amplitude of about 0.5. For this measurement, the threshold was set to –6 dB, which corresponds to an audio signal amplitude of 0.5. When the signal goes above 0.5 in level (–6 dB), the compressor attenuates it.

Compressor step responses can also vary within a single compressor plugin. For example, the Apple Logic Pro Compressor plugin has seven circuit types. So, even if all other Compressor plugin parameters are the same, switching to a different circuit type can noticeably change the step response and therefore the resulting timbre. Figure 4.9 shows step responses of the plugin's seven circuit types. Note the variation in attack and release time curves and amount of gain reduction across the various circuit types. All measurements were done with identical compressor parameters except circuit type:

- Threshold: –20 dB.
- Ratio: 20:1.
- Attack: 100 milliseconds.
- Release: 100 milliseconds.

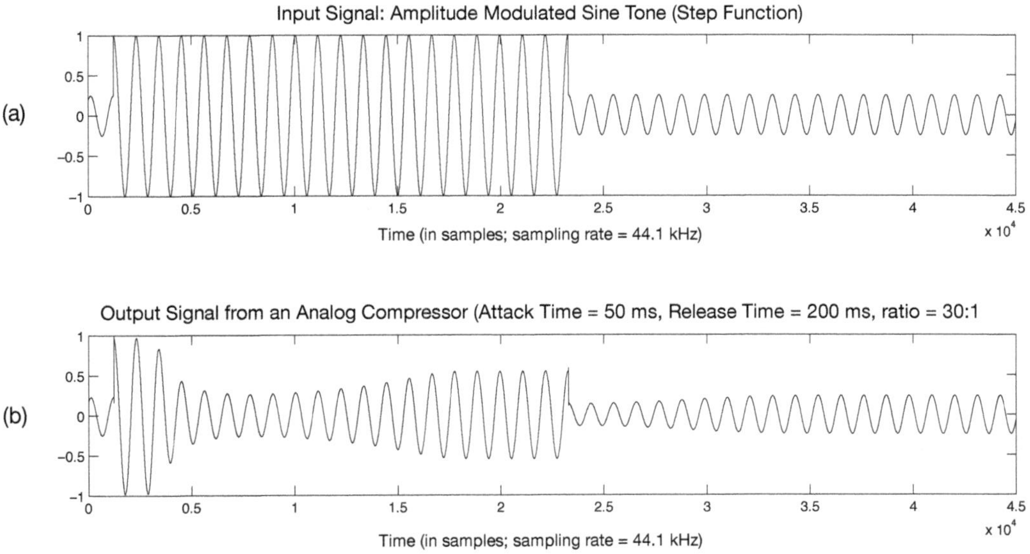

Figure 4.8 The same step function, an amplitude-modulated sine tone, was sent through a commercially available analog compressor with an attack time of approximately 50 milliseconds and a release time of 200 milliseconds. The top (a) is the input signal, and the bottom (b) is the output showing the gain that the compressor applied to the input. Note the difference between this gain curve and those found in Figure 4.7. There appears to be an overshoot in the amount of gain reduction in the attack before it settles into a constant level. A visual representation of a compressor's attack and release times such as this is not something that would be included in the specifications for a device. The difference that is apparent between Figures 4.7 and 4.8 is typically something that an engineer would listen for but could not visualize without doing this kind of measurement.

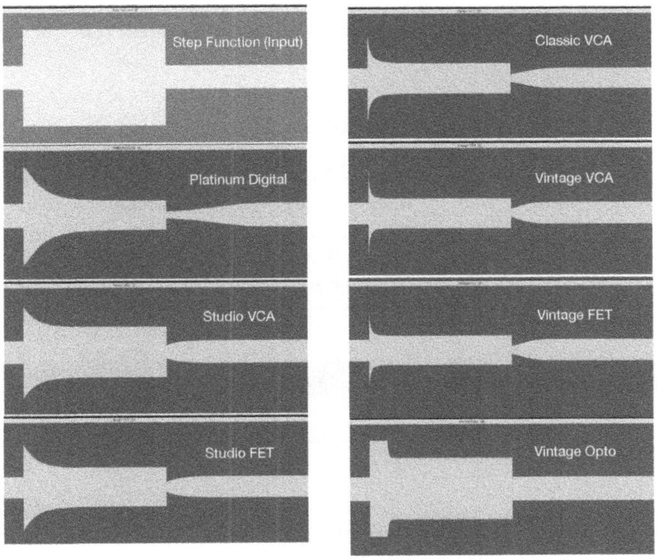

Figure 4.9 Step responses of each Apple Logic Pro Compressor plugin circuit type. The top left image is the input signal, an amplitude-modulated sine tone which was the step function. All compressor parameters were identical other than the circuit type which is labeled on each response waveform.

It can also be helpful to see the gain that a compressor applies over time compared to a signal's waveform. Figure 4.10 shows (a) an input signal on top, (b) the compressor's gain function in the middle as derived from the input audio signal and compressor parameter settings, and (c) the compressed audio signal output. The gain function shows the amount of gain reduction applied over time, which varies with the amplitude of the audio signal input. For example, a gain of 1, which corresponds to 0 dB or unity gain, results in no change in level, and a gain of 0.5 reduces the signal by 6 dB. The threshold was set to –6 dB, which corresponds to 0.5 in audio signal amplitude, in this case, so every time the signal goes above 0.5 in level (–6 dB), the gain function shows that the input signal was reduced in level.

4.2.8 *Automated Level Control Through Compression*

Dynamic range compression may be one of the most difficult types of processing for the beginning engineer to learn how to hear and use. Likely, it is difficult to hear because the goal of compression is usually to reduce wide swings in dynamic level without making it sound as though any changes to the audio signal have been made. Depending on the nature of the signal being compressed and the parameter settings chosen, the effects of a compressor can range from very subtle or inaudible to completely obvious. Perhaps another reason why novice engineers may find it difficult to identify

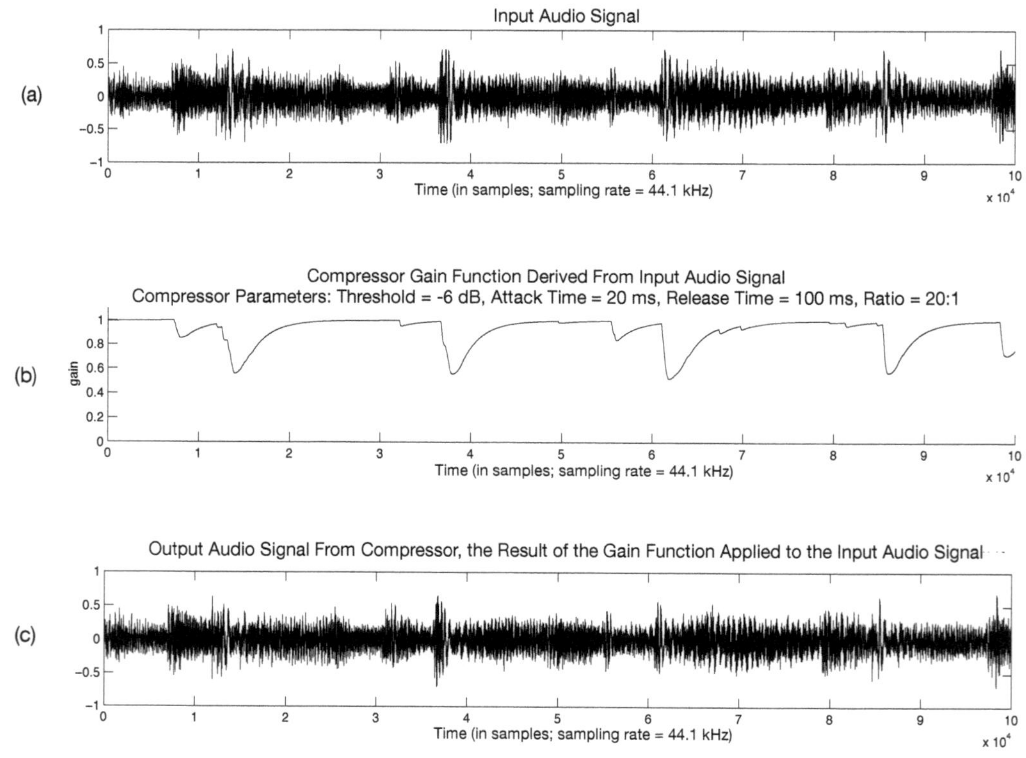

Figure 4.10 This figure shows three plots: (a) an audio signal sent to the input of a compressor, (b) the gain function of the compressor based on the input signal level, and (c) the resulting audio signal output from the compressor with the gain function (b) applied to it.

compression is that nearly all recorded sound has been compressed to some extent. Compression has become such an integral part of almost all recorded and amplified sound that perhaps listeners have come to expect it. Given the ubiquity of compression, it can help to refresh our perspective and remind ourselves what music sounds like without compression by listening to acoustic music without sound reinforcement, a listening exercise that can also help with ear training.

4.2.9 *Manual Dynamic Range Control*

Because dynamic range controllers respond to objective measures of signal level, typically in terms of peak or RMS, rather than the way we perceive loudness, the gain reduction applied by a compressor does not always suit our purposes as well as we may want it to. In other words, the automated nature of a compressor's dynamic range control may be too aggressive and not be as transparent as we would like, since objective signal level measurements do not correspond to our perceptions of loudness. As a result, a compressor may measure a signal to be louder or quieter than we perceive it to be and therefore apply too much or too little gain reduction.

When mixing a multitrack recording, we are concerned with levels, dynamics, and finding a balance for the tracks. We want to be attentive to sound sources that are masked at any point in a piece. At a more subtle level, even if a sound source is not fully masked, it is still important to find the best possible musical balance and adjust as necessary across each note and phrase of music and through the whole song or piece. Focused listening helps us find the best compromise on the overall levels of each sound source. It is a compromise because it is unlikely that every note of every sound source will be heard perfectly clearly, even with extensive dynamic range control. It's a fool's errand to turn up each sound source to be heard above all others. We will be no further ahead by doing this and will eventually run out of headroom in our mix bus and cause clipping distortion. We usually need to compromise and determine where the priorities are for any given project. In many cases, the genre dictates which elements get priority. For example, vocals in pop, hip-hop, rap, rock, country, or jazz recordings are typically the most important and therefore the most prominent element in a mix, since we want listeners to hear each word of a vocal recording. But this is not always the case and depends on the artistic vision of the band and producer. In some recordings, the vocals are mixed equally with the rest of the band. Vocals are often particularly dynamic in amplitude due to close miking techniques, so the addition of some dynamic range compression can help make each word and phrase of a performance more consistent in level.

With recorded sound, a dynamic mix can guide a listener's perspective and perception of a musical performance with level control of individual sound sources. Ideally, this is done without the listener realizing that they are being guided aurally through a song or soundtrack. Instruments and voices can be brought dynamically to the forefront and sent farther back, as the artistic vision of a performance suggests. Sound source level automation can create changing perspectives that may be obvious or not at all apparent to the casual listener. Depending on the state of the raw tracks in a multitrack recording, a mixer may need to make drastic changes in the mixing stage to create coherency and a focused musical vision. Listeners may not be consciously aware that levels are being manipulated, and, in fact, engineers often try to make level changes as transparent (i.e., inaudible) and musical as possible. Ideally, listeners should only be able to hear that each moment of a music recording is clear and musically satisfying, not that continuous level changes are being applied to a mix. Again, we often strive to make the effect of technology transparent to an artistic vision of the music we are recording. The old saying about recording and live sound engineers is that we know we are doing a good job when no one notices our work. It's only when people notice bad sound at a live event that they turn to look at the front-of-house engineer. Otherwise, that person is rightfully ignored. Listeners and musicians should be able to focus on the art and not be distracted by engineering artifacts. The work that the engineer does should support the efforts of the artist or producer and not obscure them.

In addition to learning how to identify the artifacts produced by dynamic range compression, it is also important to learn how to identify static changes in gain. If the overall level of a recording is increased or decreased by a fixed amount, it is important to be able to recognize the amount of gain change applied in decibels.

4.3 Loudness and Dynamic Range

The dynamic range of a recording can have a significant effect on our perception of its loudness. The term "loudness" refers to a sound's perceived level rather than its physical, measured sound pressure level. A few factors contribute to perceived loudness, such as a sound's frequency content, amplitude envelope, average signal level, and crest factor. If two recordings have the same peak level but different RMS levels, the one with a smaller crest factor will generally sound louder because its RMS level is closer to the peak level. The recording with a higher RMS level sounds louder because our ears respond more to average levels than to peak levels when judging the loudness of sounds. This explains why engineers used to prefer VU meters to help them judge the loudness of a track, and why VU meters were common on analog consoles.

Dynamic range compression can make a recording sound louder because it increases a recording's average level through a two-stage process. In the first stage, the peak levels of an audio signal are attenuated, which makes the overall level quieter and reduces the crest factor, since the RMS is unaffected. In the second stage, the overall level is boosted by some fixed amount with output or make-up gain. Some compressors and limiters apply automatic make-up gain at the output stage so that the gain-reduced loud sections remain at roughly the same level. Make-up gain brings up the entire signal—that is, quiet and loud parts—so if we match the attenuated audio peaks to their pre-compression levels, the net result is that the quieter audio sections are now louder. The process of compression and limiting reduces the crest factor of an audio signal, and when make-up gain is applied to restore the peaks to their original level, the RMS level is increased as well, making the overall signal sound louder. By reducing the crest factor through compression and limiting, we can make an audio signal sound louder even if its peak level is unchanged.

Another way to increase the overall level of some, but not all, signals is to normalize them. When we normalize a track in a digital audio workstation, the software scans the audio and finds the highest signal level for the entire clip. If the peak is not already at 0 dBFS, the maximum recordable level for a digital system, the entire audio clip is raised by the amount of difference between the peak and 0 dBFS. For example, if the highest peak level of a track is 6.3 dB below 0 dBFS, normalizing it will raise the entire track by 6.3 dB. But if the signal level already reaches 0 dBFS at any point, then normalizing will not change its level. This is one reason why the process of digitally normalizing a sound file does not necessarily make a recording sound louder. The only way to make a signal whose peak is already at 0 dBFS sound louder is through compression or limiting which raises the RMS level and reduces the crest factor.

Normalizing a mix is not necessarily a good idea because even if the original sample peaks are only as high as 0 dBFS, the peaks between samples, also known as inter-sample peaks, may go above 0 dBFS in the case of oversampling on playback, and cause clipping. Many mastering engineers recommend staying at least a few decibels below 0 dBFS. For recordings submitted to streaming services such as Apple Music, Apple's (2021) "recommendation is to leave at least 1 dB of headroom in order to avoid such clipping" due to processes such as oversampling during digital-to-analog conversion.

While VU and peak meters remain ubiquitous in DAWs, the best way to measure loudness and true peak levels is with a LUFS meter. A LUFS meter is like a VU or RMS meter except that it considers the way humans perceive loudness. LUFS stands for loudness units relative to full scale. LUFS is equivalent to LKFS, which stands for loudness, K-weighted, relative to full scale. The K-weighting filter is meant to mimic the frequency response of the human ear at 70 phons—that is, the loudness of a 1 kHz tone at 70 dBSPL. The K-weighting filter consists of a low-frequency cut-off at 80 Hz and a 4 dB high shelf boost at 2,000 Hz. LUFS is described in the ITU-R BS.1770 recommendation (ITU-R 2023) and the EBU R 128 recommendation (EBU 2023).

LUFS meters usually include measurement of:

- momentary term loudness averaged over 400 milliseconds,
- short-term loudness averaged over 3 seconds,
- integrated loudness averaged over the entire duration of the track or program,
- true peak level, which is a more accurate measurement of peak level to help avoid clipping that might not register on a sample peak meter, and
- loudness range, which is an indication of the dynamic range.

LUFS meters allow an engineer to match the loudness of different programs or recordings. For example, let's say the loudness of recording A is –12.3 LUFS and the loudness of recording B is –7.3 LUFS. Recording B is 5.0 dB louder than recording A, which means we can match the loudness of the two recordings by attenuating recording B by 5.0 dB.

Several loudness meters are available as plugins, including:

- Apple Logic Pro Loudness Meter (included with Logic Pro).
- Cockos REAPER Loudness Meter Peak/RMS/LUFS plugin and stereo bus metering options (included with REAPER).
- iZotope Insight 2 (paid).
- Klangfreund LUFS Meter (paid).
- PreSonus Studio One Pro Level/Loudness Meter (included with Studio One).
- Steinberg Cubase LUFS Meter (included with Cubase).
- TC Electronic LM1 Stereo Loudness Meter and LM2 Stereo Loudness Meter (paid).
- Waves WLM Plus Loudness Meter (paid).
- Youlean Loudness Meter (free and paid versions).

If a LUFS meter is not included with your DAW, I recommend the free Youlean Loudness Meter since it satisfies all requirements of the ITU-R BS.1770 recommendation.

Loudness measurements of songs in music streaming services are included in track or song metadata which allows the streaming services to match the loudness of recordings by automatically adjusting the playback level. Songs from a variety of different albums, artists, eras, and genres can be played back-to-back with minimal change in loudness if the music streaming app has the loudness matching feature turned on. The feature may be called normalize volume (for Spotify), loudness normalization (for TIDAL), or Sound Check (for Apple Music). Try listening to a playlist of songs from different artists and albums with and without loudness normalization. Do you notice a difference in loudness from one track to another when it is turned off?

4.4 Timbral Effects of Compression

When applied to a full mix, compression and limiting not only control the dynamic range but can also help the elements of a mix coalesce. Compressed musical parts can have what is known in auditory perception as the Gestalt principle of "common fate." Their amplitudes change synchronously because a mix bus compressor applies the same gain reduction to all tracks or parts at the same time. When the amplitudes of two or more instruments or voices change in the same way at the same time, our auditory systems will tend to group these elements together perceptually (Bregman 1990). This is part of the reason why dynamics processing can help blend elements of a mix together.

In addition to being a utilitarian device for managing the dynamic range of recording media, dynamics processing is also a tool for altering the color and timbre of recorded sound. Although compressors are not equalizers or filters by any stretch, they can be used to do some spectral shaping, and we should always listen for the ways that compressors affect timbre. In this section, we will look at compression as a tool to sculpt the timbre of a track.

4.4.1 *Effect of Attack Time*

With a compressor set to a slow attack time—say, in the 100-millisecond range or greater—with a low threshold and high ratio, a compressed signal's level plunges down at those points when its original level rises above the threshold. This dramatic drop in level can create an audible artifact is known as "pumping." The effect can be most audible on transient sounds with a strong pulse where a signal clearly rises above the threshold and then drops below it, such as we find in recordings of drums and percussion. Any lower-level sounds such as background noise present with the main sound being compressed will also be modulated, and this accentuates the pumping effect. Sounds that are more constant in level such as distorted electric guitar will not create such an audible pumping effect because their levels do not vary as widely above and below the threshold.

4.4.2 *Effect of Release Time*

A related effect is present if a compressor is set to have a slow release time—say, in the 100-millisecond range or greater—combined with a low threshold and high ratio. It should be apparent as the sound comes back up in level after a strong pulse that has been attenuated. The audible effect of the sound coming back up in level after significant gain reduction is called breathing because it can sound like someone taking a breath. As with the pumping effect, you may notice the effect most prominently on background sounds, hiss, or higher overtones that sustain under a strong pulse.

4.4.3 *Compression and Drums*

Although compression tends to be explained as a process that reduces the dynamic range of an audio signal, there are ways to use a compressor to accentuate the difference between transient peak levels and any sustained elements that may follow. When used this way, compression can essentially increase the dynamic range by highlighting peaks and strong pulses—that is, drum hit onsets—relative to quieter sounds, such as drum resonances or decay, that immediately follow the initial transient onset. It may seem completely counterintuitive to think of compressors as a tool to expand dynamic range. In this section, we will work through what happens when experimenting with various attack times.

A recording with a strong pulse, such as a drum, with a regularly repeating transient can trigger gain reduction in a compressor and serve as a useful sound to highlight the effect of a dynamics processing. In a typical drum sound that has not been compressed, there is a natural attack or onset, and then a release or decay, all of which are dependent on a drum's physical characteristics and tuning. A compressor can change all these properties depending on how its parameters are set.

Let us explore the sonic effect of a compressor on a drum recording. To do this exercise, use the compressor available on https://webtet.net/apcl/#/compression or add a compressor plugin to a track on a DAW and load in a drum loop (kick drum, snare drum, drum set, or percussion). Set the compressor to a low threshold, or low enough to cause significant gain reduction on each drum hit, a ratio of 10:1 or higher, an attack time of 2 milliseconds or faster, and a fast release time. If using webtet.net, set the level of difficulty to level 3 to get a wider range of possible attack and release times.

A compressor's attack time has the greatest effect in shaping drum sound onset. With a fast attack time, a compressor brings transient levels immediately down, and the excessive gain reduction makes the naturally sharp onset of the snare drum sound dull or pillowy since the rate of gain reduction matches or nearly matches the rate at which a transient signal rises in level. So, with a very short attack time along with a short release time, a compressor can nearly eliminate transients because the gain reduction brings a signal's level down at almost the same rate that the signal level rises during a transient. As a result, the initial attack of a transient signal is reduced to the level of the resonant part of the amplitude envelope—that is, the part of the sound that is below the threshold. Very fast attack times can be useful especially on limiters when used as protection against clipping. Some limiters even include a feature called "look ahead" which delays an incoming signal by a short amount of time—a few milliseconds, typically—to allow the limiter to analyze and react quickly enough to reduce the gain of sharp transient peaks before any clipping occurs. A look-ahead limiter is sometimes used in live event broadcasting to guarantee that no unexpected peaks will cause clipping. For shaping drum and percussion sounds, fast attack times are quite destructive and tend to take the life out of the original sounds. On the other hand, we can mix fast attack time compressed drums with the original, uncompressed version of the drums to maintain the original transients and use make-up gain on the compressed version of the drums to bring out the drum decay.

Experiment by increasing the attack time from 0 milliseconds to something longer (slower). As you increase the attack time to even just a few milliseconds, listen for a clicking sound to emerge at the onset of the drum's transient. The click is produced because a few milliseconds of the original audio pass through before the compressor applies full gain reduction. Listen to the way the timbre of the click changes as the attack time changes. A compressor's abrupt gain reduction reshapes the amplitude envelope of a drum hit. By increasing the compressor's attack time further, a drum's transient onset gains prominence relative to the decay portion of the sound. This happens with longer attack times because the compressor's gain reduction lags the drum attack time. Therefore, the gain reduction happens after the drum's attack and during its decay. If a drum's decay is reduced relative to its attack, the level difference increases between the loud and soft components of the sound. Thus, with compression the attack of a drum hit, the loudest part, can become even more prominent relative to the decay, the quieter part, since the quieter part gets reduced in level in this situation.

A compressor can also affect the low-frequency content in transient signals like drums. For example, when we compress low-frequency drums such as kick drum or an entire drum set and shorten the attack time, the low-frequency content is reduced. A compressor's attack time affects low frequencies because low frequencies have longer periods, and slower attack times allow more low-frequency cycles to sound before gain is reduced. Therefore low-frequency content is more audible on each rhythmic low-frequency pulse with a slow attack, compared to a fast attack. By increasing the attack time from a very fast (short) to slow (long), the amplitude of low frequencies coming from the kick drum increases. The spectral effect of increasing a compressor's attack time from near zero to 50 or 100 milliseconds is like adding a low shelf filter to boost low frequencies. We can get a similar low-frequency effect even if we maintain a fast attack time but adjust the signal detection timing (if that parameter is available to adjust). For example, the REAPER ReaComp plugin allows the RMS size to be adjusted. So, if the attack time is fixed at 0 milliseconds and RMS size is increased from 0 to 100 milliseconds or so, it has the effect of boosting low frequencies since effectively the attack time is increased when the RMS size is increased.

The release time, on the other hand, affects mostly the decay of the sound. The decay portion of a drum hit is the portion that becomes quieter after the loud transient onset. With a slow release time, a compressor's gain reduction does not quickly return to unity gain after the signal level has fallen below the threshold, since this normally happens during the decay, which is lower in level, and therefore the natural decay of the drum sound becomes significantly reduced.

4.4.3.1 Drum Compression in Recordings

Drum compression can range from transparent in its application with no artifacts to heavy-handed use that adds a nonlinear distortion effect. Drums are sometimes used as a sidechain to compress other elements in a mix such as synth pads and bass. Here are just a few examples of recordings showcasing drum compression.

"Indo Silver Club" by Daft Punk from Homework (25th Anniversary Edition) *(1997/2022)*
A sidechain compression effect like the one on this track can make the kick drum sound bigger than it would otherwise. From about 0:19 to 0:50, the pumping effect is quite prominent with a sidechain compression effect triggered by the kick drum.

"Feel It All Around" by Washed Out from Life of Leisure—EP *(2009)*
This track is heavily compressed, and there is a sidechain compression effect created by the drums. The effect is especially audible because there are many sustained, continuous sounds in the track.

"Titanium (feat. Sia)" by David Guetta from Nothing But the Beat Ultimate *(2011)*
Starting around 0:46 we start to hear some sidechain effects on the synth, and then from 1:16 to 1:31, the side chain effect triggered by the kick drum pumps or modulates the rest of the instruments (synth and bass) in the track.

"Neverender" by Justice and Tame Impala from Justice *(2024)*
There is a significant sidechain compression effect on this track.

4.4.4 Compression and Vocals

Because vocal performances tend to have a wide dynamic range, engineers often find that dynamic range control is necessary to reach the artistic goals of a recording. Compression is useful to reduce the dynamic range and to de-ess a vocal track, but unfortunately compression does not always work as transparently as desired, and timbral side effects sometimes result from the automated gain control of a compressor.

Here are a couple of simple tips to help reduce dynamic range without adding too many of the side effects that can detract from a vocal performance:

- **Use low ratios:** The lower the ratio, the less gain reduction that will be applied. Ratios of 2:1 are a good place to start.
- **Use more multiple compressors in series:** By chaining two or three compressors in series on a vocal, with each one set to a low ratio, each compressor provides some gain reduction but too much that they are creating unwanted artifacts. The overall result is more transparent (i.e., fewer artifacts) gain reduction than using a single compressor to provide all the gain reduction. Since compressors are nonlinear devices, the effect of three compressors in a row at low compression ratios will be different from a single compressor at a high compression ratio.

To help identify when compression is applied too aggressively, listen for changes in timbre while watching the gain reduction meter on your compressor. If there is any change in timbre while gain reduction happens, the solution may be to lower the ratio, raise the threshold, or both. Sometimes it may sound like the high frequencies or the low frequencies have been reduced. A compressor's gain reduction meter can also make it easier to identify compression artifacts.

A slight click or popping sound at the start of a singer's word or phrase may indicate that the attack time is too slow. Generally, a very slow attack time is not effective on a vocal since it can accentuate the attack of a vocal which can be distracting to listeners. This is especially apparent in a highly compressed vocal.

Compression of a vocal usually brings out lower-level detail in a vocal performance such as breaths and "s" sounds. A de-esser, which can reduce the "s" sound, is simply a compressor that has a high pass filtered, usually with a cut-off frequency around 5 kHz, version of the vocal as its side-chain or key input. De-essers tend to work most effectively with very fast attack and release times.

4.4.4.1 Vocal Compression in Recordings

Pop vocal recordings often provide great examples of extreme compression due to the wide dynamic range of the performances. Usually, the goal for pop song recordings is to make sure lead vocals are always front and center in a mix. To ensure they remain clearly audible and focused for the listener, pop vocals usually need heavy compression. Here are just a few examples of some recordings in which vocals have been compressed fairly noticeably.

"Let Love Rule" by Lenny Kravitz, two versions: Let Love Rule *(1989) and* Let Love Rule: 20th Anniversary Edition *(2009)*
The compression on this track is more obvious than on some other recordings. It sounds like compression or limiting has been applied to everything in the mix because the compression artifacts are so evident. The compression combined with the minimal use of reverberation and panning choices give the recording a retro aesthetic. Compare the original 1989 version to the remastered 2009 version.

"Brave" by Sara Bareilles from The Blessed Unrest *(2013)*
The main vocal on this track seems to have one dynamic level in the verses, where she seems to have sung it more quietly, and a much higher dynamic level for the choruses, where it sounds like she was belting. Nonetheless, the vocal level is constant throughout the track.

"Love on the Brain" by Rihanna from ANTI (Deluxe) *(2016)*
The lead vocal for this recording seems to have covered a wide range of dynamic levels in the original performance. With what sounds like extensive compression, the vocal level remains consistent through the song—it's never too quiet or too loud. The timbre of her voice indicates her original dynamic level changes.

"Easy on Me" by Adele from 30 *(2021)*
The vocal compression is especially apparent on the choruses in this recording where she switches between a light airy voice to a full-chest voice. Listen to the phrase "I was still a child." Although the vocal level remains constant through the phrase, it sounds like the words "I was still a" were sung much more quietly than the word "child" in her original performance yet the vocal level remains constant.

"Good Luck, Babe!" by Chappell Roan, single release (2024)
At the end of the bridge from about 2:33 to 2:40, listen to the way Chappell Roan's vocal goes from soft to loud. Compression keeps the vocal level consistent through this dramatic dynamic level change which leads into the final chorus of the song.

These are just a few examples of recordings with extensive compression, but most current popular songs have heavily compressed vocals.

4.5 Expanders and Gates

Expanders get their name from the fact that they expand the dynamic range of a signal by lowering the level of the quieter moments of a signal, thus increasing the level difference between the loud and quiet parts. A gate is simply an extreme version of an expander and is typically configured to mute a signal when its level drops below a threshold. Most of the controllable parameters on an expander are similar in function to those on a compressor except for attack time and release time. These two parameters work in the opposite way compared to a compressor. It's helpful to think of these two parameters on an expander as they relate to an audio signal's level, rather than in relation to the amount of gain reduction, as we will explore in more detail below.

4.5.1 Threshold

An expander increases dynamic range by attenuating an audio signal when its level falls *below* the threshold, thus making the quieter parts even quieter. This contrasts with compressors which attenuate a signal only when its level goes *above* the threshold. Table 4.1 shows the gain changes applied by these two types of dynamic range controllers depending on whether the signal level is above or below the threshold.

Table 4.1 This table shows the two conditions under which attenuation is applied by a dynamics processor, depending on whether it is a compressor/limiter or expander/gate.

	Compressor/Limiter	*Expander/Gate*
Signal Level	*Gain Applied*	*Gain Applied*
Above the threshold	Attenuation	No Attenuation (Unity Gain)
Below the threshold	No Attenuation (Unity Gain)	Attenuation

While attack and release times in a compressor or limiter refer to gain changes that are opposite to those of an ADSR amplitude envelope, as we saw in Section 4.2, the terms refer to the same type of signal level change with expansion and gating.

4.5.2 Ratio

The expansion ratio determines the amount of gain reduction applied once the signal drops below the threshold. It is the ratio of input level to output level in decibels below the threshold. For instance, with a 1:2 expansion ratio, the input signal below the threshold is double the number of decibels below the threshold that the uncompressed signal is below the threshold. An expander set to a ratio of about 1:10 or lower is generally considered to be equivalent to a gate. Expanders set to lower ratios apply more gain reduction when a signal goes below the threshold, making the expansion more apparent. Figure 4.11 shows the transfer function (input to output level) with ratios of 1:1, 1:2, 1:6, and 1:infinity, which is equivalent to gating.

4.5.3 Attack Time

An expander's attack time controls the rate at which a signal's level is returned to its original level after its level rises above the threshold. This is in contrast to a compressor in which the attack time controls the rate at which a signal is attenuated after its level rises above the threshold. For both,

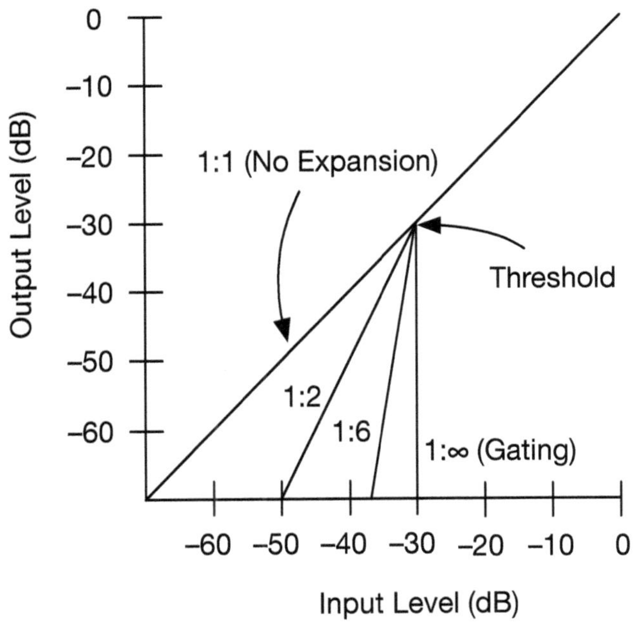

Figure 4.11 The transfer function of an expander with various expansion ratios. As input levels (x-axis) drop, output levels (y-axis) drop by a greater amount depending on the expansion ratio.

the attack time is the amount of time it takes to apply a gain change after the signal goes above the threshold. A compressor attenuates signal levels above the threshold whereas an expander boosts them.

4.5.4 Release Time

The release time on an expander is the time it takes to complete its gain reduction once the input signal has dropped below the threshold. During release time on an expander, the signal level is reduced; during the release time on a compressor, the signal level is increased.

4.5.5 Visualizing the Output of an Expander

Figure 4.12 shows the effect that an expander can have on an amplitude-modulated sine wave; the three output signals shown in the figure are also known as step responses of the expander for three different attack and release times.

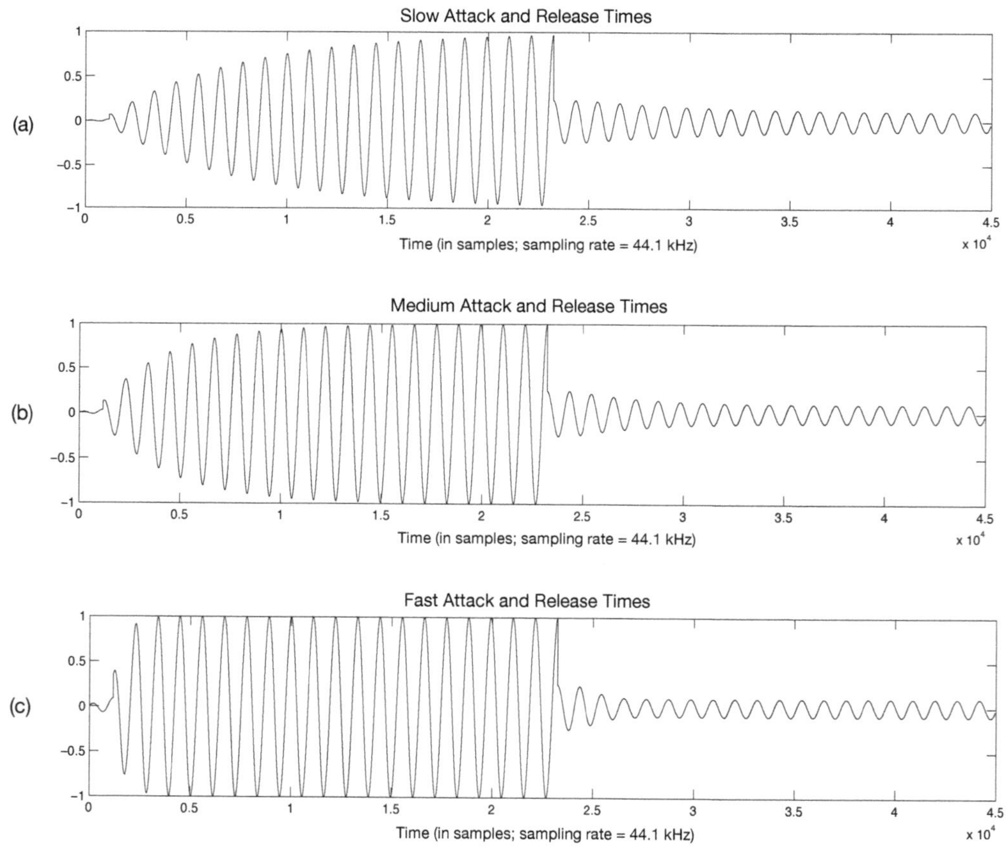

Figure 4.12 This figure shows the step response of an expander for three different attack and release times: (a) slow, (b) medium, and (c) fast. The input signal was a step function as shown in Figure 4.6.

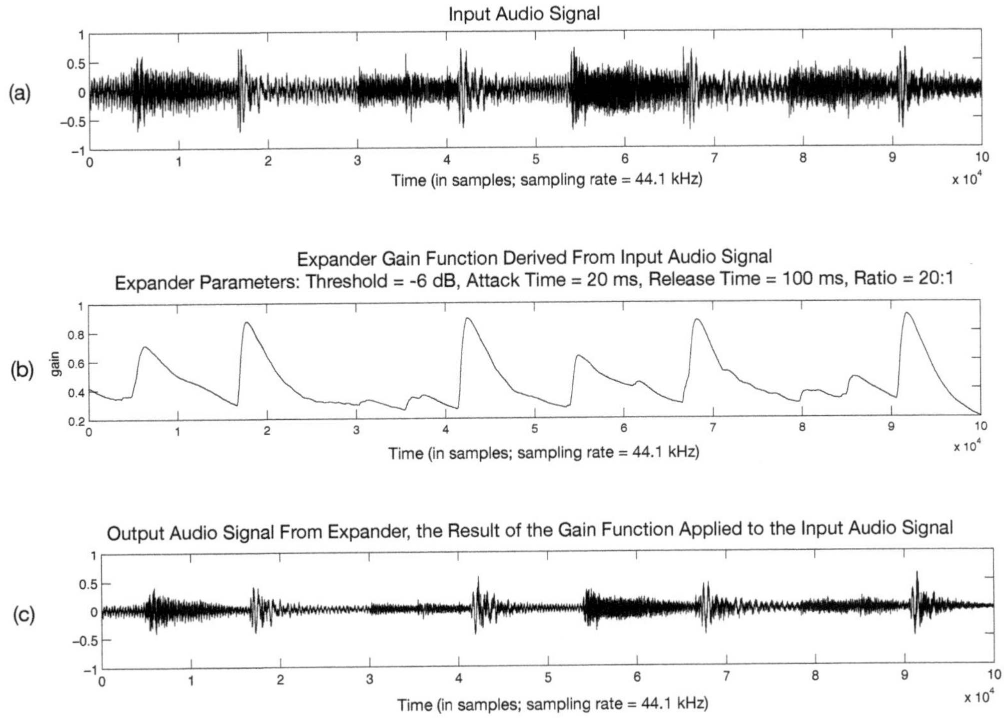

Figure 4.13 At the top (a) is the input audio signal being sent to an expander. In the middle (b) is the gain function, which is determined by the expander parameters and signal level. At the bottom (c) is the resulting audio signal output from the expander—that is, the gain function applied to the input signal.

As we saw with compressors in Section 4.2.7, it can be helpful to see the gain that an expander applies over time compared to a signal's waveform. Figure 4.13 shows a clip from a music recording with the gain function derived from the audio signal and parameter settings and the resulting output audio signal. Low-level sections of an audio signal are reduced even further in the expanded audio signal. The gain function shows the amount of gain reduction applied over time, which varies with the amplitude of the audio signal input. For example, a gain of 1, which corresponds to 0 dB or unity gain, results in no change in level, and a gain of 0.5 reduces the signal by 6 dB. For these measurements, the threshold was set to –6 dB, which corresponds to 0.5 in audio signal amplitude, so every time the signal drops below 0.5 in level (–6 dB), the gain function shows that the input signal was reduced in level.

Besides their ability to attenuate or completely mute quieter moments of a signal, expanders and gates can be used to augment a transient signal like a snare drum or kick drum with other sounds. Like compressors, many expanders have a sidechain input that can be used to control the amount of expansion applied to an audio signal. An expander can be used to add a low-frequency sine tone, at, say, 40 or 50 Hz, that matches the amplitude envelope of a kick drum. To do this, send the low-frequency sine tone through the expander and send the kick drum to the sidechain input. When the kick drum sounds, the expander will allow the sine tone to sound. Depending on the expander parameters, the envelope of the low-frequency sine tone can be shaped to match that of each kick drum hit. The two sounds—kick and sine tone—can be mixed to create a new timbre that adds low end to the drum.

4.6 Technical Ear Training Software Modules: Dynamic Range Control

The Dynamic Range Compression and Dynamic Range Expansion software modules are available on the companion website linked here: https://webtet.net/apcl/#/compression and https://webtet.net/apcl/#/expansion.

The overall functionality of the software modules focused on dynamics processing is very similar to that of the equalization module, except that there are different parameters and qualities of sound to explore.

The threshold is completely variable for all exercises and is the same for both the computer-generated question and your response. In other words, changing the threshold changes it for both the question and your response. In the compression module, the threshold level should initially be set low so that the effect of the compression is most audible. A make-up gain fader is included so that you can match the loudness of compressed and bypassed signals by ear if that is

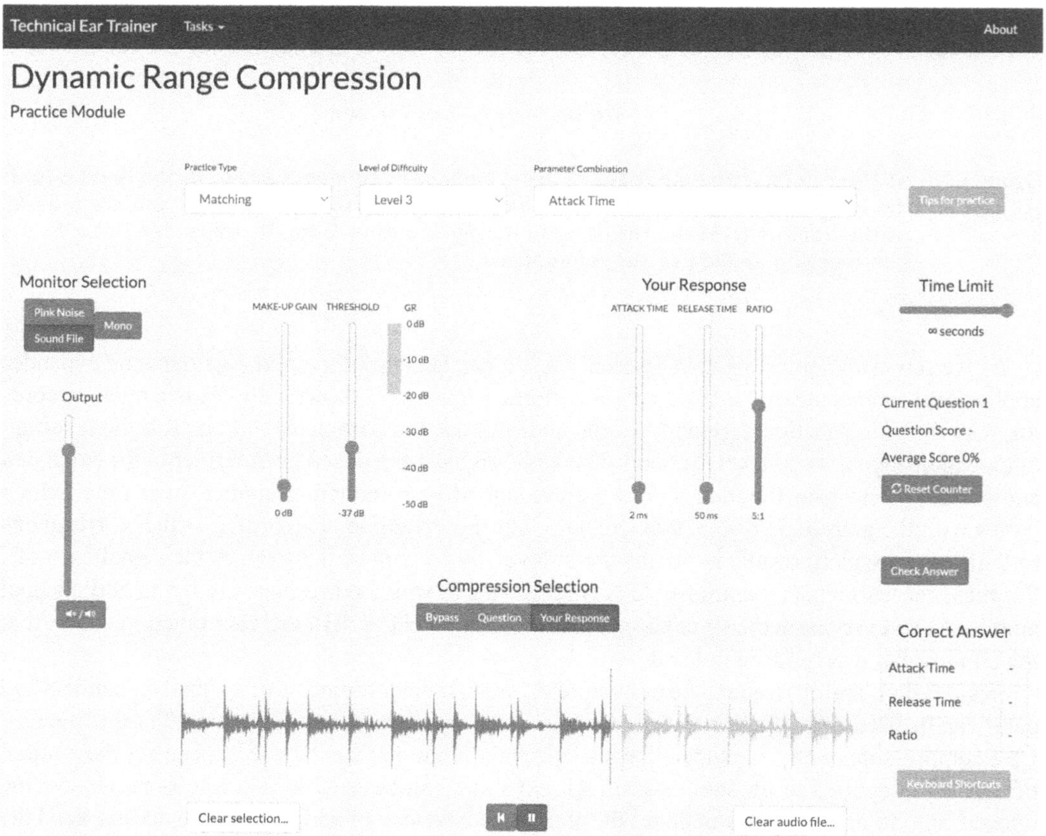

Figure 4.14 A screenshot of the software user interface for the Technical Ear Trainer practice module for dynamic range compression.

helpful. In the expansion module, it is best to start with a higher threshold which will cause the expander to produce more pronounced changes in level.

The level of difficulty determines the number of possible options from which to choose for each test parameter. With higher levels of difficulty, a greater number of options are available for each parameter.

The parameter combination determines which parameters will be included in each exercise. The dynamics modules allow you to work with attack time, release time, and ratio each on their own or in combination with one or two of the other parameters. When working with a parameter combination that focuses on only one or two parameters, the remaining user-controllable parameter(s) not being tested will apply to both the question and your response compressors.

The dynamic range control practice modules are the only ones of the entire ear training software collection in which the computer can choose "no effect" or "flat" as a possible question. In practice, this means that a ratio of 1:1 could be chosen, but only when the Parameter Combination includes "ratio" as one of the options. When you encounter a question in which you hear no dynamic range control, indicate as such by selecting a ratio of 1:1, which is equivalent to bypassing the module. If a question has a ratio of 1:1, all other parameters will be ignored in the calculation of question and average scores.

Figure 4.14 shows a screenshot of the dynamic range compression software practice module.

4.6.1 *Practice Types*

There are three practice types in the dynamics software practice module: Matching, Matching Memory, and Absolute Identification:

- **Matching:** Working in matching mode, the goal is to duplicate the dynamics processing that has been applied by the software. In this mode, you are free to switch back and forth between the question and your response to determine if the dynamics processing chosen matches the unknown processing applied by the computer.
- **Matching Memory:** Like matching, this mode allows free switching between question, your response, and bypass until one of the question parameters is changed. After a parameter is changed, the question is no longer selectable, and you must rely on your memory of its sound to determine if the response is correct.
- **Absolute Identification:** This practice mode is the most difficult and requires identification of the applied dynamics processing without having the opportunity to listen to what is chosen as the correct response. You can audition only bypass and question; you cannot audition your response.

4.6.2 *Sound Source*

Any sound recording in WAV file format rate can be used for practice. There is also an option to listen to the sound source in mono or stereo. If a sound file loaded in contains only one track of audio, as opposed to two, the audio signal will be sent out of the left output only. By pressing the mono button, the audio will be fed to both left and right output channels.

4.6.3 *Recommended Recordings for Practice*

A huge variety of loops, samples, and multitrack stems are available for purchase or free download across the web. Single drum hits are the most useful sound sources as you start training with dynamics. Then it makes sense to progress through to drum kits, as well as other solo instruments and voice, and full mixes.

A few web sites exist with free sound samples, loops, and multitrack mixes that can be downloaded for practice, such as:

- www.cambridge-mt.com/ms3/mtk
- https://freesound.org
- https://my.lewitt-audio.com/downloads
- https://reamixed.com
- https://royerlabs.com/library
- www.telefunken-elektroakustik.com/livefromthelab
- www.telefunken-elektroakustik.com/multitracks/

There are also excerpts or loops of various solo instruments bundled with Apple's GarageBand and Logic Pro, and Ableton Live that can be used with the ear training software.

4.7 Summary

This chapter discusses the functionality of compressors and expanders and their sonic effects on an audio signal. Dynamic range controllers can be used to smooth out fluctuating levels of a track, attenuate low-level sounds, or shape amplitude envelopes and modify timbre in ways that are not possible with other types of signal processing devices.

Bibliography

Apple Digital Masters Technology Brief. 2021. "Apple Digital Masters: Music as the Artist and Sound Engineer Intended." www.apple.com/apple-music/apple-digital-masters/docs/apple-digital-masters.pdf.

Bregman, Albert S. 1990. *Auditory Scene Analysis: The Perceptual Organization of Sound.* The MIT Press.

EBU. 2023. *Loudness Normalisation and Permitted Maximum Level of Audio Signals.* Technical Document R 128. European Broadcasting Union. https://tech.ebu.ch/docs/r/r128.pdf.

ITU-R. 2023. *Algorithms to Measure Audio Programme Loudness and True-Peak Audio Level.* BS.1770-5. International Telecommunication Union.

Smith, Julius O. 2007. "Why Dynamic Range Compression Is Nonlinear." In *Introduction to Digital Filters with Audio Applications.* Online book. https://ccrma.stanford.edu/~jos/filters/Why_Dynamic_Range_Compression.html.

Verfaille, Vincent, Udo Zölzer, and Daniel Arfib. 2006. "Adaptive Digital Audio Effects (a-DAFx): A New Class of Sound Transformations." *IEEE Transactions on Audio, Speech, and Language Processing* 14 (5): 1817–31.

Chapter 5

Distortion and Noise

5.1 Unwanted Distortion and Noise

Any time we work with audio equipment, technical issues may arise that can introduce noise or degrade audio quality inadvertently. If the underlying issues that create unwanted noise or distortion are not resolved, a listener's attention can get pulled towards these undesired artifacts and away from the intended artistic experience of the audio. If listeners notice a drop in audio quality because noise or distortion has been introduced when it shouldn't have or when it is unexpected, they may have difficulty enjoying the artistic experience of a recording, concert, game, or film. An audio engineer's job is to help transmit an artist's intentions to an audience, and it can be difficult for listeners to completely enjoy an artist when engineering choices add unwanted sonic artifacts that cause attention to be distracted away from an artistic experience.

Likely, most people, sound engineer or not, are familiar with the screeching sound of feedback when a microphone–amplifier–loudspeaker chain of a sound reinforcement system feeds back on itself and amplifies a signal to higher and higher levels very quickly. Although sound engineers work hard to avoid feedback, it can happen easily and quickly if a microphone on stage is pointed the wrong way (i.e., towards a loudspeaker) or turned up too loud. Feedback can be painfully loud, in the worst cases, or simply annoying, in the best cases, for both listeners and artists. Unfortunately, feedback reminds listeners that there is audio technology between them and the artist they are hearing. Feedback is so common in live sound reinforcement that film and TV sound designers often add a short feedback ringing sound at the beginning of a scene in which a character is speaking into a voice reinforcement system. With a short cue of feedback sound, listeners know a character's mic is amplified through a public address (PA) system. Feedback is probably the most extreme negative artifact produced by audio systems, but many artifacts are much more subtle than howling feedback. Even though listeners may not consciously identify sound quality problems, they can still detract from listeners' experiences. As sound engineers, we want to be as aware as possible of the sonic artifacts that can detract from an audio project or live event. As we gain experience in critical listening, we increase our sensitivity to various types of noise and distortion.

5.2 Noise

Some composers and performers intentionally use noise for artistic effect. In fact, there are whole genres of music that emphasize noise as an artistic effect, such as noise rock, industrial music, noise music, *musique concrète*, metal, sampling, and glitch, to name a few examples. Prolific Japanese noise music artist Merzbow, perhaps one of the best-known and most important

DOI: 10.4324/9781003461173-5

figures in the genre, creates soundscapes of noise and distortion in creative and expressive ways that may challenge our notions of what music is. Experimental and avant-garde electronic and electroacoustic music composers and performers often use noise to create musical effects, and they delight in blurring the line between music and noise. One of the earliest examples of *musique concrète* is by French composer Pierre Schaeffer called "Étude aux chemins de fer" [Railway Study], a piece that he composed in 1948 from his recordings of train sounds. Perhaps this is one of the earliest recordings to use samples—that is, recorded sound—as the basis for a musical piece.

Aside from recordings that use noise as their primary source material, noise is generally an unwanted signal that can get inadvertently mixed into a project, and which can distract listeners from the program material. Although noise can have innumerable spectral and amplitude envelopes, we might group them into categories based on spectral and amplitude envelopes and on the sources of the noise. I list some of the most common categories below and describe them more fully in the sections below.

- **Electronic clicks:** Transient sounds resulting from equipment malfunction or digital synchronization errors that contain primarily high-frequency content.
- **Plosives or pops:** Transient sounds resulting from plosive vocal sounds that contain mostly low-frequency sound.
- **Ground hum and buzz:** Sounds originating from improperly grounded systems.
- **Hiss or low-level white noise:** Sounds originating from analog electronics, dither, or analog tape.
- **Extraneous acoustic sounds:** Background acoustic sounds that are not intended to be recorded but that exist in a recording space, such as air-handling systems (HVAC), sound sources outside of a recording room such as noise from road, rail, or air traffic, or construction.
- **Radio frequency interference (RFI):** Audio production equipment can sometimes make an excellent, but undesired, radio receiver.

Sometimes noise exists at a low, yet still audible, level and may not register significantly on a meter, especially in the presence of musical audio signals. Therefore, in the absence of reliable visual indicators of the presence of noise, we need to use our ears to constantly track sound quality. Noises of all kinds can start and stop at seemingly random times, so we must always remain attentive.

5.2.1 Electronic Clicks

Clicks are various types of short-duration transient sounds that contain significant high-frequency energy that originate from electronic equipment. Typically, clicks originate from malfunctioning analog equipment, loose analog cable connections, digital audio synchronization errors, and any time we connect or disconnect analog cables. Clicks resulting from analog equipment malfunction can often be random and sporadic, making it difficult to identify their exact source. In this case, meters may be helpful to indicate which audio channel contains a click, especially if clicks are present even when the program material is silent. Loose or dirty analog connections may randomly break a connection causing dropouts, clicks, or sporadic noise bursts. To identify the source of intermittent unwanted clicks, it can be helpful to use a peak hold meter. A peak hold meter on a channel with clicks registers and holds the peak level if we don't happen to be looking at the meter when the click occurs. When we make analog connections in a patch bay or directly on a piece of equipment, we create signal discontinuities and therefore also clicks and pops. Breaking a phantom-powered microphone signal can make a particularly loud pop or click that can damage not only the microphone but also any loudspeakers that may reproduce the loud click.

With digital connections between equipment, it is important to ensure that sampling rates are identical across all interconnected equipment and that clock sources are consistent. Most times, interconnected equipment will not pass any audio through digital connections unless the sampling rates match. Without properly selected clock sources in digital audio, clicks are inevitable and will likely occur at some regular interval, usually spaced by several seconds. Clicks that originate from improper clock sources are often subtle, and they require vigilance to identify them aurally. Depending on the digital interconnections in a studio, the clock source for each device needs to be either internal, digital input, or word clock.

5.2.2 Plosives or Pops

Plosive consonants can produce transient thump-like sounds or pops when recorded by a microphone, and they typically have more low-frequency energy than clicks. The plosives that are particularly problematic for recording are those that are produced with both lips (technically called bilabial), namely /p/ as in plate and /b/ as in bold. When these consonants are sounded, a small burst of air is produced near the mouth opening. If you hold your hand up in front of your mouth and make a /p/ sound, you should be able to feel a puff of air coming from your mouth each time you make the sound of that letter. When the burst of air from a plosive reaches a microphone capsule, the microphone produces a low-frequency thump-like sound, and it is not normally a sound we want to record. To avoid pops, we can position a microphone slightly above, below, or to the side of the person's mouth, place a pop filter or windscreen in front of the microphone, or tape a pencil to the microphone in front of the microphone capsule. Pop filters are usually made of thin, acoustically transparent fabric stretched across a circular frame.

Low-frequency thumps from plosives do not exist when we speak or sing acoustically without a microphone since the artifact is purely a result of a microphone's response to a burst of air. Pops can distract listeners from a singer's performance or a speaker's speech because low-frequency thumps are not part of the sound of the human voice in acoustic settings. Even if a song has a kick drum in the mix, pops from a vocal usually do not line up with kick drum hits. When pops get recorded, they can be filtered out with a high-pass filter, but it is important to make sure the cut-off frequency is low enough not to affect the low harmonics in the voice. An alternative is to insert a high-pass filter only during the brief moments that pops are present.

Listen for low-frequency thumps when recording, mixing, or providing live sound reinforcement for sung or spoken voice. In live sound situations, if a microphone does not have a foam windscreen or built-in pop filter, the best way to remove pop sounds is to use a high-pass filter on the respective mixer channel or engage the high-pass filter on the microphone itself if it has one. If the mixer has a high-pass filter with a variable cut-off frequency, set it to be below the fundamental of the voice, typically not much higher than 100 Hz or so. Or if it is possible to reposition the microphone so that it is not directly in line with the person's mouth, that can help reduce the effect.

5.2.3 Hum and Buzz

Improperly grounded analog circuits and signal chains can cause noise in the form of a buzz or hum that is introduced into analog audio signals. The fundamental frequency and harmonics of the hum or buzz are determined by the frequency of the electrical power source alternating current (AC). Some places refer to this as the *mains frequency*. The frequency of the alternating current power source is either 50 Hz or 60 Hz depending on geographic location. The power distribution frequency is 60 Hz in North America, 50 Hz in Europe, either 50 or 60 Hz in Japan depending on the specific location within the country, and 50 Hz in most other countries.

When a ground problem is present, there is either a hum or a buzz generated with a fundamental frequency equal to the power source alternating current frequency, 50 or 60 Hz, with additional harmonics above the fundamental. A *hum* is identified as a sound containing primarily just lower harmonics and *buzz* as that which contains mainly higher harmonics.

We want to make sure we identify any hum or buzz before recording, when the problem is easier to solve, since removing such noises in post-production is more difficult and time-consuming. Because a ground hum or buzz often includes harmonics of 50 or 60 Hz, several narrow notch filters are needed, each tuned to a specific harmonic or multiple of the fundamental frequency, to effectively remove the offending sound in its entirety. If a ground hum or buzz has been recorded with a track inadvertently, the only option to remove the offending noise may be to use notch filters. The problem with using notch filters to remove the hum or buzz is that the filters also remove those frequencies in the program material.

Hum can also be caused by electromagnetic interference (EMI). For example, power cords carrying AC electricity placed alongside audio cables, especially those carrying microphone- or instrument-level signals, can induce hum in the adjacent audio lines, due to electromagnetic induction. An audio cable's proximity to power cords matters, so the farther away the two can be, the better. If the two types of cable do need to cross, try to make the crossing a 90° angle to reduce the strength of the electromagnetic field that crosses the audio cable.

One of the best ways to check for low-level ground hum is to bring up monitor levels with microphones on and powered but while musicians are not playing. It is helpful to turn it up and listen because tracks often get boosted in the mixing and post-production stages, and we want to know if any low-level hum or noise needs to be mitigated in the production stage before it is too late in the post-production stages. For example, if dynamic range compression with make-up gain is applied to a track that had previously inaudible low-level noise, compression could make it much more audible. If it's possible to apprehend any ground hum before getting to that stage, the result will be much cleaner.

5.2.4 Extraneous Acoustic Sounds

Despite the hope for perfectly quiet recording spaces, there are often numerous sources of noise both inside and outside of a recording space to deal with. Some sources are relatively constant, steady-state sounds, such as air-handling noise from HVAC systems, whereas other sounds may be intermittent, unpredictable, or seemingly random, such as from car horns, people talking, footsteps, dogs barking, noise from storms, nearby construction, garbage collection, subways, or when someone in the recording studio drops an item or bumps a microphone stand.

With most of the population concentrated in cities, sound isolation can be particularly challenging as noise levels rise and our physical proximity to other people increases. Besides airborne noise, there is also structure-borne noise, where vibrations are transmitted through building structures and end up producing sound in our recording spaces. Professionally built recording studios are often built with floating floors and room-in-room construction to maximize sound isolation.

5.2.5 Radio Frequency Interference (RFI)

Sometimes microphones pick up signals from terrestrial radio stations or cell phones. It's not common, but it does happen. With radio station interference, we may hear the broadcast from a local FM radio station. The audio picked up by a microphone is mainly high-frequency content, but it is annoying and distracting, nonetheless. Cell phone interference usually sounds like a series of bzzt, bzzt, bzzt sounds. If you do find yourself getting this kind of interference, it may be worth asking everyone to turn off cell phones or set them to airplane mode.

With all the noise types I describe above, the best defense is to catch them by ear when we are recording and try to eliminate the sources of the noise if possible or wait for them to pass, before doing any more recording. Noise reduction software is very sophisticated and highly effective now, but noise removal is still a manual process that takes time to do correctly. If we can avoid or minimize any extraneous noise in our recording in the first place, we save ourselves time and energy in post-production. To do that, we need to monitor our audio constantly to identify them.

5.2.6 Examples of Noise in Recordings

If noise is deliberately added to a recording, probably the most common type of noise is the vinyl crackle effect. On older recordings, analog tape hiss, which is a low-level white noise, is sometimes audible. Below are just a few examples of noise in recordings.

"Woodpecker No. 1" by Merzbow from Pulse Demon *(1996)*
This is extreme noise and distortion from a prolific noise musician.

"Teardrop" by Massive Attack from Mezzanine *(1998)*
A vinyl crackle effect is present throughout this track.

"Illusion of Time" by Daniel Avery and Alessandro Cortini from Illusion of Time *(2020)*
This track features auto-panned noise.

"Dear Alan" by Justice from Hyperdrama *(2024)*
A vinyl crackle effect is faded in during the intro and remains present for a good portion of the song, other than during a synth lead break.

Do your own survey and try to find other recordings in which noise was used for creative effect or in which extraneous noises inadvertently made their way into a recording.

5.3 Distortion

Technically, any change to an audio signal is considered distortion, but conventionally the term "distortion" refers most commonly only to nonlinear distortion, such as clipping. Such distortion usually either amplifies existing harmonics or adds new ones not originally present in a signal. There are two main kinds of nonlinear distortion: harmonic distortion and intermodulation distortion. Harmonic distortion, as the name suggests, adds frequencies that are harmonics—that is, integer multiples—of the original signal. Because harmonics added through distortion can overlap with existing frequencies in a musical instrument or voice, the added frequencies can blend with program material in a very pleasing way. By contrast, intermodulation distortion produces tones that may not be harmonically related to the original signal, and therefore this type of distortion tends to be more audible and less desirable than harmonic distortion.

Although we typically want to avoid or remove extraneous noises such as the ones I described above, distortion can be either a desired effect offering incredible expressive possibilities or an unwanted aggravation. Most audio equipment, such as digital audio workstations, audio interfaces, digital mixers, microphone preamps, and loudspeakers, is designed to be as transparent as possible—that is, to have a flat frequency response with minimal distortion. But there is a wide range of hardware and software tools available that allow us to add distortion creatively to a mix. Recording and mixing engineers seek out vintage pieces of gear because of certain types of distortion that they may add to audio signals. The sound quality of distortion from vintage gear is sometimes described as warm, full-bodied, or rich. Plugins that model vintage equipment attempt to recreate the distortion inherent in the original analog electronics and provide engineers with the sound quality of vintage equipment without the high price tag. However we describe the qualities of sound imparted by vintage equipment, they are often the result of nonlinear distortion.

Electric guitar in rock and metal genres is probably the instrumental sound in which distortion is used most often. As such, guitarists can choose from a broad palette of distortion types to shape the timbres of their instrument for ultimate levels of expression. Guitar distortion is often grouped into the three broad categories of fuzz, overdrive, and distortion, with an almost endless array of variations on these types. Within each category, there are variations and gradations that afford many timbral possibilities. Fuzz distortion seems appropriately named because it sounds fuzzy. Listen to "(I Can't Get No) Satisfaction" by The Rolling Stones (1965) and "Purple Haze" by The Jimi Hendrix Experience (1967) to hear examples of fuzz guitar. Overdrive is generally considered milder than clipping distortion, and overdrive is sometimes described as the point where the tone breaks up. Guitar distortion is usually characterized by a high-frequency boost, which can make it sound edgy or harsh. By contrast, overdrive might sound warmer because it does not boost high frequencies as much as distortion. Even "clean" guitar tones often have some amount of distortion that gives them a warm tone, especially if they are from a tube amplifier. Fuzz, overdrive, and distortion can make a guitar or any other instrument or voice sound richer, warmer, brighter, harsher, or more aggressive, depending on the type and amount used.

Even when not using distortion as an effect, we can still distort an audio signal unintentionally through equipment parameter settings, malfunctioning equipment, or low-quality equipment. We can distort or clip an audio signal by increasing its level beyond an amplifier's maximum output level or beyond the maximum input level of an analog-to-digital converter (ADC). When an ADC attempts to represent a signal level above 0 dB full scale (dBFS), it is called an over. Since an ADC can only represent signal levels below 0 dBFS, overs also get encoded as 0 dBFS, which, of course, is a distortion of the original signal level. As you may know from experience, an over can sound harsh because the distortion boosts high frequencies. More recent ADC designs include soft clipping or limiters at or just below the 0 dBFS level so that any overs that occur are much less harsh-sounding.

Fortunately, we have visual aids to help identify when a signal gets clipped. Digital meters, peak meters, clip lights, or other indicators of signal strength are present on most analog-to-digital converter input stages, microphone preamplifiers, as well as many other digital and analog gain stages. When a gain stage is overloaded or a signal is clipped, a bright red light provides a visual indication as soon as a signal goes above a clip level. The red clip light remains lit until the signal has dropped below the clip level. Some equipment has an additional red peak light, which illuminates anytime a signal is clipped. These red lights can reinforce our awareness of signal degradation and help us identify when a signal has clipped. Unfortunately, when working with large numbers of microphone signals, it can be difficult to catch every flash of a clip light, especially in the analog domain. Digital meters, on the other hand, allow peak hold so that if we do not happen to see a clip indicator illuminate during actual clipping, the peak hold function will continue to indicate that a clip did occur until we reset it. For momentary clip indicators, it is much more important to rely on what we hear to identify overloaded sounds because it can be easy to miss the flash of a red light.

In the process of recording, we set microphone preamplifiers to give as high a recording level as possible, as close to the clip point as possible, but without going over the maximum input level of the recording interface. The goal is to maximize signal-to-noise or signal-to-quantization error by recording a signal whose peaks reach the maximum recordable level (0 dBFS). In practice, we do not know the exact peak level of a musical performance until after it has happened. We set preamplifier gains based on a representative sound check, but it is wise to give ourselves some headroom in case the peaks are higher than we expect. When the actual musical performance occurs following a sound check, often the peak levels are higher than they were during sound check because musicians often perform at a more enthusiastic and higher dynamic level than they do during the sound check.

Although it is ideal to have a sound check each time we record or do live sound, sometimes we must jump right into recording or live mixing without one, make some educated guesses about preamp levels, and hope that we set the gains correctly. In these types of situations, we must be especially attentive to signal levels using our ears and level meters so that we can detect any clipped signals.

There are many forms and levels of distortion found in audio signals. Audio equipment can distort audio signals even when signal levels are not excessive. More expensive equipment usually, but not always, has lower measurable distortion. One of the problems with distortion measurements is that they do not tell us how audible or annoying the measured distortion is. Some types of distortion are pleasing, such as from tube amplifiers and audio transformers. On the other hand, one type of analog audio amplifier, known as Class B, produces offensive crossover distortion even at very low levels. Figure 5.1 shows a clean sine wave without any distortion of the waveform, and Figure 5.2 shows the sine wave with crossover distortion. Note the discontinuities in the wave

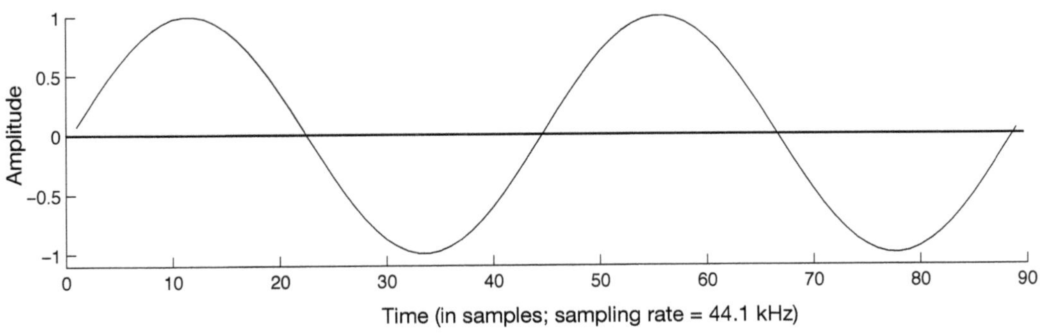

Figure 5.1 A sine wave at 1 kHz. Note that the period of 1 kHz is 1 millisecond, which corresponds to 44.1 samples at a sampling rate of 44,100 kHz.

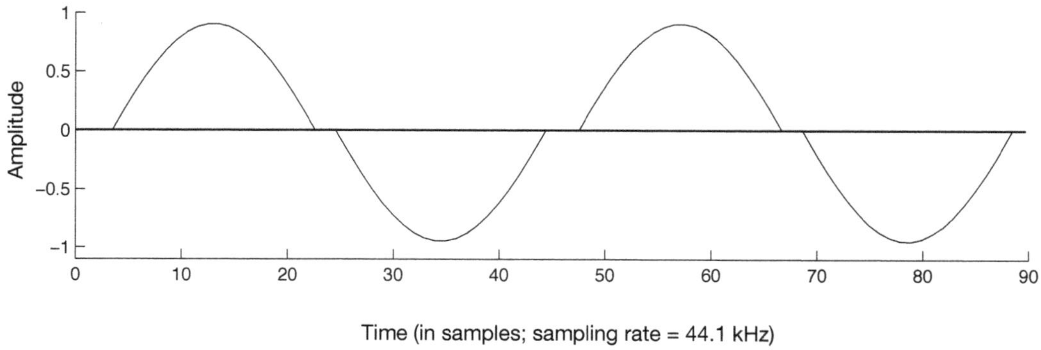

Figure 5.2 A sine wave with crossover distortion. Note the discontinuities at the points in time where the wave crosses zero.

at the points in time when it crosses over from positive to negative amplitude or from negative to positive amplitude. These short periods in which the waveform remains at 0 amplitude are characteristic of Class B amplifier crossover distortion and make this type of amplification essentially unusable for most audio applications. Even though crossover distortion may produce lower levels of measurable distortion than harmonic distortion, crossover distortion has a much more objectionable effect on perceived audio quality.

All sound reproduced by loudspeakers is distorted to some extent; however, more expensive models usually produce less distortion, or at least we hope they do. Loudspeakers are imperfect devices, and there is a wide range of quality levels across different makes, models, and price points. Equipment with exceptionally low distortion was particularly expensive to produce in the past, and therefore most of the average, less expensive, consumer audio systems used to exhibit higher levels of distortion than those used by professional audio engineers. This is becoming less true these days as the quality of inexpensive audio equipment rises. Microphones and loudspeakers, the transducers at either end of the signal chain, produce some distortion because they include a physical or mechanical component that moves, so it is worth choosing our tools carefully. Preamplifiers, power amplifiers, and other line-level signal chain components on the other hand can be designed to produce much less distortion. It seems as if the major source of distortion in most pop music is intentionally applied as an effect, as well as the side effects of heavily limited dynamic range and loudness maximization along with low bit-rate encoded versions of recordings that most consumers hear.

Most other commonly available utilitarian sound reproduction devices such as intercoms, telephones, two-way radios, and inexpensive communications headsets produce obvious distortion. For situations in which the need is simply to provide voice communication, any inherent distortion only needs to be low enough to maintain intelligibility. Some voice communication systems still produce staggering amounts of distortion.

The level of distortion found in inexpensive audio reproduction systems is usually not detectable by the untrained ear. This is part of the reason for the massive success of the MP3, Bluetooth headphones and speakers, and other perceptually encoded audio formats. Most casual listeners do not perceive audio distortion or a loss in quality in these formats. The audio formats and wireless playback devices are popular because of their convenience, low bandwidth, and portability.

There is a wide range of sound qualities that fall under the category of distortion. Here are four of the main categories of distortion within a recording, mixing, and post-production chain:

- **Hard clipping, overload distortion:** This is a harsh-sounding distortion, and it results from a signal's peaks being squared off when the level goes above a device's maximum input or output level.
- **Soft clipping, overdrive, saturation:** This is typically more pleasant and less harsh-sounding than clipping, which makes it more desirable for creative expression than hard clipping. It usually results from a specific type of circuit such as a guitar tube amplifier or analog tape.
- **Quantization error distortion or bit crushing:** This type of distortion results when a PCM digital audio signal's quantization steps are reduced. For example, converting from 16 bits per sample to 3 bits per sample without dither causes quantization error distortion. Note that we are not talking about low bit-rate perceptual encoding such as MP3 but simply a reduction in the number of bits per sample for quantization of signal amplitude.
- **Perceptual encoder distortion:** There are many different artifacts that can occur when encoding a linear PCM audio signal to a data-reduced version (e.g., MP3 or AAC), and some artifacts more audible than others. Lower bit rates generally exhibit more distortion than higher bit rates.

Whether or not distortion is intentional, as audio engineers we should be able to identify when it is present so that we can either shape it for artistic effect or remove it, according to what is appropriate for a given recording. Next, we will look at four categories of distortion: hard clipping and overload, soft clipping and saturation, quantization error, and perceptual encoder distortion.

5.3.1 Hard Clipping and Overload

Hard clipping occurs when a signal's amplitude goes above the limits of a device's maximum input or output level. A clipped signal's waveform peaks are flattened because they have surpassed the maximum level of a device, plugin, or DAW. This flattening of the waveform creates new harmonics that were not present in the original, unclipped signal. For example, if a sine wave such as shown in Figure 5.1 is clipped, the result is a square wave whose time domain waveform contains sharp edges as shown in Figure 5.3.

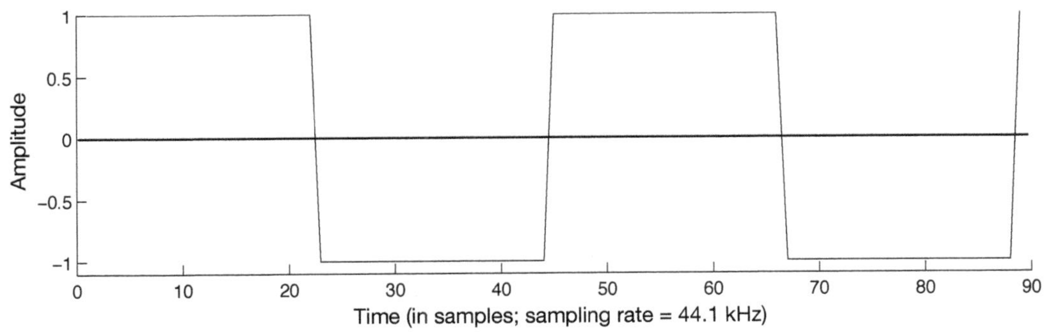

Figure 5.3 A sine wave at 1 kHz that has been hard-clipped. Note the sharp edges of the waveform that did not exist in the original sine wave shown in Figure 5.1.

The sharp corners and steep vertical portions of a clipped sine waveform like the one shown in Figure 5.3 indicate the presence of high-frequency harmonics above the 1 kHz fundamental frequency. A complex mathematical model called Fourier analysis tells us that we can decompose any complex waveform, such as a square wave, into a series of sine tones. Fourier analysis is named after Joseph Fourier, a French mathematician and physicist. Fourier proved that the opposite is also true, that we can create a complex waveform such as a square wave simply by adding together

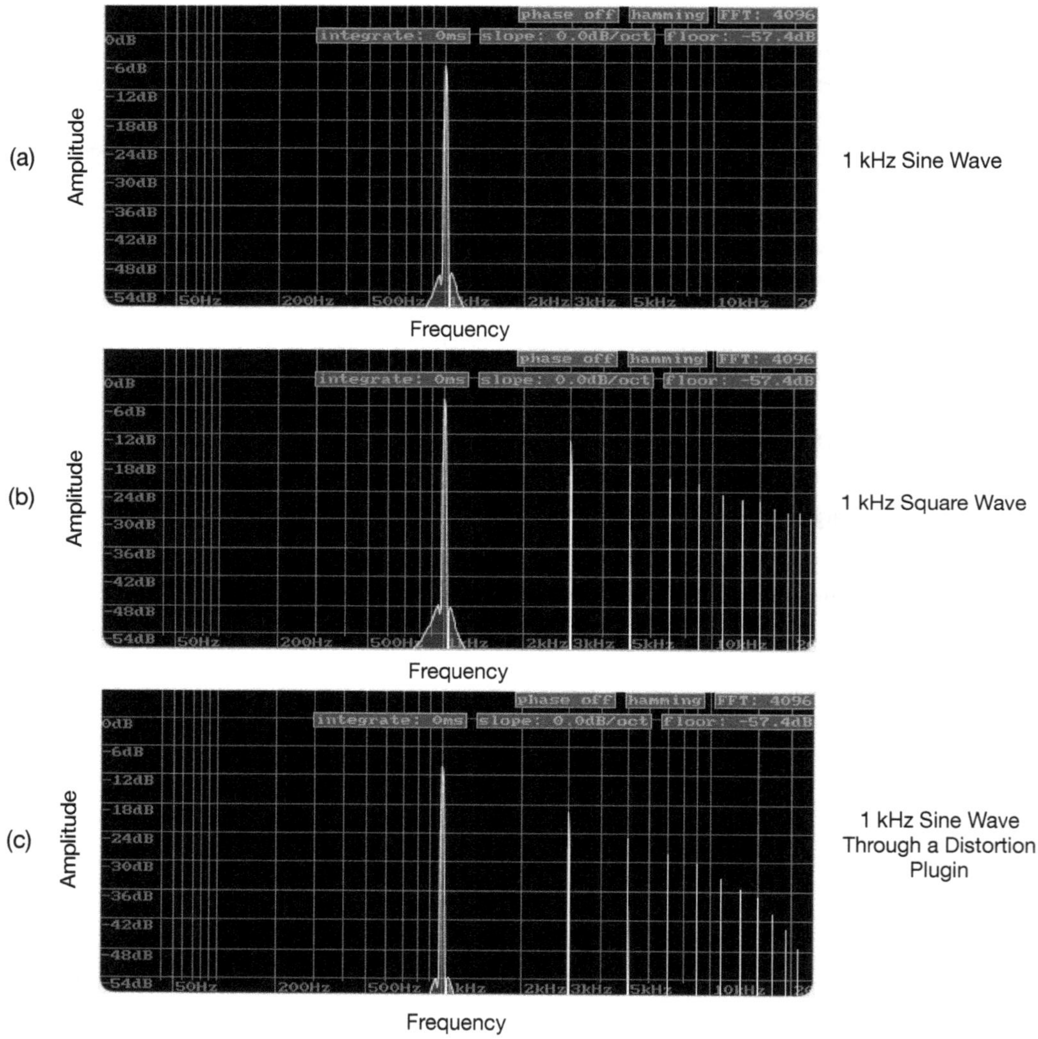

Figure 5.4 A frequency spectrum analysis of (a) 1 kHz sine wave, (b) 1 kHz square wave, and (c) 1 kHz sine wave sent through a distortion plugin. The square wave and distorted sine wave frequency spectra are similar in that they both contain only odd harmonics, 3 kHz, 5 kHz, 7 kHz, and so on. The main difference between them is that the harmonics generated by the distortion plugin appear to be lower in amplitude than those in the square wave. (The images are screenshots of the Cockos REAPER Frequency Spectrum Analyzer Meter plugin.)

sine tones at specific frequencies and amplitudes. This technique is called Fourier synthesis or additive synthesis. For example, we can construct a square wave by adding all the odd-numbered harmonics—that is, multiples of the fundamental sine tone frequency (e.g., 1st, 3rd, 5th, 7th, 9th, 11th, etc. harmonics)—at the respective amplitudes of 1/harmonic number (e.g., 1/1, 1/3, 1/5, 1/7, 1/8, 1/11, etc. amplitudes). We can confirm the composition of a square wave by doing a frequency domain analysis of it with a real-time spectrum analyzer, which is usually computed using a fast Fourier transform or FFT.

Figure 5.4 shows a frequency spectrum analysis of a sine tone, a square wave, and a distorted sine wave. As we can see from the analyzer outputs, the distortion plugin adds harmonics to the sine wave input signal, making it look similar to the square wave frequency spectrum. Because of the additional high harmonics that are added to a signal when it is distorted, its timbre becomes brighter and harsher.

5.3.2 Soft Clipping, Saturation, and Overdrive

A milder form of distortion known as soft clipping, saturation, or overdrive is often used for creative effect on an audio signal. Its timbre is often much less harsh than hard clipping. As we can see in Figure 5.5, the shape of an overdriven sine wave has flat tops but does not have the sharp corners that are present in a hard-clipped sine wave shown in Figure 5.3. The sharp corners in the hard-clipped tone indicates that high frequencies are boosted more than those in a soft-clipped sine tone.

Hard clipping distortion is produced when a signal's amplitude rises above the maximum output level of an amplifier. In gain stages such as solid-state microphone preamplifiers, timbre and sound quality change abruptly as signal levels increase from the clean, linear gain region to the maximum input level causing clipping. Once a signal reaches the maximum level of a gain stage, it cannot go any higher, regardless of any increase in input level; thus, there are flattened peaks as we discussed above. It is the abrupt switch from clean, undistorted amplification to hard clipping that introduces such harsh-sounding timbres. Some types of amplifiers, such as those designed with vacuum tubes or valves in guitar amps, exhibit a more gradual transition from undistorted linear gain to hard clipping. This gradual transition in the gain range produces a very desirable soft clipping with rounded edges in the waveform, as shown in Figure 5.5. This is the

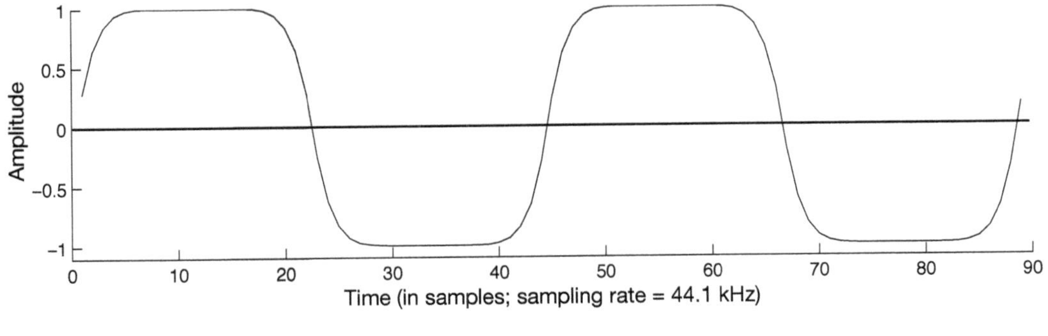

Figure 5.5 A sine wave at 1 kHz that has been soft clipped or overdriven. Note how the waveform has curved edges, with a shape that is somewhere between that of the original sine wave and a square wave.

main reason why guitarists often prefer tube guitar amplifiers to solid-state amplifiers: the distortion is often richer and warmer, and the transition from clean to distorted occurs across a wider range of levels which encourages more creative use of the effect. Soft clipping from tube-based amplifiers adds more possibilities for timbral expression than clean sounds alone. In pop and rock recordings especially, there are endless examples of the creative use of soft clipping and overdrive that enhance sounds and create new and interesting timbres. In fact, guitar amp distortion may have been the characteristic that really helped rock music differentiate itself from other music in the early days of the genre.

5.3.3 Quantization Error Distortion

When an analog audio signal is converted to a digital signal, in the standard digital pulse-code modulation (PCM) representation, signal levels for each sample are quantized to a finite number of amplitude steps. The maximum number of quantization steps available to encode a signal's level is determined by the analog-to-digital converter's bit resolution—that is, the number of bits of data stored per sample—also called the *bit depth.* An analog-to-digital converter records and stores sample values using binary digits, or bits, and the more bits available, the more quantization steps possible.

The two most common bit depths for digital audio are 16 bits and 24 bits per sample, although 32-bit floating point is also available. Usually, it is best to record, mix, and master at 24-bit or 32-bit float and set the final delivery of a recording to either 16 or 24 bits per sample. Analog-to-digital conversion using 16 bits per sample can encode 65,536 (which is equal to 2^{16}) possible steps from the highest positive voltage level to the lowest negative value, whereas 24 bits per sample can encode 2^{24} or 16,777,216 possible amplitude steps between the highest and lowest analog voltages. Even if the final product is only 16 bits, it is still better to record initially at 24 bits because any gain change or signal processing applied requires re-quantization. The more quantization steps that are available to start with, the more accurate the representation of an analog signal will be. Once audio is recorded, most DAWs process audio internally at 32-bit floating point, which helps eliminate the risk of clipping since 32-bit float has a massive dynamic range at just over 1,500 dB.

Each quantization step of linear PCM digital audio approximates the amplitude of an analog signal. There are a fixed number of digital quantization steps but, theoretically, an infinite number of analog levels. Because quantization steps are approximations of the original analog levels, there will be some amount of error in any digital representation. Quantization error is essentially a distortion of an audio signal, but at the standard bit depths of 16 or 24 bits quantization error is largely inaudible in most cases. This is because a process called dither, with or without noise shaping, is usually added during the conversion process which randomizes the error. With the random error produced by dither, distortion is replaced by very low-level noise, which is generally considered to be preferable over distortion.

In the amplitude quantization process, the signal-to-error ratio is highest when signal levels are high, and it drops as signal level drops. In other words, the error is more significant for lower-level signals. For each 6 dB a signal level is below the maximum recording level of digital audio (0 dBFS), 1 bit of binary representation is lost. For each bit lost, the number of quantization steps is halved. For example, a signal recorded at 16 bits per sample at an amplitude of –12 dBFS uses only 14 of the 16 bits available, representing a total of 16,384, that is, 2^{14} quantization steps.

Even if the signal peaks of a recording are near the 0 dBFS level, there are often other lower-level sounds within a mix that can suffer from quantization error. Wide dynamic range recordings

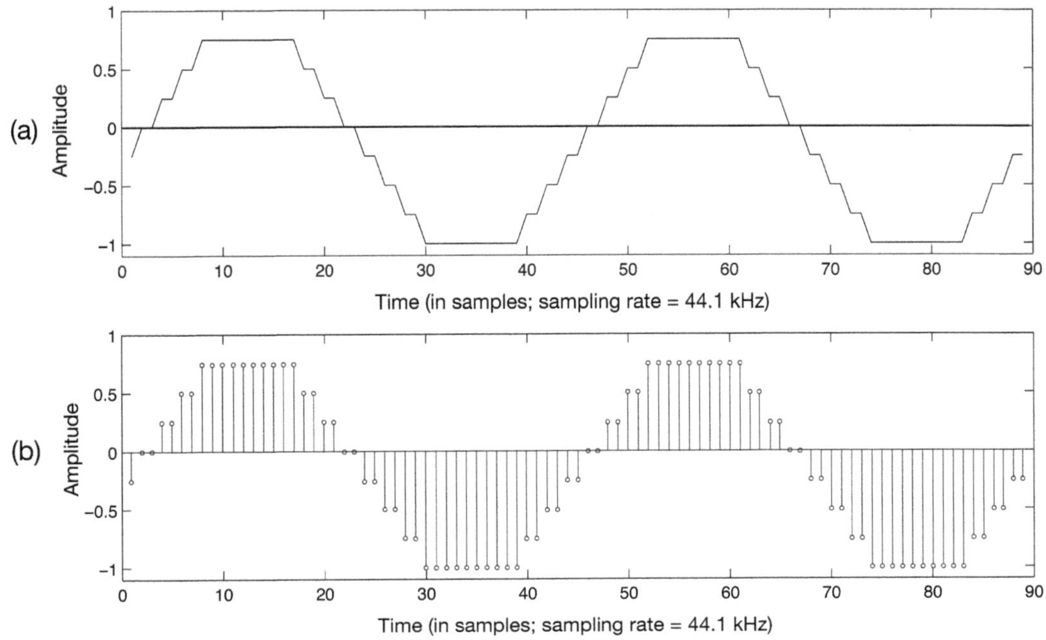

Figure 5.6 A sine wave at 1 kHz that has been quantized with 3 bits, giving 8 quantization steps. Plot (a) shows the waveform as a digital audio workstation would show it, as a continuous, if jagged, line. Since the time between each sample point is undefined, plot (b) on the bottom is a more accurate representation because we know the signal level only at each sample point but not between the points.

may include significant sections where audio signals hover well below 0 dBFS. One example of low-level sound within a recording is reverberation and the sense of space that it creates. With excessive quantization error, perhaps as the result of bit depth reduction, some of the sense of depth and width that is conveyed by reverberation is lost. By randomizing quantization error with the use of dither during bit depth reduction, some of the lost sense of space and reverberation can be reclaimed, but with the cost of some added noise.

Sometimes engineers use a bit depth reduction plugin as a distortion effect. Often called a bit crusher, this plugin simply re-quantizes an audio signal using fewer bits. Figure 5.6 shows a sine wave that has been quantized with 3 bits, giving eight–that is, 2^3—discrete amplitude steps.

If you look closely at the bit-crushed sine tone waveform, you will see that there are more negative values than positive values. This is because we have an even number of discrete amplitude steps and the mid-point must be at 0. Despite the asymmetry, it has no audible effect on an audio signal.

5.3.4 *Perceptual Encoder Distortion—Lossy Compression*

Lossy-compressed audio data is ubiquitous in streaming and downloadable music. Audio formats like MP3 were originally designed to reduce storage and bandwidth requirements of digital audio, which allowed recorded music to be more portable and convenient to listen to. Using sophisticated models of human hearing, perceptual encoders attempt to represent an audio signal with a significantly smaller amount of data while retaining as much audio quality as possible. The encoding process discards audio data to maximize data compression which reduces audio quality. The most well-known lossy codec (coder/decoder) is MP3, more formally known as MPEG-1 Audio Layer-3, but there are many other lossy formats including AAC (short for MPEG-4 Advanced Audio Coding), Ogg Vorbis, MPEG-H, and OPUS. Because of its limited bandwidth, Bluetooth uses audio codecs such as AAC, Qualcomm aptX, Sony's LDAC, and SBC (short for sub-band coding), which is the default codec for Bluetooth. As we saw in Chapter 1, wireless headphones and portable loudspeakers also use codecs for audio transmission over Bluetooth, which can further degrade audio quality.

The process of encoding a linear PCM digital audio with lossy data compression is highly complex and involves much more math than I am going to get into here. Simply put, an encoder performs spectral analyses of a signal to determine frequency content and the way it changes dynamically over time. The compression codec then adjusts the quantizing resolution in each frequency band in such a way that the resulting increased quantization noise remains under the masking threshold. In other words, it shifts the energy of the quantization noise to parts of the frequency spectrum to which the human auditory system is least sensitive. As such, it reduces the amount of data required to represent a digital audio signal by using fewer bits for quantization, and it removes components of a signal that are deemed to be inaudible or nearly inaudible based on psychoacoustic models. Inaudible components include quieter frequencies that are partially masked by louder frequencies in a recording. Whatever components are determined to be masked or inaudible are removed and the resulting encoded audio signal can be represented with less data than was used to represent the original signal. The more data compression that is applied to encode an audio signal, the more the quality is degraded relative to the original un-encoded signal.

When a linear PCM digital audio file is converted to a data-compressed, lossy format such as MP3 or AAC at typical bit rates used by music streaming services, an encoder can remove 77 to 82% of the data that was in the original audio file. Yet the encoded version often sounds very similar to the original uncompressed audio file, and to the untrained ear the two may be indistinguishable. The actual percentage of data an encoder removes depends on the target bit rate of the encoded audio. For example, the bit rate of uncompressed digital audio at the standard sampling and bit rate is 1411.2 kbits/s (44,100 samples per second x 16 bits per sample x 2 channels of audio = 1,411,200 bits per second). Typical audio codecs use bit rates between 128 and 320 kbits/s but depend on the streaming service or transmission technology as shown in Table 5.1.

For streaming services that provide different bit rates of lossy audio, it's not always clear which bit rate they may be delivering at any given moment. Based on my listening experience,

Table 5.1 This table lists the lossy audio bit rates and codecs along with lossless formats used by the major streaming services.

Streaming Service	*Lossy Bit Rates and Codecs*	*Lossless Formats*
Amazon Music	48, 192, or 320 kbits/s OPUS	FLAC
Apple Music	256 kbits/s AAC	ALAC
Qobuz	*Only lossless*	AIFF, ALAC, FLAC, WAV, and WMA
Spotify	24, 96, 160, or 320 kbits/s Ogg Vorbis	FLAC
TIDAL	96 or 320 kbits/s AAC	FLAC
YouTube Music	48, 128, or 256 kbits/s AAC and OPUS	*Only lossy*

audio quality can change midstream and can range from low bit rate with very noticeable codec artifacts to higher quality. I have heard the audio quality change midway through a streamed song, which I can only presume was due to network congestion or some other network transmission bandwidth constraint.

Satellite, terrestrial, and online radio stations sometimes use low bit rates for their audio transmissions, and the degradation of audio quality can be quite noticeable depending on the station and your network connection speed.

Although untrained listeners may not notice any difference with perceptually encoded audio, experienced sound engineers are often frustrated by the degradation in sound quality they hear in encoded versions of their work. Although lossy data compression does not maintain perfect sound quality, considering the staggering amount of data that is removed in the process, it is remarkable that it bears any resemblance to the original at all. As audio engineers, we should familiarize ourselves with the artifacts present in perceptually encoded audio so that we know what the degradations sound like.

Because audio signals are degraded during the perceptual encoding process, we consider it to be a type of distortion, but it is one that is not easily measured. Due to the difficulty in measuring the distortion and sound quality of perceptual encoders, companies and research labs that develop codecs usually employ teams of trained listeners who can identify audible artifacts that result from the encoding process. Trained expert listeners audition music recordings encoded at various bit rates and levels of quality, and in listening tests they rate audio quality on a subjective scale. Their training tells them what to listen for and where to focus their attention.

Over years of development, newer generations of codecs have become more sophisticated in the ways they remove audio data. As a result, they are more transparent at lower bit rates. That is, they produce fewer audible artifacts for a given bit rate than previous generations of codecs. The psychoacoustic models used in codecs have become more complex and the algorithms used in signal detection and data reduction based on these models have become more precise. Still, when compared side by side with an original, unaltered signal, encoded audio still can contain audible codec artifacts.

Codec distortion can create artifacts and sound quality issues such as the following:

- **Clarity and sharpness:** Listen for some loss of clarity and sharpness in transient signals such as high-frequency percussion. The loss of clarity can translate into a feeling that there is a thin veil covering the music. When compared to lossy-encoded audio, linear PCM audio should sound more direct. Some low bit-rate codecs encode at a sampling rate half of the original—for example, 22.05 kHz instead of 44.1 kHz. This means the bandwidth only extends to about 11 kHz which can explain the reduction in clarity.
- **Reverberation***:* Listen for some loss of reverberation and other low-amplitude components. The effect of lost reverberation usually translates into less depth and width in a recording, and the perceived space, whether through acoustic or artificial reverberation, around the music is less apparent.
- **Amplitude envelope:** Listen for gurgling or swooshing sounds. Held musical notes, especially prominent with piano and other solo instruments or vocals, do not sound as smooth as they should, and the overall sound can take on a tinny quality. You might hear a quick, repeated chattering effect.
- **Nonharmonic high-frequency sounds***:* Cymbals and noise-like sounds, such as audience clapping, can take on a swooshy or swirly quality.
- **Time smearing:** Because codecs process audio in chunks or blocks of samples, transient signals sometimes get smeared in time. In other words, a transient's sharp, defined attack and

quick decay in an uncompressed form can be spread across a longer time. This usually results in audible pre- and post-echoes for transient sounds.

- **Low frequencies and bass:** Are low frequencies as prominent in the encoded version? You may notice that sustain and fullness of low-frequency sound is reduced in encoded audio.

Another issue with perceptual coders is generational loss. Each time a recording is converted from one lossy format to another—say, MP3 to AAC—or from a lossy format to a lossless format and back to a lossy format—say, AAC to ALAC to AAC—its quality is reduced. The process of re-encoding an audio signal is called transcoding. Each generation of encoding reduces audio quality.

The website Perceptual Audio Codecs: What to Listen For[1] offers some technical information about audio codecs as well as audio examples that you can listen to. The site was prepared by members of the Audio Engineering Society Technical Committee on Coding of Audio Signals, most of whom are prominent researchers and experts in the field of perceptual audio coding.

It's important to note that there are also *lossless* encoding formats that reduce the size of an audio file but maintain all the original audio data without any degradation in quality. Lossless encoding is similar in concept to a computer zip file, in that the file size is reduced when encoded but all the original audio data is recovered upon decoding. The two most common lossless formats are FLAC (Free Lossless Audio Codec) and ALAC (Apple Lossless Audio Codec). Lossless codecs, whether FLAC or ALAC, have absolutely no effect on audio quality, and there is no difference in quality between the FLAC and ALAC formats. The lossless format you use depends mostly on your software or device compatibility requirements.

5.3.4.1 Exercise: Comparing Linear PCM to Encoded Audio

Sound quality deficiencies in encoded audio may not be immediately obvious unless we are tuned into the types of artifacts that codecs produce. Once we begin to explore the ways perceptual encoders degrade sound quality, we may find it easier to identify these artifacts in a broader range of situations. In other words, once we know what to listen for, we may start to hear codec artifacts almost everywhere. One of the ways to investigate sound quality degradation is to compare linear PCM sound files to their encoded versions to identify any audible differences.

According to generally accepted practices for scientific perceptual evaluation, the best way to hear differences between two audio signals is to switch back and forth between them. Immediate switching between stimuli with no pause helps us hear differences more easily than waiting several minutes or hours between stimuli. As you begin to hear the kinds of artifacts an encoder produces, you may eventually be able to hear them without doing a side-by-side comparison of encoded audio and linear PCM audio.

To create perceptually encoded files, I recommend using Cockos REAPER since it can encode any track as an MP3, AAC, or Ogg Vorbis file format at a variety of bit rates. In the "Render to File" window under the "Primary output format," you can select the codec and bit rate.

After you have converted a WAV audio track to a perceptually encoded version, compare the two versions and try to identify how your audio signal is degraded. Lower bit rates result in a smaller file size, but they also reduce the quality of the audio. Different codecs, such as MP3, AAC, and Ogg Vorbis, produce slightly different results for a given bit rate because although the general principles are similar, the specific encoding algorithms vary from codec to codec.

Switch back and forth between the original linear PCM audio version of a recording and the encoded version. Try encoding recordings from different genres of music. Note the sonic artifacts that are produced for each bit rate and encoder. Listen for the artifacts and sound quality issues in the bullet points above. Try encoding recordings of different solo instruments, different genres, and mixes of full bands. Pick a bit rate and make it the same for all codecs. Sometimes one codec will sound better for recordings of one kind of instrument, but another codec may sound better for another instrument, assuming matched bit rates.

5.3.4.2 *Exercise: Listening to the Audio Removed by a Codec*

This exercise involves listening to the audio that a codec has removed from a lossy-compressed file. To do this, we need to subtract an encoded audio file from a linear PCM version, stored in WAV file format, of the same audio file.

The first step is to render a linear PCM file to a perceptually encoded format such as MP3, AAC, or Ogg Vorbis. REAPER seems to be one of the best tools to do this task since it places the rendered file in a new track, time-aligned with the original audio. If you use another DAW or utility to render the file, import both the original WAV sound file and the encoded file into a DAW on two different stereo tracks. Zoom in to make sure the tracks are lined up in time precisely to the sample level, if possible. Reverse the polarity of one of the stereo tracks, either the original WAV file or the encoded file—it doesn't matter which one. Provided the two stereo tracks are lined up accurately in time, anything that is common to both tracks will cancel, and any remaining audio is the difference between the original audio and the audio encoded by the codec. In other words, the difference between the two is the audio that was removed by the codec. This exercise helps us hear the types of artifacts that are present in encoded audio in other situations, and once we know what to listen for, it can help us identify codec artifacts without doing this subtraction exercise.

As you listen to the difference signal, think about the following questions: What does the difference signal sound like? What is its spectral balance? Can you identify the original recording just by listening to the difference signal or is it too much like noise to identify the original recording? After hearing the difference signal and knowing what it sounds like, are you better able to identify codec artifacts in the encoded file on its own without listening to the difference signal on its own?

Do the same exercise with different bit rates and different codecs. If you end up with multiple versions of encoded audio in the DAW, remember to mute all but the original WAV file and the encoded file you want to audition.

5.3.4.3 *Exercise: Listening to Encoded Audio Through Mid-Side Processing*

Another way to hear codec artifacts is to listen to the side component of an encoded stereo mix. The perceptual encoding process relies on masking within the stereo image to hide artifacts. In many pop recordings, most elements of the mix are panned center, and this helps mask a great deal of encoding artifacts.

REAPER makes listening to the side component easy. You just need to change the mono button mode in the Mixer Master. You can do this by right-clicking on the Mono button and selecting "Mono mode: L-R." Listen to one of the encoded files you created in the previous exercise or a previously encoded file and click the mono button. You should hear left minus right, that is, the side component of the mix. The codec artifacts will not be apparent as they are in the previous exercise but should likely be more apparent than they are in the normal stereo mix.

It is possible to use the Technical Ear Trainer—Mid-Side software module to hear the effect that M-S decoding can have on highlighting the effects of a codec. The module is available on the companion website: https://webtet.net/apcl/#/midside.

5.3.5 *Examples of Distortion in Recordings*

For rock and metal guitar players, distortion is an absolute requirement to get the sound they want. Guitar distortion is one of the defining timbral qualities of rock music and its many subgenres, but we also find distortion in other genres where it may be applied to instruments, voices, or a full mix. Below are just a few examples of distortion in recordings, but distortion is present in many non-classical recordings as well as some early classical recordings. Do your own survey and try to find other recordings in which distortion was used for creative effect or in which distortion was inadvertently added due to clipping or overload.

"Zoo Station" by U2 from Achtung Baby *(1991)*
The song begins with a kind of call and response between distorted guitars and heavily distorted drums. The lead vocals are also highly distorted throughout much of the song.

"Race for the Prize" by The Flaming Lips from The Soft Bulletin *(1999)*
The drums are distorted through most of the song, except for some sections in which they are relatively clean by contrast.

"Thunder" by Imagine Dragons from Evolve *(2017)*
Distortion plays a prominent role in the timbre of the drums, vocals, and synth in this song. It is especially apparent during the choruses of the song.

"The 30th" by Billie Eilish from Guitar Songs – Single *(2022)*
This relatively quiet song gradually crescendos starting around 2:16 and peaks at 2:40. The level increase slowly pushes the song into distortion for a short section, and the effect adds an overall fuzziness to the sound.

"Gnarly" by KATSEYE from Gnarly – Single *(2025)*
From the first spoken words that start the song, to the crushed snare drum, there is no shortage of distortion in this track. Also note the snare drum resonance around 400 Hz.

5.4 Distortion and Aliasing

As we saw above, distortion creates new harmonics. Harmonics are simply integer multiples of the fundamental frequency, and most natural acoustic sounds contain harmonics. For example, a guitar playing the note A at a fundamental frequency of 440 Hz also produces harmonics at multiples of the fundamental: 880 Hz, 1320 Hz, 1760 Hz, 2,200 Hz, 2,640 Hz, and so on. When a guitar is distorted, the distortion produces harmonics that align with the frequencies produced acoustically by the guitar. The creation of new harmonics is not a problem in the analog domain, but in the digital domain things are a little different, and sometimes these new harmonics can cause issues in sound quality.

To explain the issue, we need to review sampling theory briefly. When an audio signal is converted from analog to digital, it is sampled a specific number of times per second, known as the sampling rate, and its amplitude is quantized, usually at either 16 or 24 bits per sample.

Typically, the sampling rate is 44.1 kHz, 48 kHz, 88.2 Hz, 96 kHz, or 192 kHz. The Nyquist–Shannon sampling theory states that the sampling rate must be twice the frequency of the highest frequency to be recorded. So, for example, if the sampling rate is 44.1 kHz, the highest recordable frequency is 22.05 kHz. Any frequency sampled that is above half the sampling rate—that is, the Nyquist frequency—will be incorrectly recorded as an alias frequency. Alias frequencies are recorded as new, non-harmonically related frequencies below the Nyquist frequency. Fortunately, aliasing is not an issue with analog-to-digital converters because they all use low-pass filters called anti-alias filters to eliminate frequencies above the Nyquist frequency before a signal gets sampled.

The aliasing issue can arise, however, when we use a plugin or other digital device to add distortion to an audio signal digitally. Since distortion adds high-frequency harmonics, it's possible for new harmonics to be created that are above the Nyquist frequency, since there is no upper limit to the number of new harmonics that can be created from distortion. High frequencies above the Nyquist frequency can easily be created from a distortion plugin. When new harmonics are created above the Nyquist frequency in the digital domain, aliases are created. Since aliases are not harmonically related to the original signal, they can sound harsh, out of tune, or simply out of place.

Figure 5.7 shows a sampling rate of 44.1 kHz, Nyquist frequency of 22.05 kHz, three sampled input frequencies, 30 kHz, 36, kHz, and 42 kHz, and their respective aliases at 14.1 kHz, 8.1 kHz, and 2.1 kHz. This assumes that there is no anti-aliasing filter. The highest input frequency (42 kHz, in this case) produces the lowest alias (2.1 kHz). If these three sampled frequencies happened to be upper harmonics of a 6 kHz fundamental, for example, none of the alias frequencies would align with the harmonics of 6 kHz. The harmonics of 6 kHz are 6 kHz, 12 kHz, 18 kHz, and so on.

Let's look at the harmonics that a distortion plugin can produce using a spectrum analyzer. Figure 5.8 shows three frequency spectrum analyses: (a) an 8.1 kHz sine wave, (b) an 8.1 kHz sine

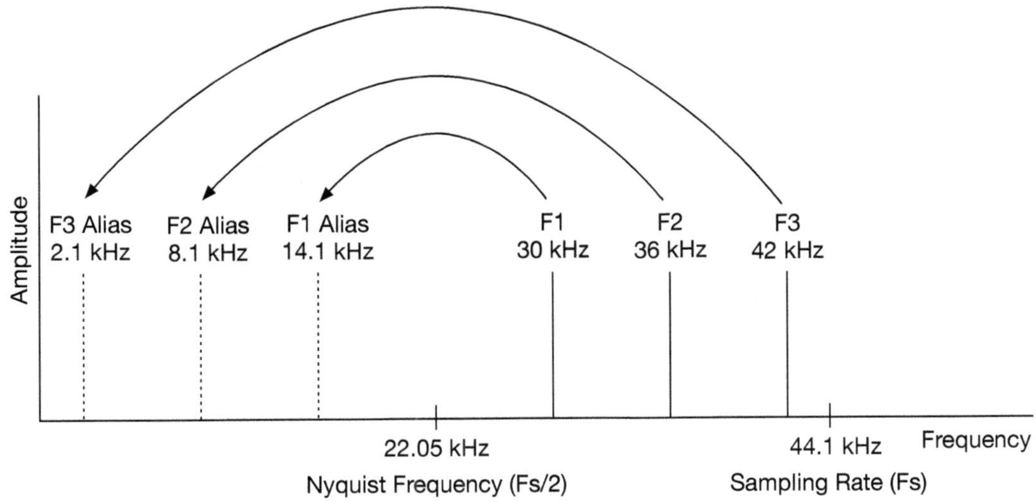

Figure 5.7 A frequency domain plot of three sampled frequencies, 30 kHz, 36, kHz, and 42 kHz, all of which are above the Nyquist frequency, and their three respective alias frequencies, 14.1 kHz, 8.1 kHz, and 2.1 kHz. In this example, the sampling rate (Fs) is 44.1 kHz, and the Nyquist frequency (Fs/2) is 22.05 kHz. Without proper anti-alias low-pass filtering at the Nyquist frequency, the sampled frequencies above the Nyquist frequency get mirrored back as aliases below the Nyquist frequency.

wave through a distortion plugin with alias frequencies present, and (c) an 8.1 kHz sine wave through the same distortion plugin, but without any alias frequencies present. All frequencies present in the middle spectrum analysis (b) but which are not present in the bottom spectrum analysis (c) are aliases.

Many distortion plugins create alias frequencies, and this may be part of the reason that many people prefer the sound of analog distortion, such as from tube amps and analog synthesizers, over distortion created in the digital domain. As we can see from Figure 5.8, it is possible to avoid aliasing in the digital domain, but it requires the plugin to perform some extra processing behind the

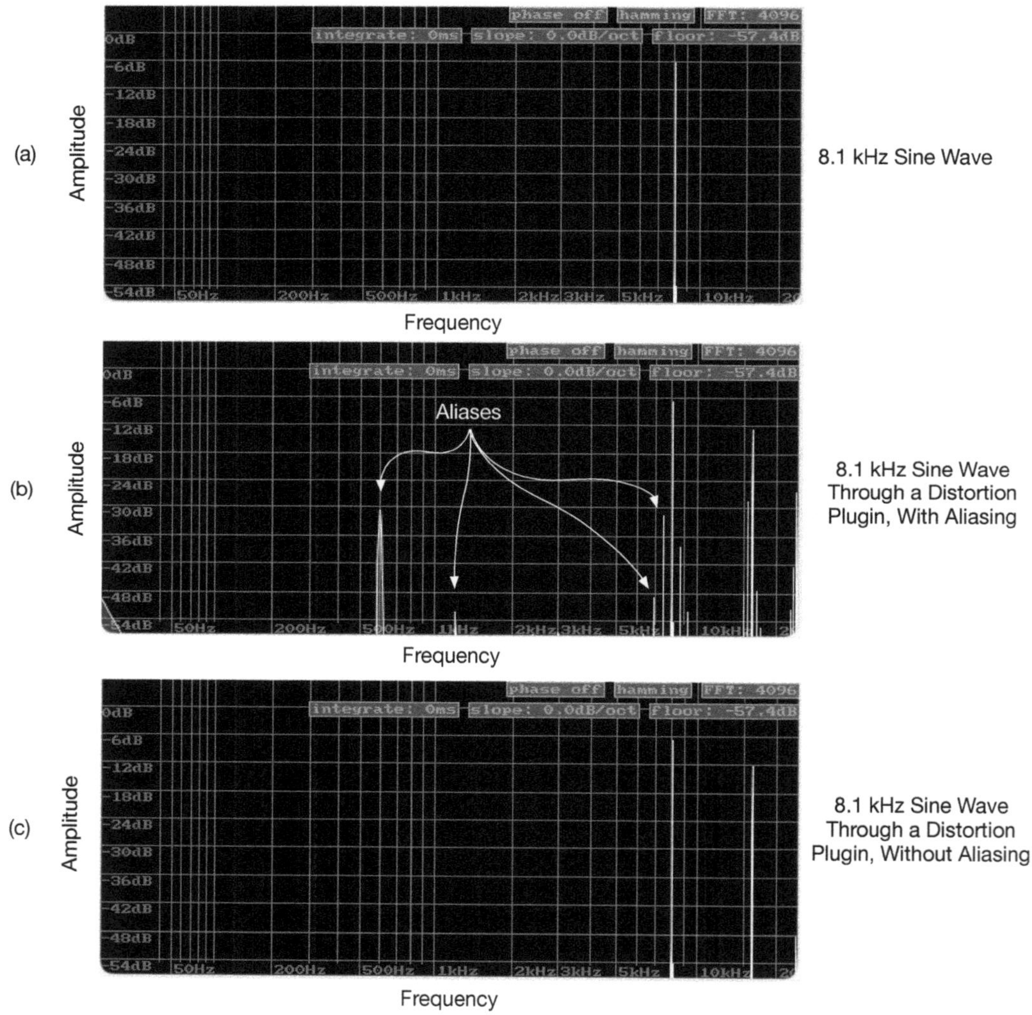

Figure 5.8 A frequency spectrum analysis of (a) an 8.1 kHz sine wave shown with a single spike, (b) an 8.1 kHz sine wave sent through a distortion plugin with numerous alias frequencies, and (c) an 8.1 kHz sine wave sent through the same distortion plugin without aliasing, showing only the first and second harmonics (8.1 kHz and 16.2 kHz). Frequencies present in spectrum (b) but not present in spectrum (c) are aliases. Note that some of the aliases are much lower in frequency than the original input signal; one is around 600 Hz, for example. (The images are screenshots of the Cockos REAPER Frequency Spectrum Analyzer Meter plugin.)

scenes. The usual way to avoid aliases from digital distortion is to resample the audio at a much higher sampling rate before applying the distortion so that any alias frequencies that may be generated are far above the range of human hearing. This is called oversampling, and it is a process in which the sampling rate may be increased temporarily to 64 or 128 times the original sampling rate. So, for example, if a sampling rate of 44,100 Hz is oversampled 128 times, the new sampling rate is 5,644,800 Hz (= 44,100 Hz x 128), and the oversampling Nyquist frequency is 2,822,400 Hz (= 5,644,800 Hz / 2). After a signal is oversampled and the distortion effect is applied, the signal is then down-sampled from 5,644,800 Hz to the system sampling rate of 44.1 kHz. But before down-sampling, it needs to be filtered with an anti-alias filter to cut all frequencies above 22,050 Hz, the Nyquist frequency of 44.1 kHz. After being low-pass filtered with the anti-alias filter, the audio is down-sampled to the original sampling rate of 44,100 Hz. The oversampling process uses more computer processing power, but it can be worth it to produce a distortion effect without aliasing. Not all distortion plugins provide oversampling, and those without it will produce aliasing. Some plugins provide an option to turn oversampling on and off. We can test distortion plugins for aliasing by sending sine tones through them and looking at the output through a spectral analyzer plugin. Another way is to send a sine tone through a distortion plugin and listen to the output while changing the sine tone frequency. If some overtones rise as you lower the sine tone frequency and vice versa, you know that the plugin produces aliasing.

5.5 Distortion and Noise in Vinyl Records

Despite the clear advantages of digital audio, sales of vinyl records continue to grow. Some people seem to prefer vinyl records over digital, whether because of sound quality, tangibility, or some other reason. Perhaps some people prefer the crackling from surface noise of vinyl records because it brings back a sense of nostalgia. Maybe the distortion inherent in vinyl is pleasing for some listeners.

These days, it's extremely rare for a recording to be made entirely in the analog domain—that is, recorded on analog multitrack tape, mixed through an analog console to analog tape, and mastered completely in the analog domain. As a result, most if not all recent vinyl record releases have been recorded and mixed entirely in the digital domain. From a production point of view, there are so many advantages to working in the digital domain compared to analog.

Here are a few advantages of linear PCM digital audio at the standard sampling rate of 44,100 Hz and bit depth of 24 bits per sample compared to recording made on analog tape and mastering to a vinyl record:

- The sound quality in digital audio is objectively better: distortion is lower, frequency response is flatter and wider, dynamic range is wider, noise is lower.
- Digital audio offers much more flexibility, control, and options for signal processing.
- Digital audio allows precise and instant recall of parameter settings.
- There is no degradation of sound quality with playback in digital audio, unlike analog tape and vinyl records which gradually wear out.
- Digital audio offers the ability to easily and instantly share mixes and tracks with anyone who has a network connection.
- The cost of digital audio is much less than analog recording. Data storage, DAWs, and plugins are much cheaper than tape, analog consoles, and racks of analog gear.

Although some listeners prefer vinyl, lossless linear PCM digital audio is objectively better by every measure than vinyl. Digital audio is perfectly able to recreate an analog waveform in its original form, even at the standard sampling rate of 44.1 kHz and bit depth of 16 bits per sample. Christopher "Monty" Montgomery, the main person behind the Ogg Vorbis codec, has a highly informative video and blog post about digital audio signals and how they behave, in which he demonstrates that the conversion of an analog signal to digital and back to analog again does not produce stairsteps in the waveform (Xiph.org and Red Hat, Inc. 2012; Montgomery 2012).

Because vinyl records often sound different from their digital counterparts, they provide an opportunity for a listening exercise if you have access to a turntable and vinyl records. If you compare a recording on vinyl to the same recording in a lossless digital audio format or even a compressed format, ideally at a high bit rate of 256 or 320 kbits/s, what differences do you hear? Listen for differences in dynamic range, high frequencies, low frequencies, and noise level, and note what you hear.

5.6 Technical Ear Training Software Module: Distortion

The included software module Technical Ear Trainer—Distortion, available at https://webtet.net/apcl/#/distortion, allows you to practice hearing three different types of distortion: hard clipping, soft clipping, and distortion from bit depth reduction or bit crushing. There are two

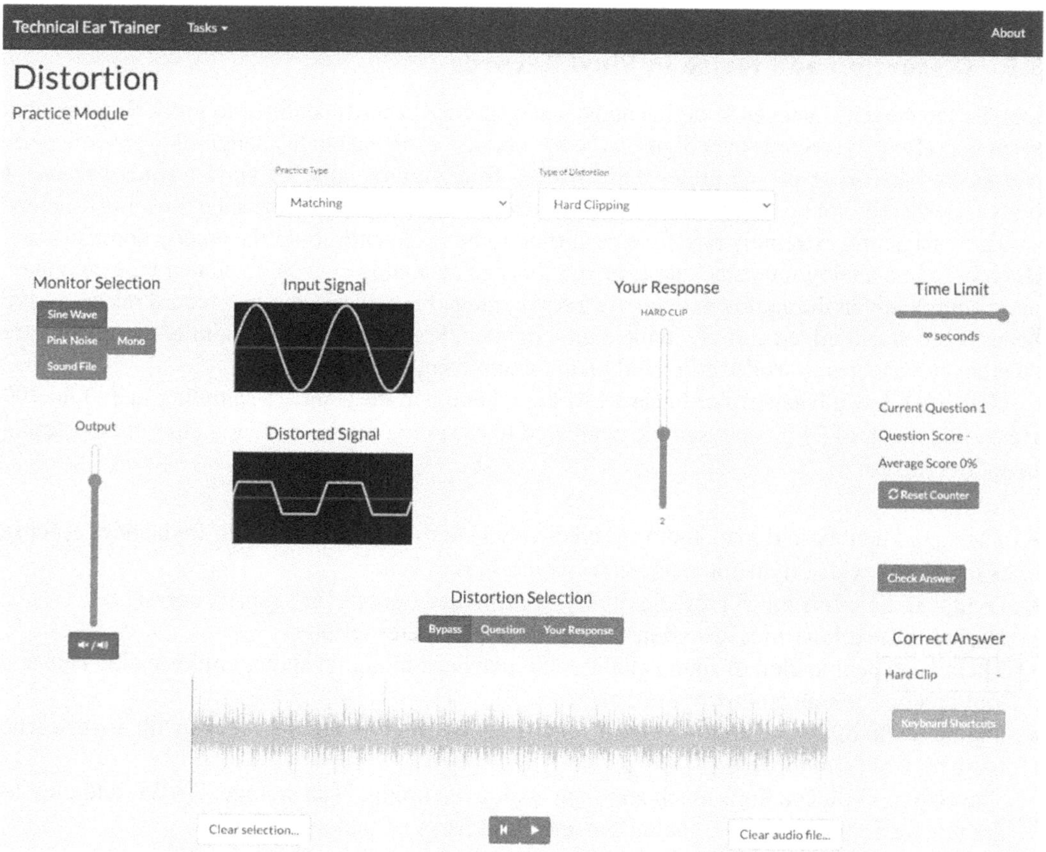

Figure 5.9 A screenshot of the software user interface for the Technical Ear Trainer practice module for distortion.

main practice types with this software module: Matching and Absolute Identification. Audio for practice can be chosen under Monitor Selection with the option to work with a 440 Hz sine tone or a sound file that you select. Since a pure sine tone contains no harmonics other than the fundamental frequency (440 Hz in this case), distortion adds new harmonics, which changes the perceived timbre of the tone and its waveform. Figure 5.9 shows a screenshot of the distortion software practice module.

5.7 Summary

In this chapter, we explored some of the undesirable sounds that can make their way into a recording. Although distortion as an effect can offer endless creative possibilities, unintentional distortion from overloading or clipping can take the life out of our audio. By practicing with the included distortion software ear training module and completing the exercises, we can become more aware of some common forms of distortion with the goal of correcting them when they occur. Although there is excellent noise and distortion reduction software available, we save ourselves time in post-production by catching noise and distortion that occurs during the recording process so we can mitigate it before it becomes a problem.

Note

1 AES. "Perceptual Audio Codes: What to Listen For." www.audiolabs-erlangen.de/content/resources/aesCodingTutorial/index.html

Bibliography

Montgomery, Christopher (Monty). 2012. "24/192 Music Downloads . . . and Why They Make No Sense." Xiph.Org, March 1. https://people.xiph.org/~xiphmont/demo/neil-young.html.

Xiph.org, and Red Hat, Inc., prods. 2012. *Digital Show & Tell.* With Christopher (Monty) Montgomery. www.xiph.org/video/vid2.shtml.

Chapter 6

Amplitude Envelope and Audio Edit Points

6.1 Digital Audio Editing: The Source-Destination Technique

Before describing the software and method for ear training, it is important to understand some digital audio editing techniques used in classical music post-production. Classical music requires a high level of precision—perhaps more so than other types of music—to achieve the level of transparency required.

Based on my experience editing classical music recordings, I have found that the process of repeatedly adjusting the location of edit points to create smooth cross-fades by ear not only results in a clean final recording, but also seems to promote improved listening skills that translate to other areas of critical listening. The highly focused listening required for audio editing, with the goal of matching edit points from different takes, can be an effective form of ear training.

Digital audio editing systems allow us to see a visual representation of our audio waveforms, and to move, insert, copy, or paste audio files to any location along a visual timeline. Source-destination editing, also known as four-point editing, offers a slightly different workflow than simply moving clips or regions around and trimming them. Not all DAWs offer the ability to do source-destination editing, but Pyramix by Merging Technologies and Sequoia by Boris FX are two notable titles that do. It is also possible to do source-destination editing on Cockos REAPER using a set of open-source scripts called ReaClassical.

Classical music recording sessions usually involve recording an entire ensemble performing together, whether it is a duo, a chamber music ensemble, or full orchestra with choir and soloists, just like a live performance. While multitrack recording is common, overdubbing is not. Because overdubbing is not usually an option, recording sessions aim to get the best performance by recording a series of takes of the same music, making sure that each bar of a piece of music sounds as good as possible.

In the editing stage following a recording session, an editor works from a musical score that has been marked to indicate the specific takes to use for each bar of music. Using source-destination editing, the editor builds a complete musical performance, called the "destination," by taking the best excerpts from a list of recording session takes, called the "source," and piecing them together. In source-destination editing, the editor identifies an edit location by listening to a recorded take while following a marked musical score and placing a marker at the appropriate point in the DAW waveform timeline. The editor listens to the same musical excerpt in the two takes which are to be sliced together and places markers in the waveforms of both takes. Once the edit point marks are placed at the chosen musical note in both the source and the destination waveforms, the next step is to audition the audio leading up to the edit point marker in each take, at which point the DAW stops playing precisely at the marker. The goal of playing up to the edit point marker is to focus on

DOI: 10.4324/9781003461173-6

the sonic characteristics of the note onset that occurs during the final few milliseconds before the marker, where the DAW stops the audio. By adjusting the precise location of the edit marker, the sound quality of the note onsets can be matched. The edit point marker may appear as a movable bracket on the audio signal waveform, as in Figure 6.1. Focusing on the final milliseconds of an audio excerpt is critical to finding an appropriate edit point. When a musical note onset is chosen as an edit point, it is important to set the edit point such that it occurs sometime during the very beginning of a note attack. Figure 6.1 shows a gate, the square bracket indicating the edit point, aligned with the attack of a note. Usually, it helps to place an edit at the onset of a musical note to help mask any artifacts that might be produced by an edit and make the transition from one take to another transparent.

When I edit music using the source-destination method, I listen through to find the musical note on which I will make a splice or edit from one take to another. I mark the rough timeline location and then narrow in on the precise placement of my edit point location by ear. Waveform views in a digital audio workstation help in the rough placement of the initial marker for an edit, but it is often more efficient and more accurate to find the precise location of an edit by ear, rather than by looking for some visual feature.

When auditioning an audio clip and stopping playback during a note's onset, only the first few milliseconds or tens of milliseconds of the note are heard. By stopping playback immediately at the note onset, it is possible to hear a transient, percussive sound at the truncated note onset. The specific timbre of the cut note varies with the amount of the incoming note that is sounded before being cut off. Figure 6.2 illustrates in block form the process of auditioning source and destination program material.

Once the audio clip cut-off timbres are matched as closely as possible between takes, the editor creates a cross-fade edit. As the name implies, a cross-fade edit fades out one take and fades in another at the edit point to help make the edit inaudible. The editor listens to the edit cross-fade to check for sonic anomalies or new artifacts that the edit may have created. Figure 6.3 illustrates the way a composite version—that is, the destination—can be created by splicing together three different source takes of the same musical program material.

When listening to a cross-fade, an editor must pay close attention to the sound quality of each cross-fade. A cross-fade length of 25 milliseconds works for most edits, but cross-fades can vary in

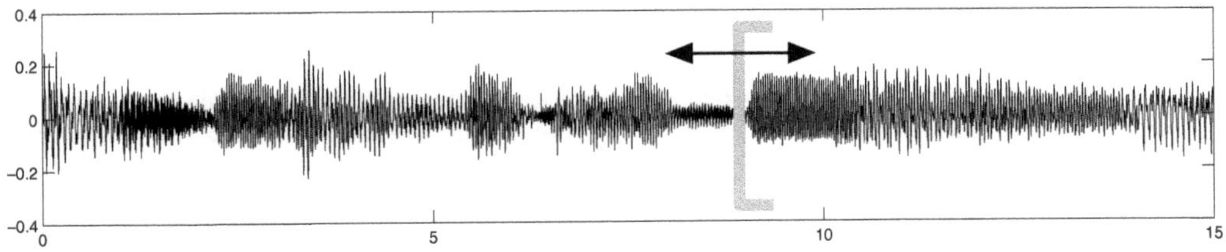

Figure 6.1 A typical view of a waveform in a digital editor with the edit point marker that indicates where the edit point will occur and the audio will cross-fade into a new take. The location of the marker, indicated by a large bracket, is adjustable in time (left/sooner or right/later). The arrow indicates simply that the bracket can slide to the left or right. We listen to the audio up to this large bracket with a predetermined pre-roll time that usually ranges from 0.5 to 5 seconds.

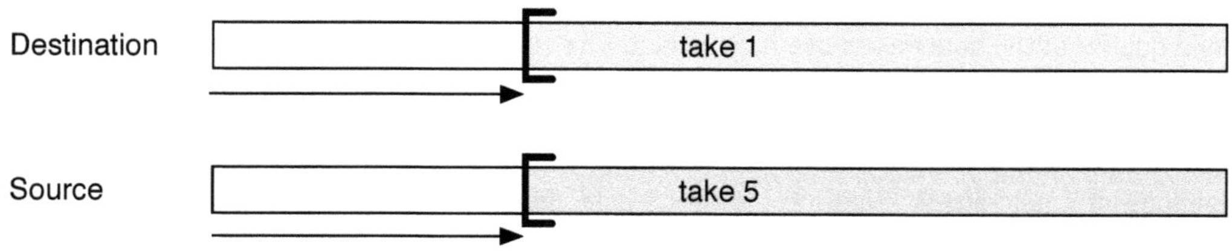

Figure 6.2 In source-destination editing, an editor auditions the source and destination audio excerpts up to a chosen edit point, usually placed at the onset of a note or strong beat. The two audio clips (source and destination) are of identical musical material but from different takes. One of the goals in placing the edit marker and auditioning each excerpt up to the edit point is to match the timbre of the source and destination clip end points. The closer the two cut-off timbres can be matched, the more successful the edit will be. The Audio Clip Edit Points ear training software module re-creates the process of auditioning a sound clip up to a predefined point and matching that end point in a second sound clip.

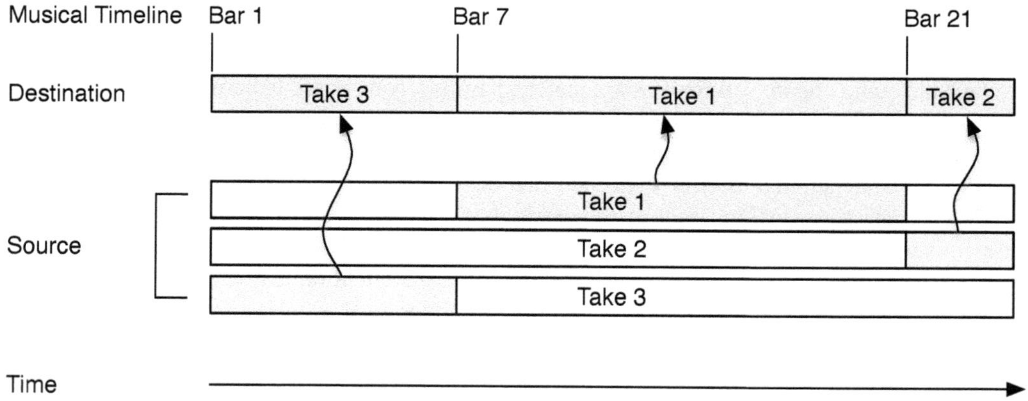

Figure 6.3 A hypothetical source and destination timeline is shown here in block form. This is an example of how a set of takes (the source) might fit together to form a complete performance (the destination). In this example, takes 1, 2, and 3 are the same musical program material and therefore a composite version is produced using the best sections from each take. In this hypothetical example, the musical piece starts with take 3, cross-fades to take 1 at Bar 7, and cross-fades to take 2 at Bar 21.

length from 10 to 300 milliseconds or more depending on the context. Typically, short cross-fades are used for transient sounds or notes with fast onsets and longer cross-fades are used for sustained or held notes.

The process of listening back to a cross-fade and adjusting the cross-fade parameters such as length, position, and shape also offers an opportunity to improve critical listening skills. For example, when doing editing for any kind of audio source material, the goal is to make each edit seamless and inaudible. Classical music recordings can contain huge numbers of edits—some recordings are in the range of ten or more edits per minute—but it is nearly impossible to hear even a single edit in a finished recording if the edits are done well.

6.2 Technical Ear Training Software Module: Audio Clip Edit Points

The Technical Ear Trainer—Audio Clip Edit Points technical ear trainer software module is available on the companion website: https://webtet.net/apcl/#/editing.

Based on source-destination editing, the associated technical ear training software module was designed to mimic the process of comparing the final few milliseconds of two short clips of identical music from different takes. The advantage of the software practice module is that it promotes critical listening skills without requiring an actual editing project. The main difference between the practice module and an editing project is that the practice software will work with only one "take"—that is, the linear PCM WAV sound file that you loaded—whereas in an actual editing project there are different takes of the same musical material. Because the ear training module uses only one audio recording, the two clips of audio it asks you to compare are the same audio signal, and therefore it is possible to find identical-sounding end points. The benefit of working this way is that the software can judge if the sound clips end at precisely the same point.

To start training with this module, load any WAV audio file that is at least 30 seconds in duration. The software randomly chooses a short excerpt or clip (which is called clip 1 or the "reference") from the recording. The goal is to determine the exact duration of clip 1 by listening to it and comparing it to clip 2 ("your answer") as many times as you want. The duration of clip 2 is displayed in the interface. The main task of the exercise is to adjust the duration of clip 2 (your answer) until its cutoff point is the same as clip 1 (reference). By focusing on the final few milliseconds before the cutoff point and listening to the amplitude envelope, timbre, and musical content, you can compare the two clips and adjust the length of clip 2 so that the sound of its cutoff point matches clip 1. By listening to each clip, comparing the timbre at the cutoff point, and adjusting the length of clip 2 until the two are matched, you can learn the sound or timbre of the clip 1 cutoff point and adjust the length of clip 2 until its cutoff point sounds the same. The software also randomizes the length of clip 1 for each question. Clip lengths range from 500 milliseconds to 2 seconds, to ensure that we are not simply being trained to identify the duration of the audio clips.

You can adjust the length of clip 2 by nudging the end point either earlier or later in time. The nudge time step sizes are 5, 10, 15, 25, 50, and 100 milliseconds. The smaller the nudge step size, the more difficult it is to hear a difference from one step to another and therefore the more difficult it is to match the reference.

Figure 6.4 shows the waveforms of four sound clips of increasing length from 825 to 900 milliseconds in 25-millisecond steps. This example shows how the end of the clip can vary significantly depending on the length chosen. Although the waveforms of the second clip (of 850 milliseconds duration) and third clip (of 875 milliseconds duration) in Figure 6.4 look similar, there is a noticeable difference in the perceived percussive or transient sound at the cutoff point. With smaller step or nudge sizes, the difference between steps is less obvious and requires more training for correct identification. When comparing clips, the focus should be on the timbral quality of the transient sound at the cutoff point of the clip to determine the one that sounds most like the reference.

For example, the 825-millisecond duration clip in Figure 6.4 contains a faint percussive sound at the end of the clip, but because the note—a drum hit, in this case—that begins to sound is almost completely cut off, it sounds like a short click. In this example, it is useful to focus on the

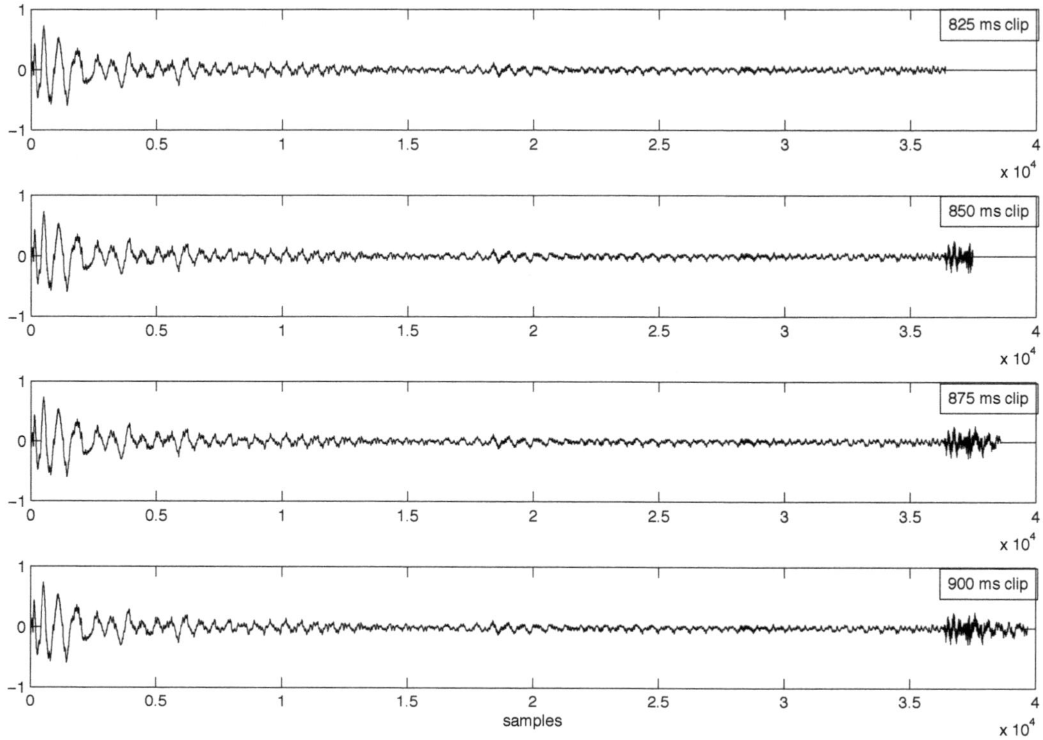

Figure 6.4 The waveforms of four clips of a music recording of different lengths: 825, 850, 875, and 900 milliseconds. These examples show how the end of the clip can vary significantly depending on the length chosen.

percussive quality, timbre, and envelope of the incoming drum hit at the clip cutoff to determine the correct sound clip length.

After matching the clip lengths, press the "Check Answer" button to display the correct duration. You can continue to audition the two clips for that question even after you know the correct duration. The software indicates whether the response for the previous question was correct or not, and if incorrect, it indicates whether clip 2 was too short or too long and the size of the error. Figure 6.5 shows a screenshot of the software module.

There is no view of the waveform as we would typically see in a digital audio editor because the goal of this training is to create an environment where we rely solely on what we hear with minimal visual information about the audio signal. There is, however, a green bar that increases in length over a timeline, tracking the playback of clip 2 in real time, as a visual indication that clip 2 is being played. Also, the play buttons for the respective clips turn gray briefly while the audio is playing and then return to white when the audio stops.

With this ear training method, the goal is to compare one sound to another and attempt to match them. There is no need to translate the sound feature to a verbal descriptor; instead, the focus lies

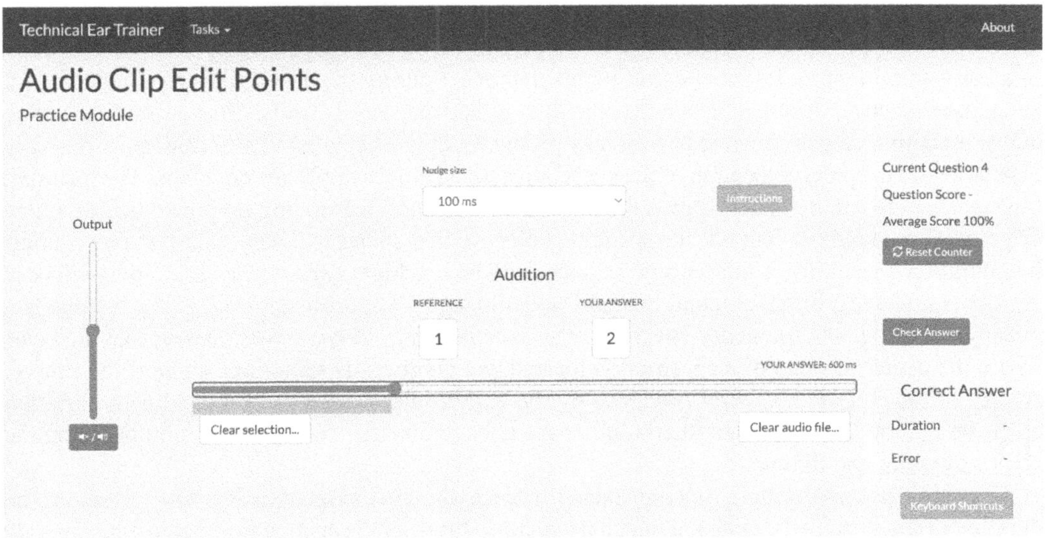

Figure 6.5 A screenshot of the Audio Clip Edit Points ear training software. The large squares with "1" and "2" are playback buttons for clips 1 and 2, respectively. Clip 1 (the reference) is of unknown duration, and the length of clip 2 must be adjusted to match clip 1. Below the clip 2 play button are two horizontal bars. The top one is a slider you use to control the duration of clip 2, from 500 to 2,000 milliseconds. The bottom green bar tracks the playback of clip 2 and it increases in length (from left to right) up to the slider position in the top bar, indicating that clip 2 is being played.

solely on our perception of the features of the audio signal. Although there is a numeric display indicating the length of the sound clip, this number serves only as a reference for keeping track of where the end point is set. The number has no bearing on the sound features heard other than for the specific excerpt in each question. For instance, a randomly chosen 600-millisecond clip will have different cutoff point features than other randomly chosen 600-millisecond clips.

I recommend starting with the less challenging exercises that use large step sizes of 100 milliseconds and progress through to the most challenging exercises where the smallest step size is 5 milliseconds. Almost any stereo recording in WAV file format can be used with the training software if it is at least 30 seconds in duration. The length of each clip is chosen randomly by the software, but clip length can also affect difficulty. Longer clips—for example, in the 2,000-millisecond range—seem to be more difficult. It is possible that with longer clips, it becomes more difficult to ignore the first several hundred milliseconds of the clip and focus only on the cutoff point at the end.

6.3 Focus of the Exercise

The main goal of the Audio Clip Edit Points ear training software is to focus on the amplitude envelope of a signal at a specific point in time—that being the end of a short audio excerpt or cutoff point. Although the audio is not being processed in any way other than that there is a fast fade-out, the location of the cutoff point determines how and at what point a musical note may get cut. In this exercise, focus on the final few milliseconds of the first clip, hold the end sound in memory, and compare it to the second clip.

Because the software randomly chooses the location of an excerpt, a cutoff point can occur almost anywhere in an audio signal. Nonetheless, I will note two specific cases where the location of a cut is important to describe: cutoff points that occur during the onset of a strong note or beat and those that occur during a sustained note—that is, between strong beats. First, a cutoff occurring at the beginning of a strong note or beat may produce a transient signal whose characteristics vary depending on the precise location of the cut relative to the note's amplitude envelope. The resulting transient timbre can be matched by adjusting the cutoff point. Depending on how much of a note or percussive sound gets cut off, the spectral content of that sound will vary with the note's modified duration. Generally, a shorter note segment will have a higher *spectral centroid* than a longer segment and have a brighter sound quality. An audio signal's spectral centroid is the average frequency of the signal's frequency spectrum. The spectral centroid is a single number that indicates where the center of mass of a spectrum is located, which gives us some indication of the timbre. A click at the end of an excerpt produced by the cutoff point can serve as a cue for the location of the end point. We can assess the timbre of the click and match the cutoff location based on the click's duration and timbre.

If a cutoff occurs during a more sustained or decaying audio signal, it is best to focus on the duration of the sustained signal and match its length. This might be analogous to adjusting the *hold time* of a gate with a very fast release time. With this type of matching, it helps to shift the listening focus to musical qualities such as tempo and timing to determine how long a final note is held before being cut off.

With any end-point location, the goal is to track the amplitude envelope and spectral content of the end of the clip. The hope is that the skills learned in this exercise can be generalized so as to increase hearing acuity, and that this facilitates our ability to hear subtle details in sound recordings that we had not noticed before spending extensive time doing digital editing. When practicing with this exercise, you may begin to hear details of a recording that may not have been as apparent when listening straight through to the entire musical piece. I have found that by listening to short excerpts out of context of an entire musical piece, I begin to hear sounds within the recording in new ways, as some sounds become unmasked and thus more audible. Listening to clips allows us to focus on features that may be partially or completely masked when heard in context (i.e., much longer excerpts) or features that are simply less apparent in the larger context. When listening to a full piece, our auditory systems are trying to follow musical lines, timbres, and spatial characteristics, so it seems as though our attention is constantly being pulled to various musical and timbral aspects through the piece, and we are not given the time to focus on every aspect of what is a complex stimulus. When we take a short clip out of context, we gain the ability to repeat it quickly while it remains in our short-term memory and therefore start to unpack details that may have eluded us while we listen to a full piece. When we repeat a clip out of context of an entire recording, we may experience a change in the perception of an audio signal. Similarly, repeating a single word many times may cause its meaning to fade momentarily as the focus shifts to the timbre of the word rather than its meaning. It is common for composers, especially of minimalist music, to take short musical phrases or excerpts of recordings and repeat them to create a new type of sound and perceptual effect, allowing listeners to hear new details in the sound that may not have been apparent before. For example, Steve Reich's piece "It's Gonna Rain" uses a recording of a person saying the words "it's gonna rain." In this piece, Reich loops those three words or portions of the three words on two different analog tape machines to create rhythmic, spatial, and timbral effects through a gradually increasing delay between the two tape machines. He takes advantage of the human auditory system's natural tendency to find patterns in sound and lose the meaning of words that are repeated many times.

The audio clip edit ear training method may help us focus on quieter or lower-level features amid louder features of a given program material. Quieter features of a program may be partly or mostly masked, perceptually less prominent, or considered in the background of a perceived sound scene or sound stage, but can be more apparent when we hear a short excerpt less than 1 second in duration, for example. Some of the features that may be more audible in this kind of listening include the following (those listed earlier are included here again):

- Reverberation and delay effects for specific instruments.
- Artifacts of dynamic range compression for specific instruments.
- Specific musical instrument sound quality: a drummer's brush sounds or the articulation of acoustic double bass on a jazz piece.
- Specific features of each musical voice/instrument such as the temporal nature or spatial location of amplitude envelope components (attack, decay, sustain, and release).
- Definition and clarity of elements within the sound image, width of individual elements.

Sounds taken out of context start to give a new impression of the sonic quality and the musical feel of a recording. Additional detail from an excerpt is often heard when a short clip of music is played repeatedly, detail that would not necessarily be heard in context.

As I was creating this practice module, I chose, perhaps arbitrarily, the jazz-bossa nova recording "Desafinado" by Stan Getz and João Gilberto (from the 1964 album *Getz/Gilberto*) as a sound file to test my software development. The recording features prominent vocals and saxophone, along with acoustic bass, acoustic guitar, piano, and drums played lightly. Through my testing and extensive listening, I have gained new impressions of the timbres and sound qualities in this recording that I was not previously aware of. Even though this might seem like a straightforward recording from a production point of view—a recording of acoustic instruments with minimal processing—I began to uncover subtle details of reverb, timbre, and dynamics. In this recording, the drums are quiet and more in the background, but if an excerpt falls between vocal phrases or guitar chords, the drum part may perceptually move to the foreground as the matching exercise changes our focus. It also may be easier to focus on characteristics of the drums, such as their reverberation or echo, if we can hear a musical part more clearly. Once we identify details within a short excerpt, it can make it easier to hear these features within the context of the entire recording and to generalize our ability to identify these types of sound features to other recordings.

In any audio project, try to listen to:

- the quality of transients—are they sharp and clear or broad and smeared?
- the shape of any cutoffs or fade-outs that are present,
- the amplitude envelope of every signal,
- lower-level and background elements such as reverb.

6.4 Exercise: Listen for Edits in Recordings

As you listen to commercially released recordings, try to see if you can hear edits or punch-ins. They are usually easier to hear in older recordings because edits using analog tape are harder to make transparent. Edits can sometimes be heard in vocals or breaths in more contemporary recordings. Here are some cross-fade artifacts that you can listen for when editing or when listening to commercial recordings:

- a slight, momentary dip in level,
- a slight, momentary increase in level,
- a note onset, speech syllable, or breath that is cut off,
- a timing mismatch—maybe the musical note following the edit feels rushed or delayed slightly,
- a sudden change in ambience or reverberation, such as when an edit is made at a cold start midway through a piece and there is no lingering reverberation in the take into which the edit goes,
- a low-frequency thump,
- a doubling, chorus, or phasing effect, especially with longer cross-fades,
- a shift in the stereo image,
- a change in timbre,
- an abrupt change in loudness after the edit,
- a click—if the cross-fade is very short.

6.5 Summary

This chapter outlines an ear training method based on the source-destination audio editing technique. Because of the critical listening required to perform accurate audio editing using this editing technique, the process of finding and matching edit points can serve as an effective form of ear training. With the interactive software exercise module, the goal is to practice matching the length of one sound excerpt to a reference excerpt. By focusing on the timbre and amplitude envelope of the final milliseconds of the clip, the end point can be determined based on the nature of any transients or length of sustained signals. By not including verbal or meaningful numeric descriptors, the exercise is focused solely on the perceived audio signal and on matching the end point of the audio signals.

Chapter 7

Analysis of Sound Recordings

7.1 Analysis of Sound Recordings

After focusing on specific features of various types of audio signal processing in the previous chapters, we can apply the knowledge and listening skills gained to explore a broader perspective of sound quality and music production. Experience practicing with each of the software modules and specific types of processing prepares us to focus on each of these sonic features in the wider context of recorded and acoustic sound analysis.

A sound recording captures an interpretation and specific representation of a single musical performance or a combination of layered musical performances. Needless to say, listening to a recording is a different experience than attending a live performance. This is true even for recordings with minimal signal processing that have as their goal to convey a concert experience. A sound recording can offer an experience that is clearer, more focused, and closer in perspective than a live performance, while also creating a sense of space. It is a paradoxical that some recordings can give the best of two perspectives – close, focused clarity with distant spaciousness. A recording can provide the clarity that we might get if seated right in front of musicians, yet at the same time it can provide the experience of listening from a more distant location through higher levels of reverberation that would not be present close to the stage. Furthermore, a recording engineer and producer often adjust levels and processing parameters over the course of a piece of music to highlight the most important aspects of a piece and guide a listener to a specific musical experience. Musicians do this to a certain extent during a performance, but the effect can be augmented in a recording.

Each audio project, whether it is a music recording, live sound event, film soundtrack, or game soundtrack, has its own unique timbral, spatial, and dynamic qualities. It is important to listen to a wide variety of recordings from different genres of music, film, and/or games, depending on our specific field of specialization, so that we can learn production choices made for each recording. We can familiarize ourselves with recording and mixing aesthetics for different genres, and the knowledge we have of others' work can inform our own work. When it comes time to record, mix, or master a project, it is helpful to have internal references for sound quality, dynamic range, tonal balance, and mix balance that we can rely on to help guide each project. The more active analytic listening we do, the stronger our internal references become.

To help expand references, it can help to look at production credits, specifically the names of the producer, recording engineer, mixing engineer, and mastering engineer, and cross-reference other recordings by the same people. While music streaming services Apple Music and TIDAL do include recording and production credits on many of their recordings, most other services do

DOI: 10.4324/9781003461173-7

not. Websites such as www.allmusic.com, www.discogs.com, and www.thelinernotesproject.com also include production credit listings. When you listen extensively to recordings by different engineers, you notice what is common across their recordings and what differentiates one person's work from another's. Furthermore, you could approach this part of technical ear training as a study of recording and production techniques used in various musical genres across the history of recorded sound.

To further develop your critical listening skills, I encourage you to actively and extensively examine, explore, and analyze sound recordings since this can help us to understand and learn the sonic signatures of artists, producers, and engineers. Through an active analytical listening process, we can learn to identify the aspects of an engineer's recordings that make them particularly successful from timbral, spatial, and dynamic points of view.

As I've mentioned before, the practice of active and critical listening does not turn us into expert listeners overnight. To develop these skills takes time and requires us to listen to hundreds if not thousands of recordings. There is no shortcut. Furthermore, it takes effort to concentrate. While you are listening, turn off visual distractions, social media, and mobile device notifications, and simply focus on what you hear in a recording. The idea is simple, but it takes significant energy to listen actively. If you only have five minutes at a time available to you to do this, it is still worth doing. As I have mentioned before, short regular listening sessions are better than infrequent long listening sessions.

The sound quality, technical fidelity, and sonic characteristics of a recording have a significant impact on how clearly the musical meaning and artistic intentions of a recording are communicated to listeners. By deconstructing the components of recording by ear, we can learn more about the way other engineers use reverberation and delays, panning, layering, balancing, dynamics processing, and equalization.

Distilled down to its most basic components, sound mixing essentially involves two main building blocks: level control and time delay. Whether level changes are made across the full band with a fader or if they are frequency selective such as with an equalizer, static or time-varying with automation, manual or through a compressor, mixing is the control of audio signal levels. Single instruments and even single notes may be brought up or down in level to emphasize and enhance the delivery of musical meaning. Frequency bands can be brought up or down in level to accentuate a musical part. The other basic component, the time delay, is the building block of artificial reverberation, delay, reflections, echo, chorus, and flanging.

The critical listening and analysis process can help us deconstruct various layers of a mix. We can listen analytically to general, overall characteristics of a full mix, or we can listen more to specific details of each sound source's timbre and signal processing effect. At a much deeper level in the analysis of a recording, an experienced engineer who is more advanced in critical listening skills may make guesses about specific models of equipment used during recording and mixing, based on the timbres and amplitude envelopes of components in a sound image.

We can analyze a stereo image produced by a pair of loudspeakers in terms of features that range from completely obvious to nearly imperceptible. A goal of ear training, as a type of perceptual learning, is to develop the ability to identify and differentiate features of a reproduced sound image, especially those that may not have been apparent before engaging in training exercises. Furthermore, by doing careful, analytical listening to commercial recordings, we develop listening skills that we can apply to our own engineering and production work.

We will now consider some of the specific characteristics of a stereo or surround image that are important to analyze. The list includes parameters outlined in the European Broadcasting Union

Technical Document 3286 titled "Assessment Methods for the Subjective Evaluation of the Quality of Sound Programme Material—Music" (European Broadcasting Union 1997):

- overall bandwidth,
- timbre, spectral balance,
- auditory image, stereo impression,
- spatial impression, reverberation, depth perspective, and time-based effects,
- dynamic range, changes in level or gain, artifacts from dynamics processing (compressors/expanders),
- noise and distortion,
- balance of elements (instruments/voices/sounds) within a mix.

7.1.1 *Overall Bandwidth*

Overall bandwidth refers to the range of frequencies in a recording—that is, how far it extends from the lowest to the highest frequencies. The goal is to estimate by ear the highest and lowest frequency (or range of frequencies) represented in a recording. In other words, what are the lower and upper end points of the frequency spectrum? Our active focus on low- and high-frequency extension in recordings can help us build internal reference points for bandwidth.

To get a feel for the lower and upper frequency ranges, try playing sine tones at various frequencies in the lower couple of octaves—for example, 20 Hz to 80 Hz—and in the upper octave—from 10 kHz to 20 kHz. You can use a DAW utility plugin such as Apple Logic Pro Test Oscillator, Avid Pro Tools Signal Generator, or Cockos REAPER Tone Generator to generate a sine tone.

Two other tools that are useful for getting to know how low or how high a recording extends are high-pass and low-pass filters. These are included as part of most parametric equalizer plugins that come with a DAW. Set each filter's slope to the maximum if the slope is an adjustable parameter. A slope of 48 dB/octave or higher is ideal.

With a high-pass (low-cut) filter, slowly sweep the cut-off frequency from 20 Hz to 500 Hz and listen to the way the low end is attenuated. At what cut-off frequency do you notice it having an effect? It is worth noting that the lowest frequencies that you hear also may depend on the frequency response of your loudspeakers or headphones. Smaller loudspeakers without a subwoofer may only extend down to 50 Hz or so, depending on the model.

With a low-pass (high-cut) filter, slowly sweep down from 20 kHz to 5 kHz and listen to the way the high end is attenuated. You may need to switch the filter on and off to home in on the frequency.

Next, sweep a high-pass filter from 20 Hz all the way up to 20 kHz to hear what the high-frequency end of the spectrum sounds like and what components of the mix are in that upper octave. Return the high-pass filter to 20 Hz and sweep a low-pass filter from 20,000 Hz down to 20 Hz and listen to only the bottom octaves—say, below 200 Hz.

While listening, ask questions such as:

- Does the recording bandwidth extend fully across the range of human hearing from 20 Hz to 20 kHz? Or is it band-limited in some way?
- How low is the lowest harmonic of instruments such as double bass, electric bass, or kick drum, or of a natural sound such as thunder?

- Are there extraneous sounds that extend below the instruments and voices in the recording, such as a thump from a microphone stand getting bumped or low-frequency rumble from an air-handling system?
- What is the highest harmonic? The highest fundamental frequencies for musical pitches do not go much above about 4 kHz, but overtones from cymbals and brass instruments easily reach 20 kHz and above. To make a judgment about high-frequency extension, we need to consider the highest overtones present in recording.

It's also worth looking at the frequency spectrum analyzer included with most equalizer plugins to see the lowest and highest frequencies in a recording.

To be able to hear these sounds requires a playback system that extends as low and as high as possible. By doing this exercise, you will run into any bandwidth limitations that your playback system may have, whether you are listening on headphones or loudspeakers. Usually, loudspeakers have more low-frequency extension than headphones, but work with what you have. Do not wait to get more gear—just start listening.

Some broadcast and communication technologies and formats deliberately limit the bandwidth of audio signals, which can provide additional references and context. For example, terrestrial FM radio broadcasts extend only up to 15 kHz, and standard telephone communication has a bandwidth of about 300 Hz to 3 kHz. A recording may be limited by its recording medium, a sound system can be limited by its electronic components, and a digital signal may be downsampled to a narrower bandwidth to reduce the amount of data transmitted. Depending on the choice of recording equipment and processing such as filters, an acoustic instrument's bandwidth may be reduced in a recording.

7.1.2 Spectral or Tonal Balance

As we saw in Chapter 2, spectral or tonal balance refers to the relative level of frequency bands across the entire audio spectrum. At a basic level, we can listen for the relative balance of high and low frequencies. For example, a recording with boosted high frequencies may sound bright or edgy. Conversely, a recording with boosted low frequencies may sound dull or dark.

As mentioned in Chapter 6, spectral centroid is an objective measure of the average frequency of a spectrum. A sound recording with more high-frequency content than low-frequency content has a higher spectral centroid. Of course, we can be much more precise in our judgment of spectral balance and identify specific frequency resonances (boosts) and anti-resonances (cuts), as we discussed in Chapter 2.

An audio signal's spectral balance measured by a real-time analyzer (RTA) helps us visualize a signal's spectral balance, and most equalizer plugins display the power spectrum in real time.

As we practice our subjective analyses of spectral balance, we can listen to recordings at two main levels:

1. Globally or overall: listening to the entire mix, noting its overall spectral balance.
2. Analytically to individual parts: listening to each instrument and voice within the mix, noting the spectral balance of each element, including any reverberation that may be present.

Where the possible combination and number of frequency resonances was simplified in Chapter 2 (e.g., up to three frequencies affected at octave or one-third-octave frequencies across an entire mix), the subjective analysis we are talking about now includes any frequency or combination of frequencies for not only the overall mix but also for individual elements. As we take a broader view of a recording or mix, we want to address questions such as:

- Are there specific frequency bands that are more prominent or deficient than others?
  - If so, do prominent frequency resonances affect specific instruments, voices, or sounds within the mix?
  - Are there specific musical notes that are more prominent than others? Another way to think about frequency resonances is to associate them with musical notes or pitches.
- Can we identify resonances by their approximate frequency in hertz?
  - Think back to the training in octave and one-third-octave frequencies from Chapter 2, and try to match the resonances with octave or possibly one-third-octave frequencies by memory.
- How prominent is each resonance?
- Are there any cuts in the spectrum? Anti-resonances or cut at frequencies are much more difficult to identify since it is harder to identify something that is missing. Listen to musical notes; some may be quieter than others.

Frequency resonances in recordings can occur for a few different reasons: the deliberate use of equalization, a microphone's placement relative to the sound source being recorded, or specific characteristics of an instrument, such as the tuning of a drumhead. The location and angle of orientation of a microphone will have a significant effect on the spectral balance of the recorded sound produced by an instrument. Because musical instruments typically have sound radiation patterns that vary with frequency, a microphone position relative to an instrument is critical in this regard.[1] Furthermore, depending on the nature and size of a recording space, resonant modes may be present, and microphones may pick up these modes depending on their positions within a room. Resonant modes can amplify certain specific frequencies, low frequencies primarily, produced by a musical instrument. All these factors contribute to the spectral balance of a recording or sound-reproducing system and may have a cumulative effect if resonances from different microphones occur in the same frequency regions.

7.1.3 Auditory Image

An auditory image, as Wieslaw Woszczyk (1993) has defined it, is "a mental model of the external world which is constructed by the listener from auditory information" (p.198). Listeners can localize sound images that occur from combinations of audio signals emanating from pairs or arrays of loudspeakers. The auditory impression of sounds located at various locations between two loudspeakers is referred to as a *stereo image*. Despite having only two physical sound sources in the case of stereo, it is possible to create phantom images of sources in locations between the actual loudspeaker locations, where no physical source exists.

Use of a complete stereo image—spanning the full range from left to right—is an important and sometimes overlooked aspect of production. Through careful listening to recordings, we can learn about the variety of panning and stereo image treatments found in various recordings. The illusion of mono sound sources positioned anywhere within the stereo image can be created by controlling interchannel amplitude differences with the standard panning function in a mixer or DAW.

As we saw in Chapter 3, interchannel time differences can also be used to position sound sources, although this technique is not widely used because its effectiveness depends on a listener's location relative to the loudspeakers. In typical stereo sound reproduction over loudspeakers, inter*channel*—that is, between loudspeakers—differences do not correspond to inter*aural*—that is, between ears—differences because sound from both loudspeakers reaches both ears. So, even though we perceive a mono sound source panned center as a single sound source, the phantom image illusion is created by two loudspeakers. Standard spaced or near-coincident stereo microphone techniques such as ORTF, NOS, and A-B were designed to provide interchannel amplitude and time differences for sources placed around the microphones. The microphone spacing of ORTF and NOS were meant to mimic the time and amplitude differences we encounter in natural listening. These stereo microphone techniques use microphone polar patterns and microphone angle of orientation to produce interchannel amplitude differences (ORTF and NOS), and physical spacing between microphones to produce interchannel time differences (ORTF, NOS, and A-B).

As we study music production and mixing techniques through critical listening and analysis, we find different conventions for sound panning within a stereo image across various genres of music. For example, popular music genres such as pop, hip-hop, country, and rock generally emphasize the central part of the stereo image. Kick drum, snare drum, bass, and vocals are almost always panned to the center of the stereo image. Guitar, keyboards, backing vocals, drum overheads, and reverb effects may be panned to the side, but overall, the stereo image will usually have a strong center component. As we saw in Chapter 3, a goniometer, correlation meter, or phase meter can confirm what we hear, since a recording with a strong center component will give a reading near +1 on a correlation meter. Likewise, if we reverse the polarity of one channel and then add the left and right channels together, a mix with a dominant center image will have significant cancellation of the audio signal. Any audio signal components that are equally present in the left and right channels—that is, monophonic sources panned center—will cancel when the two channels are subtracted or mixed with one channel's polarity reversed.

Panning and placement of sounds in a stereo image have a definite effect on how clearly listeners can hear individual sounds in a mix. There is also the phenomenon of masking, where one sound obscures another, in relation to panning. Panning sounds apart will result in greater clarity because when we pan sounds apart, we reduce masking, especially if the sounds occupy similar musical registers or contain similar frequency content. The mix and musical balance, and therefore the musical meaning and message, of a recording are directly affected by panning, and the appropriate use of panning can give us more flexibility for level adjustments.

While listening to stereo image width and the spread of an image from one side to the other, think about the following questions as a guide to your exploration and analysis:

- Taken as a whole, does the stereo image have a balanced spread from left to right with all points between the loudspeakers equally represented, or are there locations where it seems as though the image is lacking?
- How wide or narrow is the image?
  - Is the phantom image located mainly in the center, meaning that it is more monophonic, or is it spread wide across the stereo image?
- What are the locations and widths of individual sound sources in a recording?
- Are their locations stable and definite or are they ambiguous?
  - How easily can you pinpoint the locations of sound sources within a stereo image?
- Does the sound image appear to have appropriate spatial distribution of sound sources for the context?

- For classical music recordings especially, is the placement of musicians in the stereo image consistent with stage placement conventions? Can you identify a left–right reversal?

By considering these types of questions for each sound recording encountered, we can develop a stronger sense for the kinds of panning and stereo images created by professional engineers and producers.

7.1.4 Spatial Impression, Reverberation, and Time-Based Effects

Spatial processing in a recording, such as through reverberation, delay, and echo, is critical for conveying emotion and drama in music. Reverberation and echo help set the scene in which a musical performance or theatrical action takes place. Listeners are transported mentally to the space in which music exists through the strong influence of early reflections and reverberation that envelops music and sounds in a recording. Whether a real acoustic space is captured in a recording or artificial reverberation is added to mimic a real space, spatial attributes convey a general impression about the size of a space. A long reverberation time can create the sense of being in a larger acoustic space, whereas a short reverberation decay time or a low level of reverberation can convey the feeling of a more intimate, smaller space.

The analysis of spatial impression can be broken down into the following subareas:

- Apparent room size:
  - How big is the room?
  - Is there more than one type of reverberation present in a recording? Do all instruments/voices/sounds have the same reverberation, or do you hear different types?
  - Is the reverberation real or artificial?
  - What is the approximate reverberation time?
  - Are there any echoes or long delays in the reverberation and early reflections?
- Depth perspective: Are all the sounds about the same distance away or are some sounds closer and other sounds further away?
- What is the spectral balance of the reverberation?
- What is the direct/reverberant ratio?
- Are there any strong echoes or delays? Can you guess the approximate delay time of any apparent echoes? Do the echoes line up with musical tempo or are they independent?
- Is there any apparent time-based effect such as chorus, flanging, or phasing?

Classical music recordings give us the opportunity to familiarize ourselves with reverberation from a real acoustic space. Often orchestras and artists with higher recording budgets record in concert halls and churches with acoustics that are conducive to classical music performance. The depth and sense of space that can be created with microphone pickup of a real acoustic space are generally difficult to mimic with artificial reverberation added to dry sounds. Adding artificial reverberation to dry sounds is not the same as recording instruments performing in a live acoustic space. If we record dry sounds in an acoustically dead space with close microphones, the microphones pick up primarily only sound that is radiated towards the microphone, and they pick up much less of the sound that is radiated in other directions. When we record in a large, live acoustic space, usually most of our sound is from main microphones placed several feet away from even the closest instrument. Sound radiated from the back of an instrument in a live space gets reflected into the space and has a good chance of eventually reaching the main microphones. In an acoustically dry studio environment, microphones may not pick up any sound radiated from the back of an instrument. If microphones do pick up indirect or reflected sound in a dry studio, these early reflections are likely to be arriving within a much shorter time frame than those found in a large, live acoustic space. So even if high-quality convolution (impulse response-based) reverberation is applied to a dry, close-miked studio recording, it is not likely to sound the same as a recording made in a larger space.

7.1.5 Dynamic Range and Changes in Level

Dynamic range represents the variation in a recording's levels from the quietest sounds to the loudest. Over decades of collective experience listening to recorded music, listeners have developed some expectations for dynamic range. In general, classical music has the widest dynamic range. Popular music recordings tend to have some of the smallest dynamic ranges, although there are exceptions. There may be broad fluctuations in sound level over the course of a musical piece, as a dynamic level rises to fortissimo and falls to pianissimo, such as we typically find in classical music. Likewise, we can consider the microdynamics of a mix, the analysis of which is usually aided if we use a level meter such as a peak program meter (PPM), a LUFS meter (see Section 4.3), or a sample peak meter. Usually, we perceive a relatively constant level or loudness in pop and rock recordings, but we may hear and see on a meter small level fluctuations that occur from beat to beat. These are the microdynamics.

A peak level meter may fluctuate more than 40 dB over the course of one recording or as few as 2 to 3 dB for another. Large-amplitude fluctuations mean a recording has a wider dynamic range and likely has been compressed less. Amplitude levels that vary dynamically can help magnify musical intentions and enhance the listening experience. Recordings with wide dynamic ranges, such as many classical music recordings, can represent concert hall listening experiences. But there can be drawbacks with wide dynamic ranges. The main problem is that quieter sections may be partially or mostly inaudible in listening environments with significant background noise such as a car. Without manual adjustment of listening level to compensate for overall mix-level changes, a wide dynamic range may detract from the musical impact intended by an artist. Because the human auditory system responds primarily to average levels rather than peak levels in the judgment of loudness, a recording with smaller amplitude fluctuations—that is, a smaller dynamic range—will sound louder than one with larger fluctuations—that is, a wider dynamic range—even if the two have the same peak amplitude.

In dynamic range and level analysis, listen for changes in level of individual instruments and the overall mix. Level changes may be the result of manual gain changes, or they may be the result of automated, signal-dependent dynamic range control by a compressor or expander. Listen also for compression artifacts, such as pumping and breathing. Some engineers choose compression and limiting settings specifically to create timbral effects and alter the sound in some obvious way. For example, sidechain compression is sometimes used to create a noticeable pumping/pulsing effect which is common in techno, house, electronica, and pop music recordings. In this dynamics processing effect, one instrument, usually the kick drum, is used as a control signal to compress a full mix. The amplitude envelope of the kick drum triggers a compressor which then shapes the amplitude envelope of the rest of the mix, causing the level to drop every time there is a kick drum hit.

On the other hand, as we discussed in Chapter 4, compression can be one of the most difficult types of processing to hear because it's often meant to simply counter abrupt changes in level and return to unity gain when signal levels fall. When compression is used for this kind of purpose, an engineer may want the gain reduction from a compressor to be as transparent or inaudible as possible. As you listen analytically to a recording, note the amplitude envelopes of the instruments and voices. Do they sound natural or can you detect some alteration?

7.1.6 Noise, Distortion, and Edits

As we discussed in Chapter 5, many different types of noise can disrupt or degrade an audio signal in one way or another and can come in different forms such as 50 or 60 Hz buzz or hum from an electrical grounding issue, low-frequency thumps from a microphone or stand being bumped, external noises such as car horns or airplanes, clicks and pops from inaccurate digital synchronization, clipping of signal levels that go over the maximum level of an audio system, and drop-outs—that is, very short periods of silence—resulting from defective recording media or network connectivity problems. Generally, the goal is to avoid accidental instances of noise and distortion since they can detract from the artistic intentions of a project.

Unless intentionally distorting a sound, engineers try to avoid clipping any of the stages in a signal chain. So it is important to recognize when it is occurring and reduce a signal's level appropriately. Sometimes it is unavoidable or it slips by those involved and is present in a finished recording. When you analyze recordings, listen for extraneous sounds that do not seem to fit the artistic intentions of the mix.

Most recordings contain extensive edits as engineers and producers create composites of the best performances of an artist. Usually, these edits are transparent and inaudible, but their artifacts sometimes do slip through and remain in a released recording. Are there any sounds such as breaths, notes, or reverberation that seem to be cut off abruptly? If so, you may be hearing an edit or punch in.

7.1.7 Balance of the Components Within a Mix

Finally, in the analysis of recorded sound, consider the overall mix—that is, the balance of the elements within a recording. The relative balance of instruments, voices, and sounds can have a highly significant influence on the artistic meaning, impact, and focus of a recording. The amplitude of one element within the context of a mix can also influence the perception of other elements within the mix. Sometimes level adjustments even as small as 1 or 2 dB on a single instrument can have a noticeable effect on the overall perception of musical intent and meaning.

When you are listening for mix balance, think about these questions:

- Are the amplitude levels of the instruments, voices, and sounds balanced appropriately for the music, film, or game genre or style?
- Is there an element in the mix that is too loud or another that is too quiet?
- Can you hear what you need to hear to make sense of the recording?

Mix balances can change from moment to moment based on the natural dynamics of sound sources, changes in distance between a microphone and a sound source (performers sometimes move, and thus their recorded levels may change proportionally), and fader movements that an engineer makes during mixing.

We can analyze the entire perceived sound image as a whole, but we may also analyze less significant features of a sound image and consider these secondary elements as a subgroup. Some of these sub-features might include the following:

- Specific features of each component, musical voice, or instrument such as the temporal nature or spatial location of amplitude envelope components (e.g., of a note's attack, decay, sustain, and release).
  - In other words, are the note onsets of a particular instrument fast or slow? Are notes sustained or does the sound decay quickly?
  - Are the notes' attacks located in the same place as the sustained portion of the sound, or are the attack and sustain portions of a sound spread across the stereo image?
- Definition and clarity of each element within a sound image.
  - Are notes clearly articulated or do they blend like those in a legato musical passage? Musicians are often deliberate about the way notes are phrased and articulated, but the recording technique can influence the way it sounds to a listener.
- Width and spatial extent of each element.
  - How wide is each sound source? Some sources may be narrow and focused whereas others may extend the full width. What do you think is contributing to a source's width? Reverberation (real or artificial), delays or reflections, microphone techniques, double tracking (recording the same part twice and panning the two out wide) can all affect the width of a mix element.

For an untrained or casual listener, specific features of recordings may not be obvious or immediately recognizable. As trained listeners, we are more likely to be able to identify and distinguish specific features of reproduced audio that are not apparent to an untrained listener. An example of this is perceptual encoder algorithm development, which we saw in Chapter 5. Expert listeners are trained to identify shortcomings in codecs since the artifacts and distortion produced during perceptual encoding are not necessarily immediately apparent until one learns what to listen for. After learning to identify audio artifacts in codecs, it can become difficult *not* to hear them.

Music recordings, at least those that are audio-only and accompanied by video, are distinct from a live music concert listening experience since music recordings require us to rely entirely on the sense of hearing. There is no visual information to help follow a musical soundtrack, unlike an in-person live performance where visual information helps to fill in details that may not be as obvious in the auditory domain. As a result, recording engineers sometimes exaggerate certain sonic features of a sound recording, through level control, dynamic range processing, equalization, and reverberation, to help engage us as listeners.

7.2 A Survey of Stereo Recordings

In this section, we survey several commercial recordings and highlight some timbral, dynamic, spatial, and mixing choices that are apparent from listening. I have included key engineering and production credits as well.

Meters can help guide our listening, so I have also included the measured loudness in terms of LUFS Integrated, averaged across the full program, and loudness range in Loudness Units or LU for each track. I measured integrated loudness and loudness range with the Cockos

REAPER Loudness Meter Peak/RMS/LUFS and confirmed the measurements with the Youlean Loudness Meter 2. I have also included screenshots of the Cockos REAPER Goniometer plugin to show variations in stereo image width from track to track and between different sections with a song.

In choosing these tracks, there are some themes with respect to recording and mix engineers and record labels. Three of the tracks in this list are mixed by Serban Ghenea, one of the top mix engineers for pop music these days. Listen to the quality of his mixes, especially in terms of low-frequency presence and control, vocal timbral quality, dynamic range control, stereo image, and reverberation. In the classical and acoustic music recording world, producer and recording engineer Morten Lindberg has been making stunningly beautiful recordings which are released on his 2L Lindberg Lyd record label. I have included two of his recordings below. I discuss two jazz recordings from iconic jazz and contemporary music label ECM Records. This label has some distinct aesthetic approaches to mixing and sound quality across many of their recordings, which result in timbral clarity and prominent use of reverberation. George Massenburg is a highly respected engineer, producer, and equipment designer, and I include two of his recordings. Prolific classical recording engineer Richard King continues to make exceptional recordings, and I include one recording that he engineered.

This list is, of course, just a tiny sample of the many millions of recordings that can be analyzed.

7.2.1 *"My My My!" by Troye Sivan from* Bloom *(2018)*

- Producer: Oscar Görres. Mixing Engineer: Serban Ghenea. Engineer: John Hanes. Mastering Engineer: Randy Merrill.
- Loudness: −10.4 LUFS Integrated.
- Loudness Range (LRA): 6.7 LU.

This track has many little details that are fun to pick out and that make the track interesting sonically. Through their use of equalization, reverberation, and mixing choices, the engineers support the musical performance and songwriting as the song progresses.

The song starts with Troye Sivan's voice layered and unaccompanied, with some subtly glitchy effects, pitch shifting, and what sounds like backwards audio effects. Similar kinds of glitchy, reversed audio effects return later in the song. Programmed drums and percussive synth harmonies enter with the first verse. The kick drum has a nice low end, and the percussive parts fill out the mid-range and provide good support for the lead vocal. Different drum parts have different reverberation effects on them. At least one reverb is mono, but others are stereo. The percussive synth reverb takes on a different quality between vocal phrases where it sounds as if it is gated. The reverb decay sounds as though it is cut short when the vocals are not present. The snare drum, panned center and relatively dry, has sixteenth-note repeating delay that occurs every second bar, giving it a dub-like feel. Vocals are doubled during the verse, and the reverb sounds like a medium room. There is a sense of space around the vocals but there is not a noticeable reverb tail per se during the verses. In the pre-chorus, a synth pad enters, and a low-pass filtered vocal overdub echoes the lead vocal words "my baby." Some additional high-frequency sounds are present in the choruses, including glitchy synths, additional backing vocals, and long reverb tails.

The reverb time during the second verse is shortened again, like the first verse, which provides a nice contrast to the chorus that just ended. The mix is drier again but still has a sense of space just without the long reverb tails. New synth pads and backing vocals are present in the second verse that were not present in the first verse.

This mix uses the full stereo image, especially with the vocals and their associated reverb, as we can see in the Lissajous pattern in Figure 7.1. Timbrally, the low frequencies from the kick drum and synth bass are beautifully controlled and even. They support the many vocal layers, programmed drums, and synth parts. There is still a nice sparkle to the high end, primarily from the vocal sibilance, glitchy synth sounds, and the snare drum. Reverb and delays are used creatively to contrast song sections, emphasize musical moments, and create space. The song ends with a long, lush reverberation tail decay following the last vocal note.

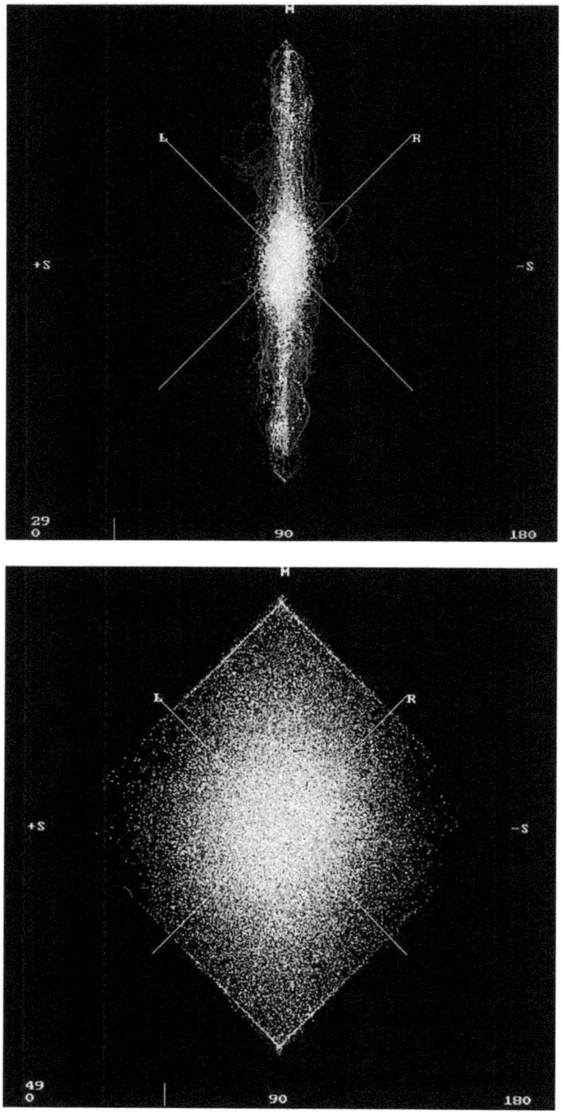

Figure 7.1 Screenshots of a goniometer during the verse (top) and chorus (bottom) of "My My My!" by Troye Sivan showing the variation in stereo image width of the mix from one section to another.

7.2.2 *"The Fate of Ophelia" by Taylor Swift from* **The Life of a Showgirl** *(2025)*

- Producers: Taylor Swift, Max Martin, and Shellback. Recording Engineers: Max Martin, Shellback, and Lasse Mårtén. Mixing Engineer: Serban Ghenea. Mastering Engineers: Randy Merrill and Ryan Smith.
- Loudness: –8.1 LUFS Integrated.
- Loudness Range (LRA): 7.5 LU.

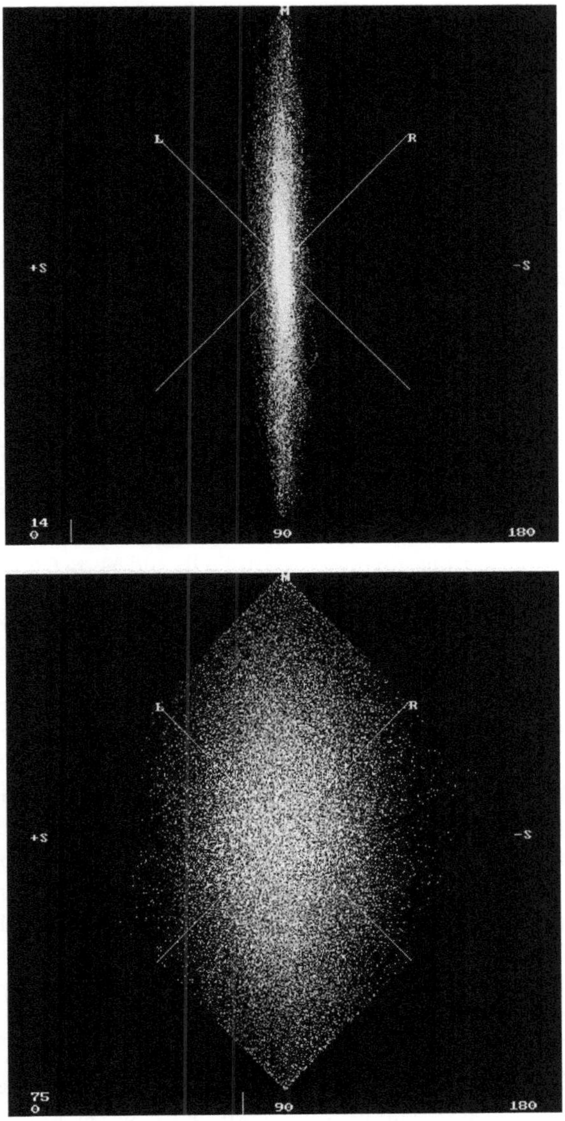

Figure 7.2 Screenshots of a goniometer during the verse (top) and chorus (bottom) of "The Fate of Ophelia" by Taylor Swift showing the variation in stereo image width of the mix from one section to another.

A snare drum fill with a mono spring reverb, both panned center, along with a kick drum to round out the low end for the opening of this song. Following the snare fill, a light, dry hi-hat taps eighth notes to accompany a sparsely voiced, reverberant piano part. Swift's vocals in the verses are lush and reverberant, which contrasts with the relatively dry drum set and synth bass. During the choruses, a few new parts are added to the mix including reverberant, wide-panned backing vocals, wide-panned by dry hand claps, shimmery synth parts, and electric guitar. The lead vocals are doubled and panned wide in the choruses, and they seem a little drier than the verse vocals. In the second pre-chorus, reverberant hand claps are added to the mix. For the bridge, everything drops out except the vocals, piano, and shimmer synth. The main vocal is panned center, and words at the end of each phrase are doubled. Additional vocal layers are panned wide and reverberant.

Listening to just the side component—that is, L–R—during the verse makes it easier to hear the multiple layers of delays on the vocals. The vocal delays repeat and the delay timing matches the song's tempo, so each delayed echo lines up with drum beats. The echo onsets are masked by drum beats, making them difficult to hear as separate echoes when not listening to the side component on its own. The side component also unveils a modulation effect on the vocal delay effect.

Overall, this mix has excellent timbral balance. The low, mid, and high frequencies are all appropriately level matched, and no frequency sticks out. Dynamic range control is excellent, especially on the vocals and low end. Both are controlled and supported but never excessively compressed or constrained. The stereo image width also varies with a relatively narrow stereo image for the verses and much wider choruses. Width differences are evident in the Lissajous patterns shown in Figure 7.2. Reverberation and delays create contrasts between section and musical parts.

7.2.3 *"Worst of Me" by Cynthia Erivo from* I Forgive You *(2025)*

- Vocal Producers: Cynthia Erivo, Will Wells. Mixing Engineer: Serban Ghenea. Mastering Engineer: Randy Merrill. Recording Engineers: Will Wells and Myra Choo.
- Loudness: –7.2 LUFS Integrated.
- Loudness Range (LRA): 5.2 LU.

This track begins with a wonderfully wide stereo image of Erivo's ethereal, reverberant vocals layered and panned wide. A muted piano joins the backing vocals, and both accompany the lead vocal through the first verse. When the first verse starts, the high frequencies of the backing vocals are attenuated somewhat, leaving room in the frequency spectrum for the clear lead vocals to soar above the backing vocals. In the pre-chorus, drums, bass, and additional backing vocal parts enter. The entrance of the bass fills out the low end of the track nicely while providing support for the lead vocal without overwhelming the track. The drums in the pre-chorus consist mainly of cymbal rolls, kick drum on beat one, and a bright reverberant snare on beat four. A short electric guitar line enters near the end of the first chorus for a couple of bars.

The second verse varies from the first verse since it includes the drums playing time, bass, and strings. In the second pre-chorus, I'm hearing a little bit of fluttering in the track and I'm not sure of the source of that effect.

The mix is full in this track. Once we get out of the first verse, there seems to be a lot packed into the mix. There is significant compression overall, but it serves to control signal levels in a way that supports the music. The lead vocal dynamic levels are controlled perfectly in the recording, allowing us to hear each word even though the timbral changes in Erivo's voice indicate that her original performance had a wide dynamic range. The panning of the background vocals helps create a nice wide stereo image that blends with the strings after the strings are added. Figure 7.3 shows snapshots of the goniometer during verse and chorus sections, confirming that the verses are nearly as wide as the choruses.

The reverb decay times are longer on the vocals and strings, which helps create the sense of being in a large space. We hear the vocal reverb decay at the very end of the song. Other than the splashy, reverberant, high-frequency snare drum present in the pre-choruses, the drums are drier than the vocals, which helps keep them from getting too muddy.

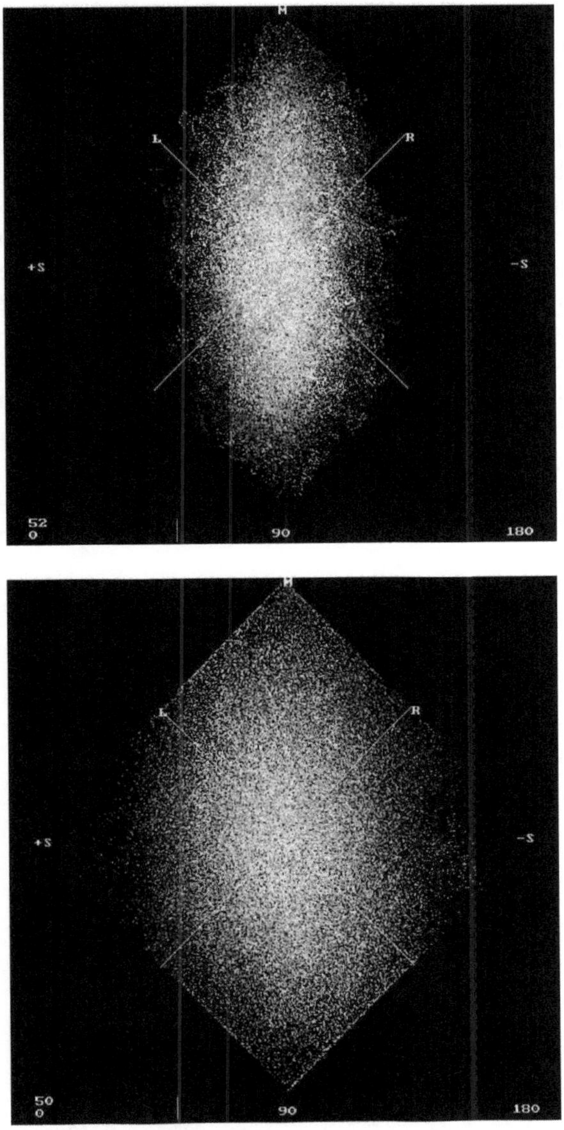

Figure 7.3 Screenshots of a goniometer during the verse (top) and chorus (bottom) of "Worst of Me" by Cynthia Erivo showing the variation in stereo image width of the mix from one section to another.

7.2.4 *"Stille, Stille Kommer Vi" by Hoff Ensemble from* Quiet Winter Night: An Acoustic Jazz Project *(2012)*

- Producer, Recording Engineer, Mixing Engineer, Mastering Engineer: Morten Lindberg. Recording Technician: Beatrice Johannessen.
- Loudness: –18.3 LUFS Integrated.
- Loudness Range (LRA): 14.5 LU.

It's unusual to record an acoustic ensemble with drums like this one in a reverberant space such as a church, but that's just what they did for this recording. The natural acoustics of the recording space are apparent from the first notes, but surprisingly the longer reverb and more distant microphone technique complement the music very nicely and do not detract from it. The slower tempo of this track and the style of the music lend themselves to a longer reverberation time than we would expect for music with a faster tempo played by similar instrumentation, such as in a jazz recording.

The track starts with the drum set played with brushes. The room acoustics are clearly audible in each beat, especially on the snare drum hits on beats two and four. The piano enters next, followed by acoustic, upright bass and trumpet. The singer enters after about a minute with the repeated phrase, "stille, stille, kommer vi," and the vocal level is mixed to match the trumpet level.

There seems to be no compression at all on this track, or if there is, it is very minimal, and this is reflected in the loudness range of 14.5 LU. The track sounds much quieter than a typical pop music recording, and this is reflected in the LUFS value of –18.3. The musical performance is enveloped and supported by the reverberation, which sounds like the natural reverberation of the space, although it is possible that some artificial reverberation was added. The reverberation does not get in the way of the music, and while the reverberation seems most noticeable at the beginning of the track since it is so prominent, it seems to blend effortlessly with the music. In terms of perspective, the cymbals and auxiliary metallic percussion sounds seem closest, which probably helps avoid

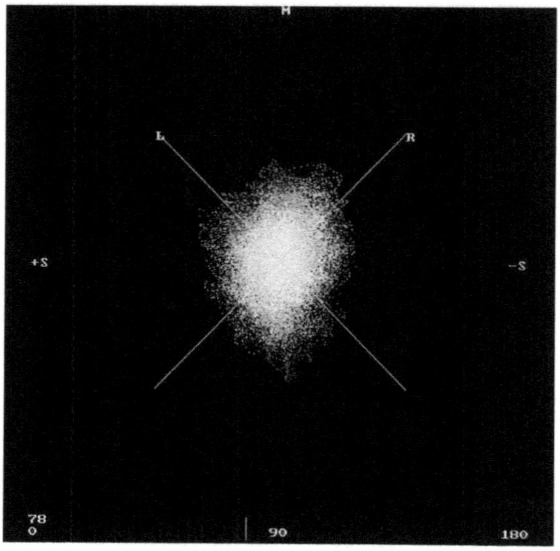

Figure 7.4 A screenshot of a goniometer of "Stille, Stille Kommer Vi" by Hoff Ensemble showing the typical stereo image width of the mix.

making the reverberation too washy with transient signals. The low end of the kick drum and the acoustic bass provide a solid foundation upon which the rest of the instruments sit. Timbrally, the high-frequency clarity of the cymbals, voice, and trumpet complement and are balanced with the low and mid-range from the other instruments.

Photos from the recording session showing can be found on the 2L recording label website: www.2l.no/pages/album/087.html.

7.2.5 *"Allegro" from Mozart: Violin Concerto No. 4 in D Major, K. 218, Performed by TrondheimSolistene and Marianne Thorsen from* Mozart: Violin Concertos *(2006)*

- Producer: Morten Lindberg. Balance Engineers: Hans Peter L'Orange and Morten Lindberg.
- Loudness: –19.6 LUFS Integrated.
- Loudness Range (LRA): 14.1 LU.

With impeccable clarity, timbral balance, instrumental balance, stereo width and depth, liveness, and spaciousness, this recording is an excellent classical music recording. The positions of each orchestral section—that is, first violins, second violins, viola, cello, double bass, oboes, and horns—on stage are clearly audible. First violins and cellos sound closest in the stereo image, and wind instruments are further back in the stereo image, consistent with the stage layout of an orchestra. The solo violin is focused front and center, making its nuances easy to hear, but its level is not excessive, and it does not sound artificially boosted. The timbre of the solo violin is balanced, which gives it clarity but without any harshness. The solo violin sounds as if it was recorded with at least two microphones that are panned wide, since its location seems to vary subtly and because it sounds wider than I would expect from a single microphone signal panned center. Despite

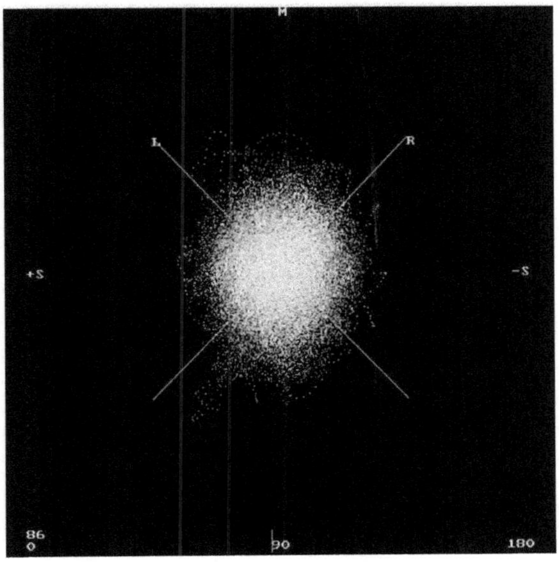

Figure 7.5 A screenshot of a goniometer of "Allegro" from Mozart: Violin Concerto No. 4 in D Major, K. 218 performed by Trondheimsolistene and Marianne Thorsen showing the typical stereo image width of the mix.

the subtle movement and image quality, the solo violin width sounds natural. The acoustics of the recording space provide reverberation which envelops and supports the musical performance. The reverberation time and amount of reverberation in the mix sound natural, and there is an appropriate amount of liveness in the acoustic space for the music. As is typical for classical music recordings, I do not hear any compression on this recording; as a result, it has a wide dynamic range which is reflected in the measured LUFS and loudness range values. Figure 7.5 shows the goniometer reading at one moment in the mix, but this is representative of the stereo width for the

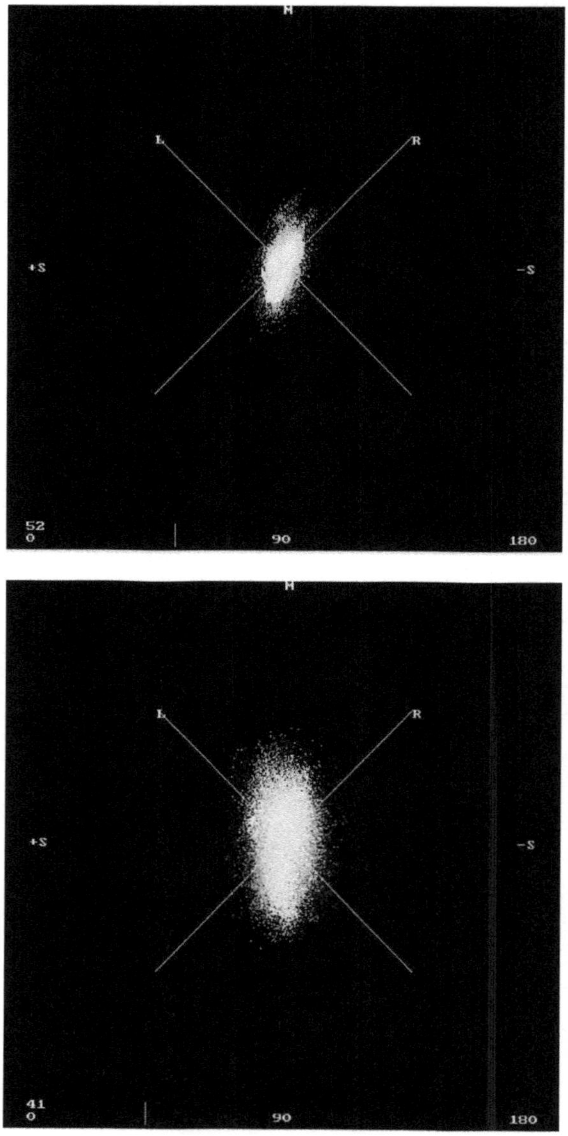

Figure 7.6 Screenshots of a goniometer from two points in time of "Sweet Jane" by Cowboy Junkies showing the stereo image width of the mix.

entire recording. Overall, this is a stunning classical music recording that seems to do everything right without ever being excessive in any one area.

7.2.6 *"Sweet Jane" by Cowboy Junkies from* The Trinity Session *(1988)*

- Producer: Peter Moore. Recording Engineers: Peter Moore and Perren Baker.
- Loudness: –21 LUFS Integrated.
- Loudness Range (LRA): 7.7 LU.

Using an interesting and unusual microphone setup and recording process, Cowboy Junkies' *The Trinity Session* album was recorded in a church in downtown Toronto with a single Calrec Soundfield Ambisonic microphone on a single day. According to a *Sound on Sound* magazine article about the recording, all the musicians were positioned in a circle around the microphone (Doyle 2015). The lead singer, Margo Timmins, was positioned outside of the circle, but her vocals were sent through a Klipsch Heresy monitor that was in the circle with the other musicians.

There is obvious echo or reverb from the natural acoustics of the space, especially from the side stick sounds of the snare drum, but the reverb is slightly less noticeable on the kick drum and guitar. The drums are positioned mostly on the right side of the stereo image and the guitar on the left. The bass is centered in the stereo image. The lead vocal has a light and airy timbre and seems to lean just to the right of center. The vocal has a fairly prominent level of sibilance—that is, in the 5 to 8 kHz range—especially on the "s" sounds, which sometimes come across as whistles, especially when Timmins sings the word "sweet."

This recording is relatively quiet as reflected in the low LUFS value, but the dynamic range is not very wide, accurately reflecting the band's performance in this acoustic space. The goniometer screenshots in Figure 7.6 show the width of the recording, and the smaller plots in the center indicate low recording levels.

There are very few if any recordings that sound like this one. It was a remarkable feat to achieve the mix balance, tonal balance, and reverberation they did with a single microphone in a highly reverberant space. It sounds both intimate and spacious due to the close-sounding vocals and the more reverberant-sounding drums.

7.2.7 *"No. 5, Arc-en-ciel" from* Ligeti: Etudes, Book 1, *Performed by Khatia Buniatishvili from Labyrinth (2020)*

- Producer: Nicolas Bartholomée. Recording Engineer: Hugo Scremin.
- Loudness: –17.5 LUFS Integrated.
- Loudness Range (LRA): 26 LU.

This recording of solo piano is exquisite, and it invites the listener into the music with its spaciousness. The reverberation immediately helps reinforce the feeling of this moving performance. In terms of imaging, the piano extends the full width of the stereo image, which is entirely appropriate for a solo piano recording. Timbrally, this piano sounds full and rich; the upper melodic lines soar, and the middle and lower register lines are clearly articulated. Since this recording does not sound as though it has any compression, it has a wide dynamic range which is reflected in the measured LUFS and loudness range values. Most of the recording is quiet except for two significantly louder sections in the middle. Figure 7.7 shows the goniometer reading at one moment in the mix, and this is representative of the wonderfully wide stereo image for the entire recording.

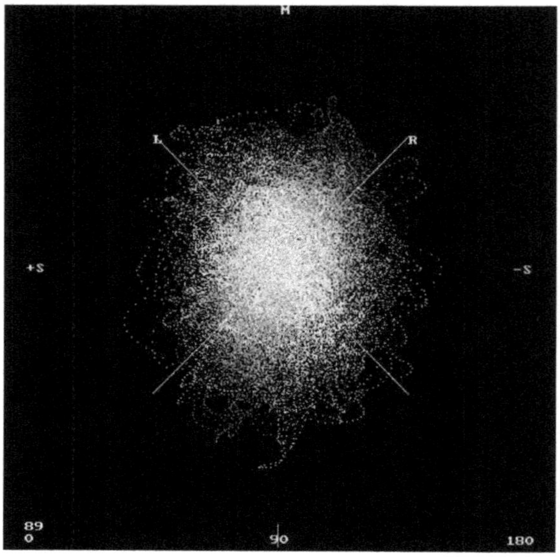

Figure 7.7 A screenshot of a goniometer of "Arc-en-ciel" from Ligeti: Etudes, Book 1 performed by Khatia Buniatishvili showing the typical stereo image width of the mix.

7.2.8 *"In Your Eyes" by Peter Gabriel from* So—25th Anniversary Edition *(1986/2012)*

- Producers: Daniel Lanois and Peter Gabriel. Recording Engineers: Kevin Killen and Daniel Lanois. Mastering Engineer: Ian Cooper.
- Loudness: –11.8 LUFS Integrated.
- Loudness Range (LRA): 8.9 LU.

A classic hit from the 1980s, this track is a study in successful layering of sounds that work together to create a timbrally, dynamically, and spatially exciting mix. The music starts with chorused piano, synthesizer pad, and auxiliary percussion. Bass and drum kit enter soon after, followed by Gabriel's lead vocal. There is an immediate sense of space on the first note of the track but without an obvious reverberation decay in the beginning, mainly because the sustained piano and synth pad cover the reverb tail. Reverberation decay is more audible during the pre-chorus and after the chorus especially on the snare drum. The combination of instrument and voice sounds with their associated reverb and delay effects creates a spacious, open, and enveloping feeling.

Despite multiple layers of percussion such as talking drum and triangle, along with the full rhythm section, the mix is pleasingly full yet remains uncluttered. The various percussion parts and drum kit occupy a wide area in the stereo image, helping to create a space in which the lead vocal sits. Listen closely to the timbre of the triangle, which taps on the off beats in the introduction and through the verses. The triangle is mostly consistent timbrally, but note that there is a very slight change in its timbre for a few beats here and there. These changes in timbre might be the result of edits or punch-ins during the recording sessions.

Gabriel's lead vocal features prominently in the mix, and its timbre is warm yet somewhat gritty, with a slight emphasis on the sibilance. The lead vocal is doubled for some words and phrases. It is supported by the variety of drums, wide-panned auxiliary percussion, bass, percussion, and synthesizers through the piece. The drums, played by one of the most famous drummers in the world, Manu Katché, have a natural dynamic range with limited compression and emphasize his amazing stick work on cymbals and snare drum. There is less emphasis on the low end of the kick. The electric bass seems to provide most of the low end, and it is punchy and articulate, sounding as though it was compressed heavily. The bass contributes significantly to the rhythmic foundation of the piece, especially with the grace notes and rhythmic accents in the stunning performance that bass players Tony Levin and Larry Klein provide. Listen for the vocal and synthesizer layering, especially during the pre-chorus. The electric guitar in the pre-chorus and chorus sections has a bright, thin timbre and sounds compressed, but it provides an excellent complement to the low end

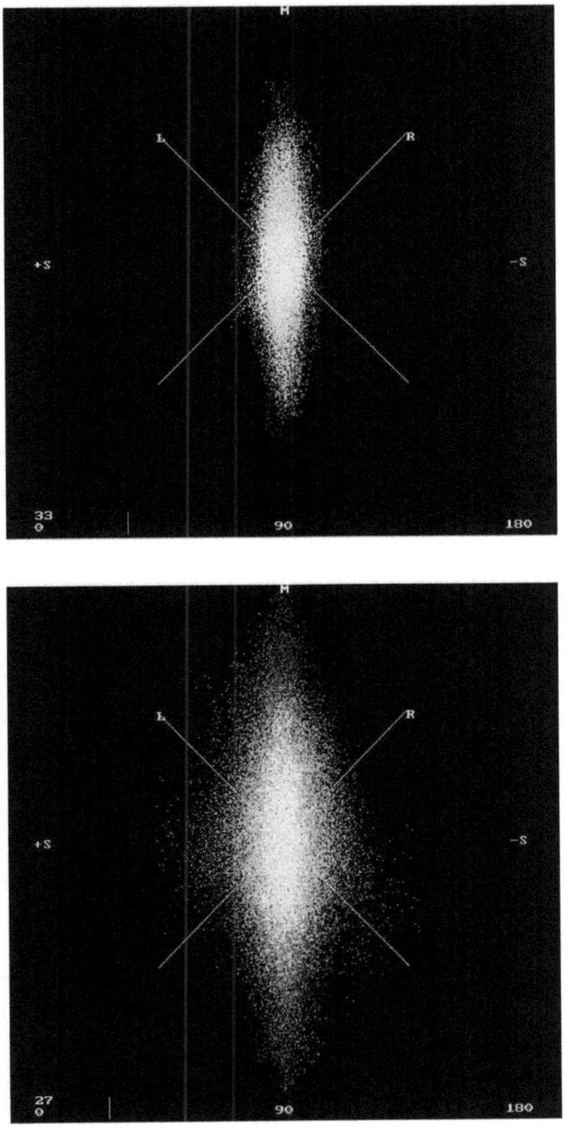

Figure 7.8 Screenshots of a goniometer during the verse (top) and chorus (bottom) of "In Your Eyes" by Peter Gabriel showing a slight variation in stereo image width of the mix from one section to another.

from the bass and drum kit. Senegalese singer Youssou N'Dour enters towards the end of the piece, and his vocal is layered with backing vocals that are panned out to the sides.

Distortion is certainly present in this recording, starting with the slightly crunchy drum hit, which sounds like a floor tom, on the downbeat that starts the piece. The very first notes of the piano and synth play the pickup (beat four) to start the tune, and then the floor tom hit establishes beat one.

Other sounds are slightly distorted in places and compression effects are audible. This is certainly not the cleanest recording we can find, yet the distortion and compression artifacts work to add life and excitement to the recording.

The stereo image for this track has a strong focus towards the center as shown in the goniometer readings in Figure 7.8, but with some auxiliary percussion, toms, guitar, backing vocals, and reverberation panned wide.

Overall, this recording demonstrates a fascinating use of many layers of sound, including acoustic percussion and electronic synthesizers, which create the sense of a large open space in which a musical story is told. In the credit listing for this recording on the compact disc, the drums and percussion are listed first, followed by the bass. I have heard that this is intentional because Gabriel feels that these are the most important elements of his recordings.

7.2.9 *"Demons" by Imagine Dragons from* **Night Visions** *(2012)*

- Producer: Alex Kid. Recording Engineer: Josh Mosser. Mixing Engineer: Manny Marroquin. Mastering Engineer: Joe LaPorta.
- Loudness: –5.7 LUFS Integrated.
- Loudness Range (LRA): 12.0 LU.

This track is a study in distortion. The song opens with the lead singer alone while a keyboard accompaniment gradually fades in. There are at least two echoes or delays on the vocal: one is a shorter slap-back echo, and the other is a longer delay timed to the tempo of the song. The keyboard that fades in under the lead vocal begins to sound noisy or distorted as it gets louder, as though it was processed with a bit-crusher plugin—that is, significant bit depth reduction. In the few beats before the chorus, the drums enter but they are low-pass filtered, giving them a dark, distant sound. The drum kit low-pass filter is removed exactly on the first beat of the chorus, which has the effect of moving the drums immediately to the forefront, precisely with the start of the chorus.

During the choruses, the listener is blasted with distorted kick drum and snare drum. The drums sound fuzzy and excessively distorted. Also during the choruses, the backing vocals sound highly compressed and distorted. It also sounds as though there is a slightly modulated, high-frequency noise during the chorus. This noise could be due to a bit-crusher plugin, but it is not clear what has been bit-crushed. With the distortion and compression/limiting on the choruses of this song, the sound image seems overly full, as though there is no room for one more instrument or voice. The tension created by the compression and distortion is released when the next verse starts and everything drops out except the lead vocal and keyboard accompaniment.

While the integrated loudness for the entire track is high (–5.7 LUFS), the recording has a relatively wide dynamic range. For example, the first verse has a loudness of –16.1 LUFS compared to the chorus which is much louder at –3.6 LUFS.

The stereo image seems to be center-heavy, except for backing vocals, reverb, and delay, most of which are panned out to the sides. This is another interesting track for a mid-side processor to hear just the side (or difference) component of the mix. In the side component, high frequencies from the distortion are more apparent, and the delays are also easier to hear. Regardless of your opinion of the sound quality of this recording, it was a hit and, as such, it is worth analyzing for features of the production and recording. Figure 7.9 shows contrasting goniomenter displays from two different moments track.

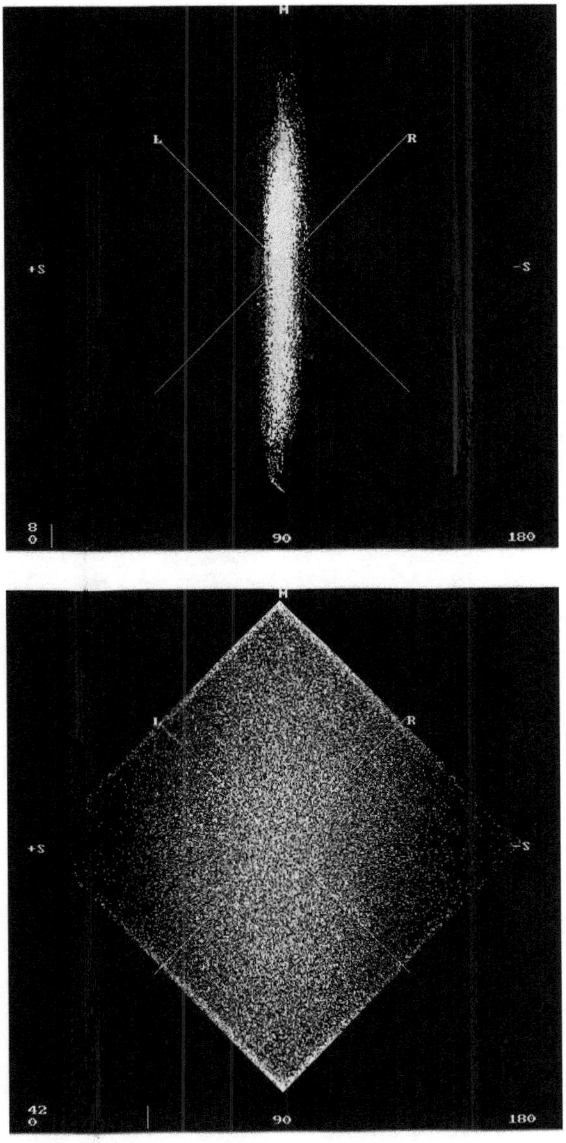

Figure 7.9 Screenshots of a goniometer during the verse (top) and chorus (bottom) of "Demons" by Imagine Dragons showing the contrast in stereo image width from one section to another.

7.2.10 *"Church" by Lyle Lovett from* Joshua Judges Ruth *(1992)*

- Producers: George Massenburg, Billy Williams, and Lyle Lovett. Recording Engineers: George Massenburg and Nathan Kunkel. Mastering Engineer: Doug Sax.
- Loudness: –17.3 LUFS Integrated.
- Loudness Range (LRA): 9.2 LU.

Lyle Lovett's recording "Church" represents contrasting perspectives, wide dynamic range, timbral clarity and tonal balance, and excellent mix balance. The track begins with piano giving a gospel choir a starting note, which they subsequently hum. Lovett's lead vocal enters immediately with handclapping from the choir on beats two and four. The piano, bass, and drums begin some sparse accompaniment of the voice and gradually build to more prominence. One thing that is

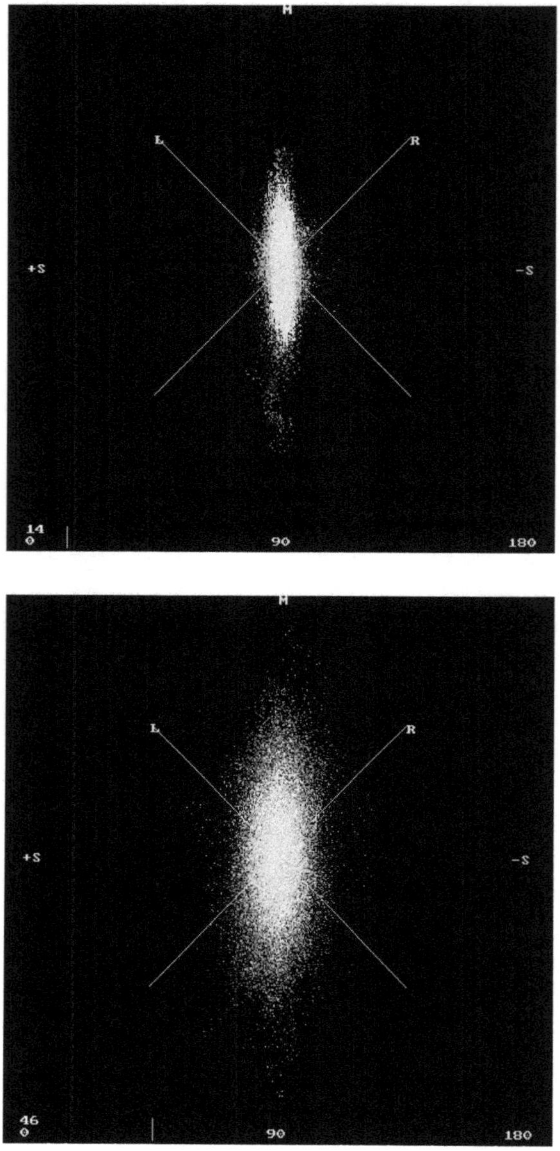

Figure 7.10 Screenshots of a goniometer during the verse (top) and chorus (bottom) of "Church" by Lyle Lovett. Because of the wide dynamic range and microdynamics in this recording, goniometer snapshots cannot convey the rich stereo image, spatial impression, and panning.

immediately striking in this recording is the clarity of each sound. The timbres of instruments and voices represent evenly balanced spectra, coming forth from the mix as natural-sounding.

Lovett's vocal is up front with very little reverberation, and its level in the mix is consistent from start to finish. The drums have a crisp attack with just the right amount of resonance. Each drum hit pops out from the mix with toms panned across the stereo image. The cymbals are crystal clear and add sparkle to the top end of the recording. In terms of perspective, the drums sound quite close and big within the mix.

The choir in this recording accompanies Lovett and responds to his singing. Interestingly, the choir sounds as though it is set in a small country church, where the reverberation is especially highlighted by hand claps. The choir and associated handclaps are panned widely across the stereo image. As choir members take short solos, their individual voices come forward and are particularly drier than they are when with the choir.

The lead vocals and rhythm section are presented in a dry, up-front way, and this contrasts with the choir, which is clearly in a more reverberant space or at least more distant.

Levels and dynamic range of each instrument are properly adjusted, presumably through some combination of compression and manual fader control. Each component of the mix is audible, and none of the sounds is obscured.

Goniometer snapshots in Figure 7.10 show moments in the verse (top) and chorus (bottom). Because of the wide dynamic range and microdynamics in this recording, goniometer snapshots cannot convey the rich stereo image, spatial impression, and panning.

Noises and distortion are completely nonexistent in this recording, and obviously great care has been taken to remove or prevent any extraneous noise. There is also no evidence of clipping, and each sound is clean. I do hear what sounds like a subtle chorus effect on the lead vocal in the first verse during the words "it came time to let us out." Perhaps another vocal take was mixed in momentarily.

This recording has become a classic in terms of sound quality and is often used as program material to audition loudspeakers. It is an excellent example of George Massenburg's engineering style, which puts sound quality and timbral clarity first, while minimizing distortion, such that the recording medium remains transparent to the musical intentions of the artist.

7.2.11 "Suite" by Tord Gustavsen Quartet from The Well *(2012)*

- Producer: Manfred Eicher. Recording Engineer: Jan Erik Kongshaug.
- Loudness: –19.6 LUFS Integrated.
- Loudness Range (LRA): 12.7 LU.

Jazz recordings from ECM Records tend to have a distinctively spacious sound. Partly, this is due to their choice of musicians to record and the types of pieces that the musicians play, but it is also due in large part to the engineering choices. ECM recordings typically exhibit a lot of clarity, minimal dynamics processing, high sound quality, and substantial amounts of reverb. The ECM production style has evolved slightly over the decades, with artificial reverb becoming less prominent than in early recordings by the label. This recording by the Tord Gustavsen Quartet is a good example of current ECM recording and production aesthetics. The piece begins with piano alone, played by Gustavsen. The introduction is slow with brief pauses. The reverberation supports the feeling of space and peaceful contemplation created by the music. The piano sound extends to the full width of the stereo image, but it seems anchored in the center of the image. In other words, there is good continuity of the stereo image from left to right. Listening closely, we can hear the piano dampers lifting from the piano strings. The upright bass played *arco* (with a bow) enters in the far right-hand side of the image at about 1:20. At around 2:40, the piano settles into a slow, consistent tempo and the saxophone and drums enter. The bassist also switches to *pizzicato* (plucked) playing at this point.

The ride cymbal on the drum set is dry, and it seems to be the closest sound in the image. The saxophone becomes the lead instrument after it enters, and it sounds slightly further back than the ride cymbal. The sax has a bright and clear sound, and its level is high enough in the mix so that we can hear it clearly, but it does not overpower the other instruments.

The piano, saxophone, and snare drum have quite a bit of reverb on them. The reverb tail is long, and it creates a sense that the group is in a large space. At the same time, the clarity and closeness of the piano, saxophone, and especially the ride cymbal make it sound as though we are quite near the instruments. The bass plays a less prominent role than it did during the *arco* playing at the beginning, and even though its level seems lower in the mix, we can still hear its articulation. The kick drum sounds big and round, but it is mixed low enough so as not to be obtrusive.

The piano sounds compressed, and there is some compression or limiting on the overall mix, mostly triggered by the bass and kick drum, which subtly affects high frequencies from the cymbals. Overall, the spectral balance seems even. The low frequencies from the kick drum and bass blend well but remain distinctive and provide a solid foundation for the piano and saxophone. High frequencies remain clear but not harsh. Figure 7.11 shows a goniometer display at one moment of this track.

Some listeners may not be fond of this much use of reverb in a jazz recording, but it is worth exploring recordings by ECM. The label has produced a huge catalog of jazz recordings, with Manfred Eicher as producer and the late Jan Erik Kongshaug as recording engineer on most of them.

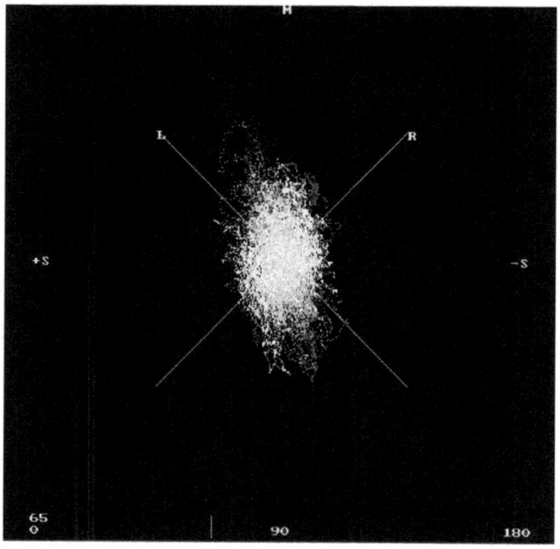

Figure 7.11 A screenshot of a goniometer of "Suite" by the Tord Gustavsen Quartet showing the typical stereo image width of the mix.

7.2.12 *"Spiel" by Julia Hülsmann Quartet from* **In Full View** *(2013)*

- Producer: Manfred Eicher. Recording Engineer: Jan Erik Kongshaug.
- Loudness: –21.1 LUFS Integrated.
- Loudness Range (LRA): 11.3 LU.

This recording has the classic reverberation, timbral clarity, and sparseness found on many releases by ECM Records. The track begins with a minimalist piano motif mixed with an

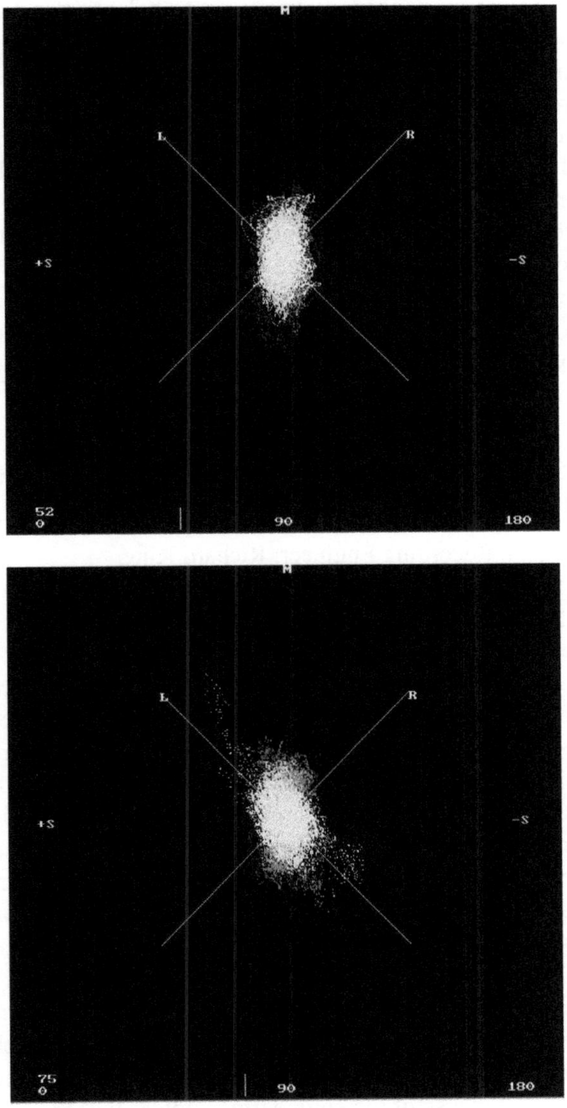

Figure 7.12 A screenshot of a goniometer of "Spiel" by the Julia Hülsmann Quartet showing the typical stereo image width of the mix in the top image. The bottom image shows the stereo image during a drum fill which is panned left.

undercurrent of low-pass filtered reverberation. The piano timbre seems dark but not muddy, emphasizing mid and low frequencies. Drums enter soon after with brushes, allowing us to hear the reverb from its transient sounds. There is plenty of detail in the brush work on the cymbals and snare drum. The drums sound close but at the same time are mixed with a layer of reverberation running underneath. The drum set seems to occupy mostly the left side of the stereo image, with the floor tom panned hard left and the ride cymbal panned slightly left. The drum set panning gives it a good sense of space. Trumpet, panned center, joins for the main melody (the head) and sits prominently in the mix. It is quite a bit louder than the piano and drums. At certain moments, the trumpet and snare drum sound louder and closer than the piano. Partway through the head, the upright bass enters, panned slightly right to balance the drums positioned to the left. The bass drops out on occasion, which leaves a hole in the lower end of the audio spectrum that is occasionally punctuated by the kick drum. The piano is brought forward in the mix for its solo and occupies its rightful place in the spotlight.

The dynamic range on this recording is naturally wide although some compression is audible on the piano, trumpet, and drums. The double bass seems as though it may also have been compressed since the notes are even, but the compression is less audible on it.

Overall, the long reverberation decay time mixed low creates warmth and space and contrasts with the clarity of the cymbals, drums, and trumpet. While the reverberation is audible as a separate sound, it creates a welcoming atmosphere that supports the music, helping to situate it in a large, beautiful space. The instruments blend perfectly, but we can still hear inner parts and subtle details. Figure 7.12 shows goniometer displays during two different moments in this track.

7.2.13 "Attaboy" by Yo-Yo Ma, Stuart Duncan, Edgar Meyer, and Chris Thile from The Goat Rodeo Sessions *(2011)*

- Producer: Steven Epstein. Recording Engineer: Richard King.
- Loudness: –12.3 LUFS Integrated.
- Loudness Range (LRA): 12.4 LU.

The pristine recording quality of this track stands in stark contrast to the Imagine Dragons track discussed above. Thile's mandolin opens the first tune on this bluegrass/classical crossover album. The mandolin sound is detailed and present while a gentle wash of reverb creates a space around the instrument. Duncan's fiddle, Ma's cello, and Meyer's double bass enter soon after, playing sustained notes that create a moving harmony under the mandolin melody. The timbres on all these string instruments are warm yet crisp. It sounds as if the instruments were recorded in a live room with reverb added. The reverb, although it does sound like artificial reverb primarily, is never obtrusive but simply helps support the direct sounds from the instruments as they trade roles playing melody and harmony throughout the piece. This recording is very clean, detailed, warm, spacious, dynamic, clear, and it places the instruments in ideal positions across the stereo image. We can hear subtle details in the sound of each instrument, but the instruments also blend beautifully with each other and with the acoustic space. The rich spaciousness of the recording is reflected in the goniometer display shown in Figure 7.13. The music from this recording comes alive in part because of the engineering choices that Steven Epstein and Richard King made.

Steven Epstein and Richard King make an amazing team of producer and engineer, and their work stands among the best-sounding recordings in the classical and crossover genres. *The Goat Rodeo Sessions* is no exception. Fortunately for us, they shot some nice video from the Goat Rodeo

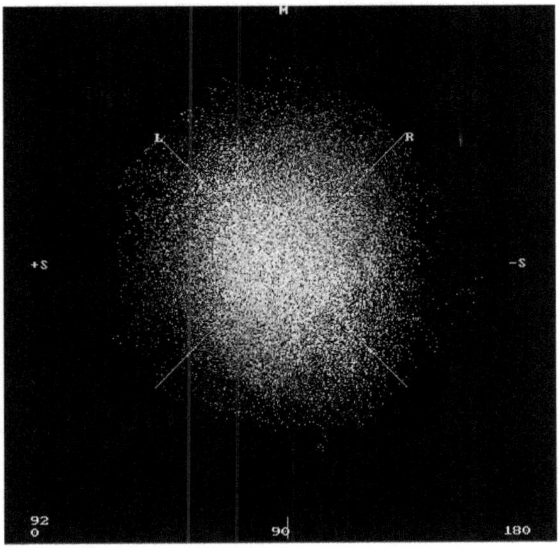

Figure 7.13 A screenshot of a goniometer of "Attaboy" from *The Goat Rodeo Sessions* performed by Yo-Yo Ma, Stuart Duncan, Edgar Meyer, and Chris Thile showing the typical stereo image width of the mix.

recording sessions, so if you are curious about microphone placement and technique, you can find the video on YouTube.[2]

7.2.14 *"Aria" from J. S. Bach:* **Goldberg Variations,** *BWV 988, Performed by Víkingur Ólafsson from* **J. S. Bach: Goldberg Variations** *(2023)*

- Producer, Recording Engineer, Editing Engineer, Mixing Engineer, Mastering Engineer: Christopher Tarnow.
- Loudness: –25.8 LUFS Integrated.
- Loudness Range (LRA): 9.5 LU.

The quality that strikes me first is the natural spaciousness of the recording. Each note has a wonderfully airy placement and is supported by the natural acoustics of the recording space. In terms of the stereo image, higher notes generally originate more from the left side, and lower notes originate more from the right, as they would if we were listening in front of the piano on stage, although this placement is not strict. The stereo image has a diffuseness that makes precise localization of individual notes difficult, but this helps create an immersive feeling. The reverberation sounds natural, giving the recording a suitable sense of space without an excessive reverberation decay time. Timbrally, the piano has great clarity in the high notes as well as defined articulation in the middle and lower registers. The measured loudness level (–25.8 LUFS) and loudness range (9.5 LU) reflect the overall quiet dynamic level of the piece and more limited dynamic range than the other solo piece mentioned above composed by Ligeti. The goniometer snapshot shown in Figure 7.14 confirms the spaciousness of the stereo image and the lower dynamic level of the recording. The correlation meter fluctuates widely in this recording, reflecting the richness of the stereo image. Overall, the timbral clarity, hall acoustics, and stereo image working together make this an excellent recording.

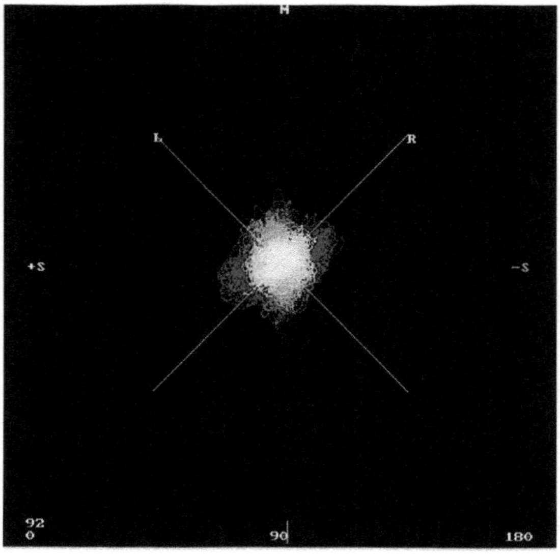

Figure 7.14 A screenshot of a goniometer of "Aria" from J. S. Bach: *Goldberg Variations*, BWV 988 performed by Víkingur Ólafsson showing the typical stereo image width of the mix.

7.2.15 *"In the Country" by Jon Randall from* Walking Among the Living *(2005)*

- Producers: George Massenburg and Jon Randall. Recording Engineers: George Massenburg and David Robinson. Mastering Engineer: George Massenburg.
- Loudness: –12 LUFS Integrated.
- Loudness Range (LRA): 6.1 LU.

The fullness and clarity of this track are present from the first note. Acoustic guitar and mandolin start the introduction, followed by Randall's lead vocal. The rhythm section enters in the second verse, which extends the bandwidth with cymbals in the high-frequency range and kick drum in the low-frequency range. Various musical colors, such as dobro, fiddle, Wurlitzer, and mandolin, rise to the forefront for short musical features and then fade to the background. It seems apparent that great care was taken to create a continually evolving mix that features musically important phrases.

The timbres in this track sound naturally clear and completely balanced spectrally. The voice is consistently present above the instruments, with a subtle sense of reverberation to create a space around it. Notice the consistency of the vocal level from word to word. We can hear every word effortlessly. The drums, while they sound amazing, are not as prominent as they are on the Lyle Lovett recording (also produced and recorded by Massenburg) discussed above, and in fact they are a little understated. The cymbals are present and clear, giving a rhythmic pulse and accents, but they certainly do not overpower other sounds in the mix, and the snare has more mid-range than crack. The bass is smooth and full, with enough articulation for its part, and provides consistent support for the rest of the group. The fiddle, mandolin, and guitar sounds are all full-bodied, crisp, and warm, with a nice sense of space around them. The high harmonics of the strummed mandolin and guitars blend with the cymbals' harmonics in the upper frequency range. Further to the timbral integrity of the track, there is no evidence of any noise or distortion, as we expect with Massenburg's engineering work.

The stereo image is used to its full extent with mandolins, guitars, and drum kit panned wide. The balance on this recording is impeccable and makes use of musically appropriate spatial treatment (reverberation and panning), dynamics processing, and equalization. Goniometer displays in Figure 7.15 show moments in the verse (top) and chorus (bottom).

Unfortunately, the lossless encoded version of this track that I can access on Apple Music seems to contain watermarking artifacts or some other type of subtle distortion that is not present

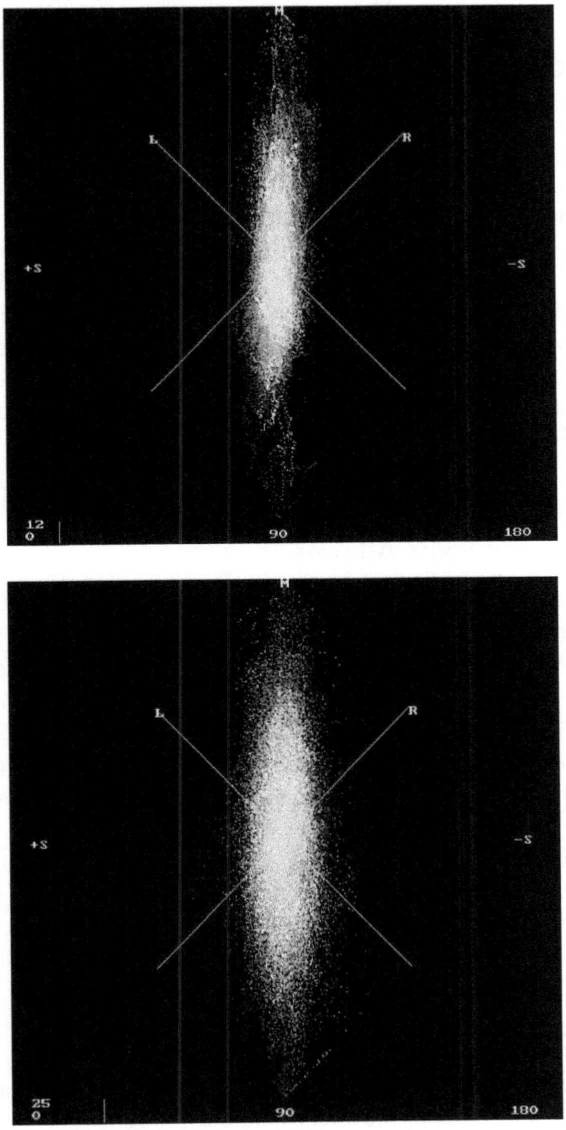

Figure 7.15 Screenshots of a goniometer during the verse (top) and chorus (bottom) of "In the Country" by Jon Randall. The goniometer snapshots show the prominence of center-panned sounds including voice, bass, kick drum, and snare drum, but it does not convey the panned acoustic guitar, mandolin, and cymbals.

in the original CD audio. I confirmed my suspicion by mixing the lossless streamed version with the CD audio version, aligned them down to the sample, and reversed the polarity on one of the versions. When the two version are subtracted, a distorted mid-range frequency artifact remains that might possibly be a result of audio watermarking. I discuss audio watermarking in Section 7.11 below.

I encourage you to listen to a wide range of different recordings from a variety of different genres and styles. Using a goniometer can help focus on the stereo image width.

In the following sections, I offer suggestions for listening exercises to do.

7.3 Comparing Original and Remastered Versions of Recordings

Many recordings have been remastered and re-released several years after their original release. Remastering an album usually involves returning to its original stereo mix and applying new equalization, dynamics processing, level adjustments, and maybe also mid-side processing and reverberation. Comparing an original release of an album to a remastered version is a useful exercise that can help highlight timbral, dynamic, and spatial characteristics typically altered by a mastering engineer. Streaming services do not always make the original version of an album available if a remastered version exists, so it can take some effort to locate different versions. Sometimes the only way to hear an original version is on CD (compact disc) if you can find it and have a device to play it. Different versions may also exist on vinyl records, but there is no guarantee that the audio from a vinyl disc will be without distortion or surface noise, so it may be difficult to know if audible differences between versions are due to the vinyl or from the mastering process.

7.4 Comparing Taylor Swift Albums

It is rare that an artist re-records an album let alone several of them, but that's just what pop superstar Taylor Swift has done. She has re-recorded four of her albums, note for note, recreating the sounds of the original versions. There are now an original version and a "Taylor's Version" of the albums *1989* (original 2014, Taylor's Version 2023), *Speak Now* (original 2010, Taylor's Version 2023), *Red* (original 2012, Taylor's Version 2021), and *Fearless* (original 2008, Taylor's Version 2021). That two versions of these albums exist offers a unique opportunity for a listening exercise since each pair of versions are musically identical and very similar, but not identical, in terms of timbre, dynamics, and spatial processing.

Compare the original version with the Taylor's Version recording of each album and note the similarities and differences in panning and placement of sounds, timbre or equalization, loudness and dynamics processing, mix balance, reverberation, and any other differences that may be apparent. Which one do you like better and why?

7.5 Comparing Peter Gabriel Mixes

It is also unusual for an artist to release two different stereo mixes plus a spatial audio mix of an album simultaneously. Peter Gabriel's most recent album, *I/O* (2023), includes three versions: *I/O (Bright-Side Mix)* in stereo, *I/O (Dark-Side Mix)* in stereo, and *i/o (In-Side Mix*) in Dolby Atmos. This album release offers an opportunity to compare the different versions as a listening exercise. The main musical parts seem to be identical among the different versions, but a mix may bring forth auxiliary musical parts that are not present in the other versions.

Compare the different mixes and note the similarities and differences in panning and placement of sounds, timbre or equalization, loudness, mix balance, dynamics/compression, reverberation, or any other differences that may be apparent. Which one do you like better and why?

7.6 Comparing Snare Drum Recordings

Another useful exercise is to pick one instrument and compare its timbre and processing across a range of different recordings. For this example, I have chosen the snare drum as the instrument to compare. Despite the relative consistency of physical form from one snare drum to another, many factors come into play to affect a snare drum's timbre including its depth, drum head material and tuning, shell material, a drummer's technique, the acoustics of the room in which it is played, the choice and placement of microphones, and any processing applied in the mixing and mastering processes. Listening to snare drum sounds across a variety of recordings, we find that there are endless variations in timbre, reverberation, dynamics processing, and mix placements. It's common for contemporary pop recordings to use programmed drums or to replace acoustic drum recordings with samples.

To make it easier to hear the differences between snare drum sounds, try importing a variety of tracks into a DAW and trimming each recording down to a single snare hit. Figure 7.16 shows a series of snare drum hits from a variety of different commercial recordings, as an example of what you might do.

Below are a few examples of recordings in which I briefly highlight some timbral, dynamic, and spatial qualities of snare drums. They are listed in chronological order by release date. I encourage you to listen to these recordings. When you do, notice that the earlier tracks on this list have subtle variations in snare drum timbre, timing, and loudness from one drum hit to the next, which is characteristic of drums played by a drummer as opposed to those which are programmed or replaced with samples.

"When the Levee Breaks" by Led Zeppelin from Led Zeppelin IV (Remastered) *(1971)*
The drum sound on this song is iconic. Listen for the natural echo from the acoustic space in which it was recorded and the resonance around 500 Hz.

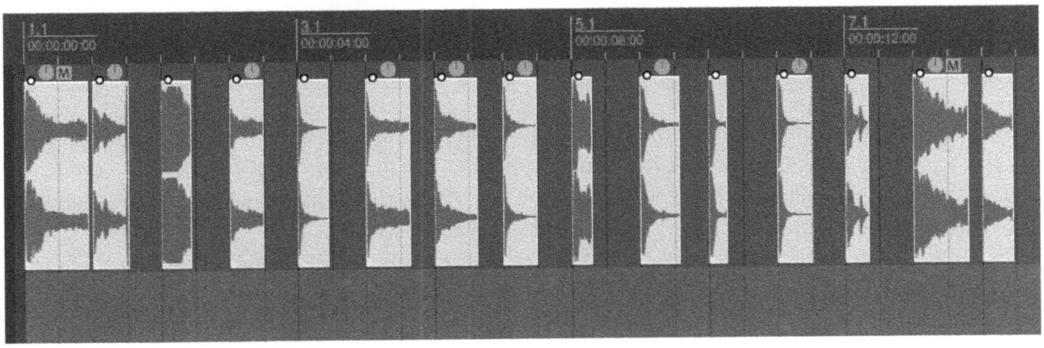

Figure 7.16 A screenshot of a collection of snare samples taken from a variety of different recordings and placed in sequence in a DAW. Playing the samples back to back brings their differences into stark contrast. The variety of amplitude envelopes is apparent just by looking at the waveforms.

"In the Air Tonight" by Phil Collins from Face Value *(1981)*
Programmed drums keep this song's pace until about 3:43, at which point the drum set enters with a dramatic drum fill. This drum recording is often cited as an example of the 1980s gated reverb sound. Listen for the high-frequency focus of the snare and the short splash of reverb following each drum hit.

"Valerie" by Steve Winwood
There are at least three different versions of this track on music streaming services and two distinctly different snare drum sounds.

> *"Valerie" from* Talking Back to the Night *(1982)*
> This sounds like programmed drums. The snare drum is dry and balanced nicely with the rest of the track.
>
> *"Valerie (Remix)" from* Chronicles *(1987)*
> The snare sound on this version is very different from the original. It sounds extremely compressed and has a splashy stereo reverb on it. The reverb seems to have more high-frequency content than the snare itself.
>
> *"Valerie (Remastered)" from* Revolutions: The Very Best of Steve Winwood *(Deluxe) (2010)*
> This is a remastered version of the original 1982 version. The overall mix is louder, and the snare timbre is brighter than in the original version. As a side note, the compression and limiting from the remastering process on this version brings out the delay effect on the vocal.

"Owner of a Lonely Heart" by Yes from 90125 *(1983)*
There are two main snare drum sounds on this track. The intro snare drum is bright and has longer reverb. Following the intro, the main snare drum part enters with a darker timbre and a short, gated reverb. It seems dry compared to the much longer reverb tail on the vocals.

"Smells Like Teen Spirit" by Nirvana from Nevermind *(1991)*
There is a natural-sounding room reverberation on the drums which sounds as though it was augmented with artificial reverb.

"Umbrella (feat. JAY-Z)" by Rihanna from Good Girl Gone Bad: Reloaded *(2007)*
The drums sound as if they are primarily samples of acoustic drums. The snare drum has some room sound on it, but it is otherwise dry and has a bright timbre.

"Radioactive" by Imagine Dragons from Night Visions *(2012)*
Clearly a programmed drum sound, this snare drum enters in the first verse. It is bright, aggressive, and mixed loud. The snare decay spreads out to the left and right channels, and the reverberation decay on the snare sounds as if the high frequencies are boosted.

"1-2-3" by Café Tacvba from Jei Beibi *(2017)*
This snare drum sounds like it was played by a drummer as part of a drum set. It is set back in the mix and seems to have a boost in the low mid-frequency range.

"I Like That" by Janelle Monae from Dirty Computer *(2018)*
This sounds like a classic drum machine sound. It has a crisp, dry sound and is placed forward in the mix.

"Last Train Home" by John Mayer from Sob Rock *(2021)*
With a full bar of acoustic drum set on its own to start the song, it is easy to isolate the snare drum sound on this recording. The snare drum has a natural resonance and has been treated with some artificial reverberation.

"Man I Need" by Olivia Dean from The Art of Loving *(2025)*
The snare drum in this track has a retro sound. It's dry and has a good amount of low- and mid-range. It sounds as though there is a little noise burst during the attack portion of each snare hit which gives it a bit of brightness.

7.7 Classical Music Recordings

7.7.1 Comparing Recordings of the Same Piece of Music

Many of the great pieces of classical music have been recorded multiple times. In fact, some orchestras have recorded the same pieces in the same hall years later. For example, major symphony orchestras such as the Berlin Philharmonic, the Cleveland Orchestra, Chicago Symphony Orchestra, London Philharmonic Orchestra, London Symphony Orchestra, and Vienna Philharmonic have each recorded Brahms's Symphony No. 3 more than once.

For this listening exercise, pick a well-known piece of classical music, find different recordings of the piece, and compare the recording quality of them. When you listen, make note of aspects such as:

- stereo image width,
- locations of instruments in the stereo image,
- timbral balance,
- reverberation, spaciousness, reflections or echoes, and hall sound,
- noise and distortion, and
- dynamic range.

Tempos and interpretation will likely vary from one recording to another, but try to focus only on the bullet points above. Even solo instrument recordings can vary widely.

7.7.1.1 Johann Sebastian Bach's Goldberg Variations, *BWV 988*

Because of the iconic nature of J. S. Bach's *Goldberg Variations*, many different recordings of it exist for piano. Listen to different piano recordings of the "Aria" movement from the *Goldberg Variations* and compare them. Pay attention to the stereo image width of the piano, reverberation and sense of space, timbre, loudness, and distance or perspective. Below is a list of suggested recordings of the piece, including the one by Víkingur Ólafsson described in Section 7.2.

- Glenn Gould, *Bach: The Goldberg Variations, BWV 988 (The 1955 & 1981 Recordings)* (Sony Classical, 2013). [Listen to both the 1955 mono recording and the 1981 stereo recording.]
- Lang Lang, *Bach: Goldberg Variations* (Deutsche Grammophon, 2021).
- Víkingur Ólafsson, *J. S. Bach: Goldberg Variations* (Deutsche Grammophon, 2023).
- Murray Perahia, *Bach: Goldberg Variations* (Sony Classical, 2000).
- András Schiff, *Bach: Goldberg Variations* (Decca, 1983).

- Peter Serkin, *Bach: Goldberg Variations* (Sony Classical, 1986).
- Rosalyn Tureck, *Bach—Goldberg Variations & Well Tempered Clavier* (Deutsche Grammophon, 2022).

Perhaps the most famous recordings of this piece are by Glenn Gould, who can be heard humming in the 1981 recording. Some other recordings have low-level background noise such as low-frequency traffic rumble or air-handling noise, which can be heard if you listen closely.

7.7.1.2 *Igor Stravinsky's* The Firebird Suite

Another piece that has been recorded many times is Igor Stravinsky's *The Firebird Suite.* Below are short descriptions of sound quality of five recordings of the same movement from five different decades.

- Leonard Bernstein and New York Philharmonic, "V. Infernal Dance of King Kashchei" (1919 version) from the album *Stravinsky: The Rite of Spring & Suite from "The Firebird"* (Sony Music, 2004, originally released 1958).

There is plenty of nice detail in the percussion and string pizzicato. The brass instruments' early reflections arrive with a short delay after the direct sound, but from opposite sides than the direct sound. At around 45 seconds in, the first violins on the left seem to get louder and move further left. The reverberation tail at the end decays to the center of the stereo image. Despite the age of this recording, it is quite clean overall.

- Leopold Stokowski and London Symphony Orchestra, "4. Infernal Dance of King Kaschei (1919 Version)" from the album *Mussorgsky: Night on the Bare Mountain/Stravinsky: The Firebird Suite* (Decca, 1968).

There is significant distortion on the loudest moments of this recording, unfortunately. Nonetheless, the raw power of the orchestra comes through, and the recording is louder than many other recordings of the piece.

- Robert Shaw and the Atlanta Symphony Orchestra and Chorus, "IV. Infernal Dance of King Kastchei (1919 Version)" from the album *Stravinsky: The Firebird Suite (1919 Version); Borodin: Overture & Polovetsian Dances from Prince Igor* (Telarc, 1978).

One of the first orchestral recordings to be made digitally, this recording has excellent clarity and spectral balance, and a natural sense of space in the hall. Instruments have focused locations in the stereo image and there are no strong reflections or echoes. Despite the use of early digital recording technology, the sound quality is remarkable.

- Pierre Boulez and Chicago Symphony Orchestra, "XVI. Danse infernale de tous les sujets de Kastchei" from the album *Stravinsky: L'Oiseau de feu; Feu d'artifice; 4 Etudes* (Deutsche Grammophon, 1993).

This recording has a wide stereo image, good front/back depth, and a natural spaciousness. High frequencies seem to be more present in this recording compared to the version by Shaw and the Atlanta Symphony.

- Gustavo Dudamel and Los Angeles Philharmonic, "Tableau I. Infernal Dance of Kastchei's Subjects (Live)" from the album *Debussy: La mer; Stravinsky: The Firebird* (Deutsche Grammophon, 2013).

One of the first things I noticed in this recording when listening with one headphone model in particular (e.g., Sony MDR-7506), is the amount of low end in the bass drum hits right from the start of the movement. Other headphones (e.g., Sennheiser HD 650) do not seem to emphasize the low end of the bass drum as much. Low frequencies from the bass drum hits seem louder in this recording than in other recordings. There is also significant detail in the strings, woodwinds, and timpani, which all sound as if they were close-miked.

7.7.2 Comparing Classical Music Record Labels

As you listen to classical recordings, pay attention to the record labels. Do you hear any characteristics that are common to a given record label and that distinguish it from other labels? Some labels may have a more reverberant sound, while others may have a closer sound. It can help to audition different recordings of the same piece of music for this comparison. Classical music record labels usually include their logos on their album covers, so this can be an easy way to identify the label of most recordings.

Some historically important classical music recordings come from the Mercury Living Presence series, which began issuing recordings in the early 1950s. Orchestral recordings in this series generally used only three omnidirectional microphones. Produced by Wilma Cozart and engineered by C. Robert (Bob) Fine, the Mercury Living Presence recordings still sound remarkably good, especially considering the period in which they were made and the technology available at that time. While they do have a noticeable noise floor, which is expected from recordings of that period, the spaciousness and stereo image clarity has stood the test of time.

Some of the bigger, highly respected record labels include Decca, Deutsche Grammophon, and Sony Classical. Smaller labels such as 2L, BIS, Channel Classics, ECM New Series, Harmonia Mundi, and Telarc, many of which now exist as part of larger record labels, also have excellent quality recordings and may offer timbral and spatial characteristics that differ from the big labels. Labels used to have staff recording engineers, which meant there would be a lot of consistency in microphone setup and approaches to mixing. Note the record labels of the recordings you audition and listen for similarities among recordings from a specific label.

7.8 Graphical Analysis of Sound

In research on sound image perception produced by car audio systems, researchers have used graphical techniques to elicit listeners' perceptions of the locations and dimensions of sound images (Ford et al. 2002, 2003; Mason et al. 2000). Work done by Wieslaw Woszczyk and John Usher (Usher and Woszczyk 2003; Usher 2004) has sought to visualize the placement, depth, and width of sound images within a multichannel reproduction environment, to better understand listeners' perceptions of sound source locations in an automotive sound reproduction environment. In listening experiments, participants were asked to draw sound sources using elliptical shapes on a computer graphical interface.

When listeners translate what they hear to a visual, two-dimensional diagram, they can achieve a level of analysis distinct from simply using verbal descriptions. Although there is no standard method for visually illustrating an auditory perception, the exercise of doing so is very useful for sonic analysis and exploration. One of the classes offered through the Graduate Program in Sound Recording at McGill University is called Analysis of Recordings. When I took this class, I was introduced to graphical analysis of stereo images to document my perceptions, fine-tune my critical listening, and create a more concrete document of a stereo image for further discussion and analysis. I give credit to Wieslaw Woszczyk at McGill for introducing me to this idea and I present it here for you to explore.

7.8.1 *Graphical Analysis Exercise*

Using a template such as in Figure 7.17, try to draw what you hear coming from a sound system. Your listening location relative to a pair of loudspeakers and the speaker placement will have a direct effect on the localization of phantom images. For example, sitting slightly to the left of the ideal listening location will make it sound as though most of the stereo image is in the left speaker. Section 1.8 illustrates the ideal listening location for stereo sound reproduction that will give accurate phantom image locations. Sound reflected from nearby surfaces such as walls, windows, or objects near a loudspeaker will also affect the stereo image that you hear. Headphone listening will give the best isolation from the effects of seating position and room acoustics, but we perceive the stereo image from headphones differently than loudspeakers due to in-head localization and the lack of crosstalk in headphones.

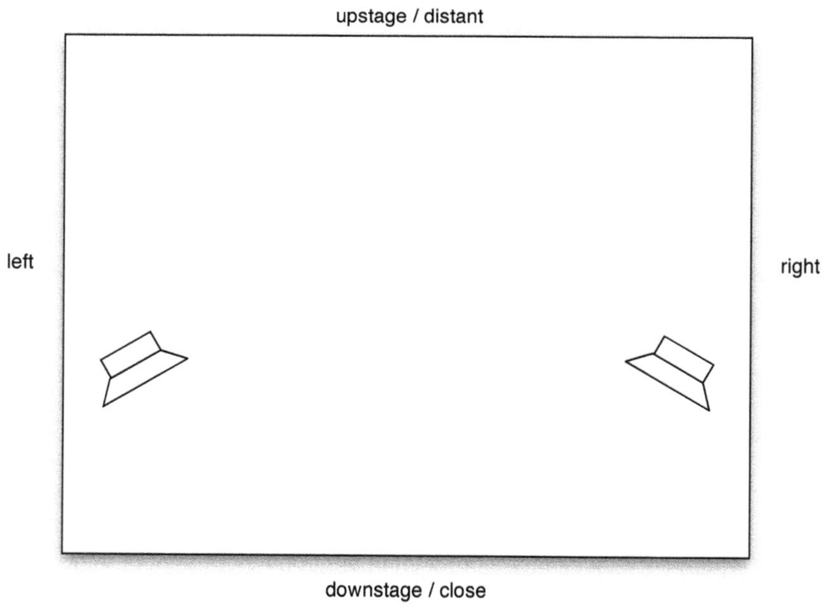

Figure 7.17 A template for the graphical analysis of a sound image. Use this template as a guide to visualize and draw the perceived locations and spatial extents of sound images within a sound recording. Try to translate the perceived locations of sound sources in a mix in terms of their left/right and front/back positions to this two-dimensional layout.

Loudspeaker locations are included in the template because we may perceive that some sound sources are positioned in front of the loudspeakers and others are positioned behind the loudspeakers.

When you draw images of the stereo image, try to represent the width and depth of the sound images that you perceive from your loudspeakers or headphones. The shapes that you draw on the template should not resemble musical instrument images. For example, do not draw a guitar to represent the sound of a guitar. Rather, translate the dimensions of the guitar sound image to a visual representation. Likewise, a solo piano stereo image will likely be different from the image of a piano playing within an ensemble; thus the images we draw for an instrument of each of these different types of ensembles will also look significantly different. The shapes also do not need to be of any particular geometry. The sound images you draw may look more like blobs than circles, ovals, triangles, or squares.

I recommend labeling your stereo image drawings to indicate how the visual forms correspond to the perceived aural images and include the name of the recording that you analyzed. Without labels, this kind of drawing may appear too abstract to be understood by someone else or by you later.

This is a challenging exercise, and some questions may arise such as:

- How do we translate our aural impressions of a stereo image into a visual image? There is no right or wrong way to represent sounds visually. Each person who draws the stereo image of a recording will come up with a slightly different interpretation. There may be commonalities among drawings of the same recording, especially having to do with sound source placement from left to right. The actual shapes and textures that we use to represent each sound will vary widely from person to person and that is fine.
- For some recordings, timbres, processing, panning, and mix levels change over the course of a track. How do we indicate movement or changes in spatial extent?
- How do we draw the variety of timbres that we hear, such as "round" low-frequency sounds or "sparkling" high-frequency sounds?
- Sound sources in a stereo image may not have clearly defined dimensions. What kinds of outline textures best represent the outer limits of a source's extents? Are the outlines jagged? Are they represented by a gradient?

Graphical analysis allows us to focus on the location, width, depth, and spread of sound sources in a sound image. A visual representation of a sound image should include not only direct sound from each sound source but also any spatial effects such as reflections and reverberation present in a recording. Try to draw everything that you hear within the stereo image. Drawing sound images challenges us to focus more closely on sound source locations in a stereo image. Figure 7.18 shows an example of one way we might draw the stereo image of a recording. Use your ears and have fun with it.

7.9 Spatial Audio

Conceptually, the term "spatial audio" encompasses a range of other terminology including multichannel audio, surround sound, 3D audio, and immersive audio. It allows the most realistic reproduction of an enveloping sound field, especially for recordings of purely acoustic music in a concert hall setting. This type of recording and reproduction can leave listeners with the impression that they are seated in a hall or virtual environment, completely enveloped by sound.

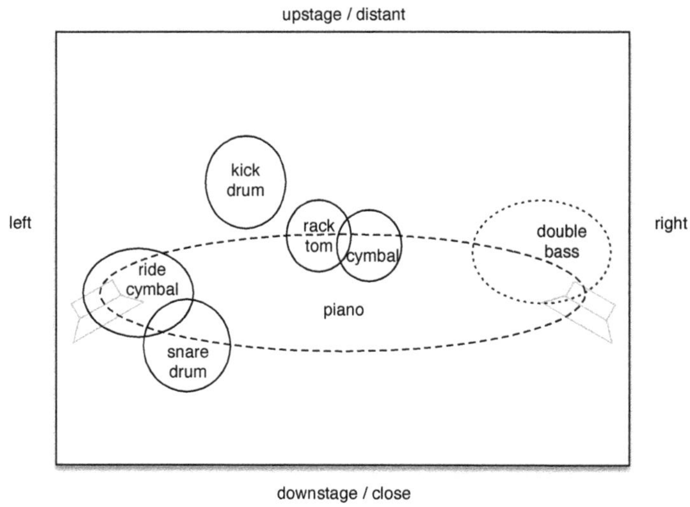

Figure 7.18 This is an example of a graphical analysis of a stereo image of a jazz piano trio recording. Your shapes do not need to look like the shapes in the drawing. The shapes of the sources may not be round or oval but may be more like blobs. The main goal is to identify left/right and front/back positioning and size of each source. (Recording analyzed in this example: Tord Gustavsen Quartet. (2012). "Playing" from the album *The Well*. ECM Records. Produced by Manfred Eicher. Engineered by Jan Erik Kongshaug.)

Conversely, spatial audio is also a format that can offer entirely unrealistic sound fields because it allows an engineer to position sound sources around a listener. Typically, there are no musicians placed behind audience members at a concert, other than perhaps antiphonal organ, brass, or choir that we sometimes hear in classical music performances, but multichannel audio reproduction allows a mix engineer to place direct sound sources to the rear and sides of the listening position. Certainly, spatial audio has many advantages over two-channel stereo, but there are still challenges to be considered and opportunities for critical listening to help with these challenges.

The term "multichannel audio" typically refers to audio reproduction formats with four or more loudspeakers arranged around a listening area. Dolby Atmos is probably the most widely used multichannel or spatial audio format, but other formats include Auro 3D, DTS, and Sony 360 Reality Audio.

The term "3D audio" typically refers to headphone reproduction that uses binaural processing to position virtual sound sources at any location around a listener, such as used in virtual reality applications.

The conventional 5.1-channel surround-sound format found in recommendation ITU-R BS.775 (International Telecommunication Union 2022) includes three loudspeakers in front and two behind (see Figure 1.4). With this arrangement, there is a wide space between each front loudspeaker and its nearest surround loudspeaker—for example, the wide space between the left front loudspeaker at 30° and the left surround loudspeaker at 120°. The wide space to the side between front and rear loudspeakers makes lateral sound images difficult to produce, at least with any stability and locational accuracy.

Dolby Atmos loudspeaker layouts can include height and overhead channels as well as surround and side channels. Loudspeaker setups for Dolby Atmos range from 5.1.2 channels—that is, five main channels (typically three front and two surround loudspeakers), one subwoofer, and two height/overhead loudspeakers—to 11.1.8 channels—that is, 11 main loudspeakers, one subwoofer, and eight height/overhead loudspeakers. Dolby Atmos recordings can also be auditioned over headphones. By convolving multichannel signals with the appropriate head-related transfer functions (HRTFs), a binaural renderer can convert a 7.1.4-channel Atmos mix, for example, into a two-channel binaural (left/right) signal for playback over headphones. While binaurally rendered Dolby Atmos mixes played back over headphones may not always produce the same sense of envelopment and localization accuracy as a multi-loudspeaker playback system, the advantage of binaurally rendered spatial audio is that it offers us an opportunity for ear training to compare traditional stereo to spatial audio over headphones.

7.9.1 Front Sound Images

A distinctive feature of most multichannel reproduction systems is the presence of a center speaker, something that is not present in a standard two-channel stereo system. The advantage of a center channel is that it can help solidify and stabilize center sound images. Center phantom images in a stereo loudspeaker setup by contrast appear to come from the center only when we are seated in the ideal listening location and equidistant from the loudspeakers (see Figure 1.2). When we move to one side of the ideal listening position, center phantom images appear to move to side as well. When seated to one side, we are no longer equidistant from the two loudspeakers, and therefore sound arrives first from the closest speaker. Because of the law of first-arriving wavefront, also known as the precedence effect or Haas effect, we localize center-panned sounds to the closest loudspeaker when seated off-center.

Soloing or muting the center speaker of a surround mix helps give an idea of what is being sent to the center channel. When listening to the center channel and exploring how it is integrated with the left and right channels, think about these questions:

- Does the presence or absence of the center channel make a significant difference to the front image?
- Are lead instruments or vocals the only sounds in the center channel?
- Are any drums or components of the drum kit panned to the center channel?
- Is the bass present in the center channel?
- If it is a classical recording with a soloist, is the soloist in the center channel?

If a recording has prominent lead vocals which are panned only to the center channel, then it is likely that some of the reverberation, echo, and early reflections are panned to other channels. In such a mix, muting the center channel can make it easier to hear the reverberation without any direct sound.

Sometimes phantom images produced by the left and right channels are reinforced by the center channel. Duplicating a center phantom image in the center speaker can make the central image more stable and solid. Often the signal that is sent to the left and right channels may be delayed or altered in some way so that it is not an exact copy of the center channel. If all three channels produce the same audio signal, the listener can experience comb filtering with changes in head location as the signals from three different locations combine at the ears (Martin 2005).

The spatial quality of a phantom image produced between the left and right channels is markedly different from the solid image of the center channel reproducing the same audio signal on its own. A phantom image between the left and right loudspeakers may still be preferred by some listeners despite its shortcomings, such as a phantom image perceived location that corresponds to listener location. A phantom image produced by two loudspeakers will generally be wider and more full-sounding than a single center loudspeaker producing the same sound, which we may perceive as narrower and more constricted.

It is important to compare different channels of a multichannel recording and start to form an internal reference for various aspects of multichannel sound images. By making these comparisons and doing close, careful listening, we can form solid impressions of what kinds of sounds are possible from various loudspeakers in a surround environment.

7.9.2 Side, Rear, and Overhead Sound Images

In our analysis of surround recordings, it is useful to focus on how well a recording in 5.1-, 7.1.4-, 11.1.4-channel, or binaural audio achieves a smooth spread from front to rear to overhead and if side images exist. Side images are difficult to produce without an actual loudspeaker positioned on the sides because of the nature of binaural hearing, which is far more accurate at localizing sounds originating from the front.

When you listen to a multichannel or spatial audio recording, try to localize the various elements in a mix and consider the placement of sounds around the listening area with these questions:

- How are different elements in the mix panned?
- Do they have precise locations, or is it difficult to determine the exact location because a sound seems to come from many locations at once?
- What is the nature of the reverberation and where is it panned?
- Are there different levels of reverberation and delay?

In a 5.1-channel surround playback system, the rear channels are widely spaced. The wide loudspeaker spacing, coupled with our forward-facing outer ears (pinnae) that have less spatial acuity in the rear, makes it challenging to create a cohesive, evenly spread rear image. It is important to listen to the surround channels soloed, with the other channels muted. When listening to the entire mix, the rear channels may not be as easy to hear because of the human auditory system's predisposition to sound arriving from the front.

7.9.3 Stereo and Spatial Audio Mixes Comparisons

Music streaming services including Amazon Music Unlimited, Apple Music, and TIDAL offer spatial audio versions of many songs in their catalogs, in either Dolby Atmos or Sony 360 Reality Audio format. With both stereo and spatial audio versions available on music streaming services, it is easier than ever to compare spatial audio and stereo mixes over headphones.

Some headphones, particularly wireless models, offer headtracking and personalized binaural audio features which can improve the binaural experience. These features help with localization accuracy and externalization of binaural sound images, and reduce timbral coloration. Headphones with a headtracking feature update binaural spatial images in real time based on a person's head movement and position. This helps lock spatial audio images so that it sounds as if they remain fixed in place, which mimics the way real acoustic sources stay in place even if we move around or turn our heads. Because generic HRTFs used for most binaural rendering can cause localization errors, in-head localization, and timbral coloration, companies have found ways to approximate a person's HRTFs to improve the binaural experience. For example, custom binaural processing is provided with Apple's Personalized Spatial Audio feature, which uses images of a listener's head and ears to select HRTFs that most closely match a listener's HRTFs. Headtracking and customized HRTFs both improve the listening. If you have access to either of these features, try to compare the sound with and without them engaged, and note the differences in sound localization and timbre.

Regardless of the headtracking and personalized binaural audio features on some models, any headphone can be used for this exercise. In fact, it is preferable to use wired headphones, which usually do not have either feature, because they do not undergo Bluetooth codec distortion (as we saw in Chapter 1). It's also worth mentioning that the binaural spatial audio formats available on music streaming services are not lossless, so you may hear codec artifacts on spatial audio versions even when listening on wired headphones.

When you compare the stereo and spatial audio versions, listen for and note differences and similarities between them in terms of:

- **Panning and spatial location of sound sources:** Are mix elements in roughly the same locations or are they different? If they are different, how are they different?
- **Reverberation decay time and spaciousness:** Is the reverberation the same in the two versions? Is there more reverberation in one version? Is one more spacious and immersive than the other? Given that the format is called spatial audio, it may be tempting to think that it will automatically sound more spacious, but this may not be the case, depending on the track. Compare the two and decide based on what you hear rather than what you know about each version. See Section 7.15 on bias in the listening and evaluation process.
- **Image width:** This may seem counterintuitive, but some spatial audio mixes can sound narrower than their stereo versions when listening over headphones. This is due to the crosstalk inherent in a binaural rendering of a multichannel audio mix compared to the distinct left/right channel separation of a stereo mix over headphones.
- **Timbre and equalization:** One drawback of binaural processing is that it can create unwanted timbral coloration. Timbral effects result mainly from the use of generic HRTFs in the binaural rendering process that do not match a listener's own HRTFs. Headtracking and customized binaural processing both improve localization and mitigate timbral coloration. Do you notice timbral coloration when compared to the stereo versions?
- **Dynamic range and compression/limiting:** Are there any differences in compression?
- **Loudness:** Usually, Dolby Atmos mixes are quieter than standard stereo mixes because Dolby requires that Atmos mixes must not exceed –18 LUFS.
- **Mix balance:** Since the spatial audio format allows sounds to be panned anywhere around a listener, mix elements may get balanced differently than in the corresponding stereo version. Because spatial audio allows more separation between sound sources, there may be less masking in a spatial audio mix, and some elements may be more audible in the spatial audio version. Because of the possibilities for reduced masking, elements may be included in a spatial audio mix that are not included in the stereo version.

Below are some tracks that utilize the spatial audio format quite effectively:

- "Yoshimi Battles the Pink Robots, Pt. 1" by the Flaming Lips from *Yoshimi Battles the Pink Robots* (2002), Dolby Atmos.
- "Boom" by Tiësto & Sevenn from *BOOM—Single* (2017), Dolby Atmos.
- "Come Together (2019 Mix)" by the Beatles from *Abbey Road (2019 Mix)*, 1969/2019, Dolby Atmos.

- "III. Poco allegretto" of Symphony No. 3 in F Major, Op. 90 by Johannes Brahms, recorded by Ben Gernon and the London Philharmonic Orchestra from *The Immersive Experience* (2020), Dolby Atmos.
- "Diso" by William Orbit from *Starbeam EP* (2021), Dolby Atmos.
- "Oxytocin" by Billie Eilish from *Happier Than Ever* (2021), Dolby Atmos.
- "Sharks" by Imagine Dragons from *Mercury—Acts 1 & 2* (2022), Dolby Atmos.
- "Road to Joy" by Peter Gabriel from *i/o (In-Side Mix)* (2023), Dolby Atmos.

7.10 High-Res Audio

There have been debates about the advantages of high sampling rates in digital audio, but it is not clear that recording at high sampling rates brings any audible benefits, and in fact there may be some drawbacks to high-resolution audio. Standard-resolution digital audio has a sampling rate of 44,100 Hz or 48,000 Hz and a bit depth of 16 bits per sample. As recording technology, data storage capacity, and network bandwidth have evolved, they have allowed recording and distribution of audio at much higher sampling rates. There is no question that 24 bits per sample is preferable to 16 bits during the recording, mixing, and mastering stages, since processing done in the digital domain benefits from the higher resolution. After the recording, mixing, and mastering stages, a bit depth of 16 bits per sample is acceptable for delivery unless the recording has an unusually wide dynamic range. As an exercise, compare a 24-bit recording to a dithered-down 16-bit version of the same recording and try to see if you can hear any differences in spatial characteristics, dynamics, or timbre.

As we saw in Chapter 5, the sampling rate determines the highest frequency that can be recorded and therefore the bandwidth of a recording. Sampling theorem states that the highest frequency we can record is equal to half the sampling frequency, known as the Nyquist frequency. Higher sampling rates allow a wider bandwidth for recording. But do ultra-high sampling rates improve audio quality?

The difference between a high sampling rate of 96 kHz or 192 kHz, and a standard sample rate of 44.1 or 48 kHz is subtle, there is no question. It is still up for debate whether listeners can reliably and consistently hear any difference between 96 kHz and 44.1 kHz sampling rates, for example. Well-respected recording and mastering engineers report being able to hear differences in ideal listening environments, but controlled double-blind listening tests have failed so far to provide conclusive evidence that listeners can hear any difference between 44.1k and 96k sampling rates. Does bias play a role in our perception of sampling rates? In other words, if we know what we are listening to, does the higher sampling rate always sound better? There may be some advantage to recording at a high sampling rate for subsequent mixing and processing, but for now we do not seem to have any firm scientific evidence to support it.

There are now several sites that sell recordings at high sampling rates. Conduct your own listening tests and see if you can hear the differences between 44.1 kHz and 96 kHz or 192 kHz. Some of the high-res audio download sites include:

- **2L:** www.2l.no
- **HDtracks:** www.hdtracks.com
- **ProStudioMasters:** www.prostudiomasters.com

In the late 1990s, Sony and Philips Electronics introduced a high-resolution audio format called DSD (Direct Stream Digital) that specified a 2.8224 MHz sampling rate—that is, 64 times the sampling rate of 44.1 kHz, at one bit per sample. They released DSD recordings for a few years on a medium called Super Audio CD (SACD). Some engineers say that recordings from SACD offer a greater difference than 96 kHz or 192 kHz when compared to CD-quality audio of 44.1 kHz/16 bit. One of the differences they say has to do with improved spatial clarity. The panning of instruments and sound sources within a stereo or surround image can be more clearly defined, source locations are more precise, and reverberation decay is generally smoother. Again, it does not appear that double-blind listening tests support this as conclusively, but more work is needed.

Although it is becoming difficult to find SACD discs and players, there are some websites that allow you to download or stream DSD audio from sites such as:

- **2L:** www.2l.no
- **Blue Coast Records:** http://bluecoastrecords.com
- **Native DSD Music:** www.nativedsd.com

To play back DSD properly, you will likely need appropriate software and hardware as specified on the download and streaming sites. Try comparing audio at different sampling rates. With any of these comparisons, it is easier to hear differences when the audio is reproduced over high-quality loudspeakers or headphones. Lower-quality reproduction devices do not allow full enjoyment of the benefits of high sampling rates.

A sampling rate of 44.1 kHz is capable of recording the entire audible range up to 20 kHz, the limit of human hearing, and as long as our converter has good anti-aliasing filters, that sampling rate is good enough. As we discussed in Section 5.4, aliasing can occur as an artifact of digital audio processing if new harmonics are generated in the digital domain without any kind of mitigation such as oversampling. In that case, high sampling rates do make sense. The main problem with recording and distribution of high sampling rate audio, such as 96 kHz or 192 kHz, is that ultrasonic frequencies (those above 20 kHz) are present. Depending on the playback system and its capability to handle the ultrasonic frequencies that may be present in a high sampling rate recording, intermodulation distortion may occur which can produce frequencies in the audible range that were not present in the original recording. Besides the possible introduction of distortion products in the audible range, high sample rate audio also requires more data storage capacity and faster networks (Montgomery 2012).

7.11 Audio Watermarking

On occasion, audio from sources like terrestrial radio or streaming audio contains strange swooshy, fluttery, or warbly artifacts in some recordings. If you notice these kinds of artifacts on a lossless or high-bitrate stream (e.g., 320 kbits/s), in which codec artifacts are nonexistent or likely inaudible, then the artifacts may be due to audio watermarking and not due to a codec. Matt Montag (2012) wrote in his blog about the audibility of watermarking on audio releases from Universal Music Group and their subsidiary record labels, such as Interscope, The Island Def Jam, Universal Republic, Verve, GRP, Impulse!, Decca, Deutsche Grammophon, Geffen. Audio watermarking involves adding a known audio signal to recordings so that copyright may be enforced more easily as these recordings move about the internet. Unfortunately, the added watermark is audible, and it degrades the audio quality quite noticeably in some cases. Based on some listening I have done, solo piano recordings seem to be affected the most, but other recordings of acoustic music also suffer. Pop and rock recordings with limited dynamic range seem to suffer the least. If you stream the music you listen to, take a moment and listen more closely and see if you can hear these artifacts. Test a track by recording your audio stream into a DAW and then line it up with a WAV copy of the track from a CD or other known lossless source, which we assume should be free of watermarking, and subtract the two versions by flipping the polarity on one version and mixing the two together. Lossless encoded audio, such as ALAC and FLAC, and high-quality coded audio, such as 320 kbps Ogg Vorbis, streams should be perceptually identical or very, very close to the original uncompressed CD versions, but the artifact, presumably from watermarking, is noticeable. Furthermore, it is much worse than the artifacts we would expect from coded audio at bit rates above 128 kbps. Streaming media, especially lossless, offers amazing possibilities for accessing millions of sound recordings for study and enjoyment, but for those of us concerned with high-quality audio, the presence of audio watermarking artifacts means that we cannot even count on lossless streaming audio for the highest-quality listening at least for now. While the presence of watermarking seemed to be widespread when I was writing the second edition of this book, record label catalogs seem to have been largely updated with clean audio for most tracks. Unfortunately, it seems to be still present in some tracks on some streaming services. Listen closely to see if you can hear it.

7.12 Comparing Loudspeakers and Headphones

Each model of loudspeaker or headphone has a unique sound. Objective characteristics such as frequency response, power response, total harmonic distortion, and intermodulation distortion contribute to the sound that we hear and thus influence our decisions during recording and mixing sessions. While we may not know the specifications for headphones or loudspeakers or how their specifications relate to sound quality, we can still listen and compare.

For this exercise, do the following:

- Choose either two different pairs of loudspeakers, two different headphone models, or a pair of loudspeakers and a pair of headphones.
- Select several familiar music recordings.
- Document the make/model of the loudspeakers/headphones and listening environment.
- Compare the sound quality of the two different sound reproduction devices.
- Describe the audible differences with comments on the following aspects and features of the sound field:
    - Timbral quality and tonal balance:
        - Describe differences in frequency response and spectral or tonal balance.
        - Is one model deficient in a specific frequency or frequency band?
        - Is one model particularly resonant in a certain frequency or frequency band?
    - Spatial characteristics:
        - How does the reverberation sound?
        - Does one model make the reverberation more prominent than the other?
        - Is the spatial layout of the stereo image the same in both?
        - Is the clarity of sound source locations the same in both? That is, can sound sources be localized in the stereo image equally well in both models?
        - If comparing headphones to loudspeakers, are there differences in the elements that are panned center?
        - How do the central images compare in terms of their location front/back and their width?

  - Overall clarity of the sound image:
    - Which one is more defined?
    - Can details be heard in one that are less audible or inaudible in the other?
  - Preference: Which one do you prefer overall and why?
  - Overall differences: Describe any differences beyond the list presented here.
- Sound files: It is best to use only linear PCM files (AIFF or WAV) or lossless encoded audio that has not previously been converted from MP3 or AAC.

When comparing loudspeakers to headphones, one main difference is stereo image width. Because there is crosstalk involved with loudspeaker listening but not with headphone stereo, stereo images over loudspeakers tend to sound a little narrower.

Each sound-reproducing device and environment has a direct effect on the quality and character of the sound we hear, and it is important for us to know our sound reproduction system—that is, the loudspeaker/room combination—and to have reference recordings that we know well. Reference recordings do not have to be perfectly pristine recordings, although that helps, but it is more important that the recordings be familiar. Be aware that listening level affects the perception of quality and timbre. Even a small level difference can affect the perception of timbre.

7.13 Evaluating a Sound Enhancer

Some software music players offer so-called enhancement controls. One example is the "Sound Enhancer" feature found in the Apple Music player found on macOS computers (but not iOS devices), and the Apple iTunes player on Windows computers. In Apple Music/iTunes, the Sound Enhancer is turned on by default. You can toggle it on or off and adjust the amount of enhancement, which ranges from low to high, in the Playback Preferences.

This type of processing offers another opportunity for critical listening, when we compare the audio quality with sound enhancement turned on to the audio quality with it turned off. Pick a recording, turn the Sound Enhancer to "high" and toggle it on and off.

When you turn on the Sound Enhancer, listen for issues of sound quality and consider questions such as:

- How does the Sound Enhancer processing affect different recordings?
- How does it affect the stereo image? Is the overall image width affected? Are the locations of sound sources in the stereo image altered in any way?
- Is the reverberation level affected?
- Is the timbre altered in any way? Try to identify as precisely as possible how the timbre is altered by this processing. Identify if any equalization has been added and what specific frequencies have been altered.
- Is there any dynamic range processing occurring? Are there artifacts of compression present? Does the enhanced version sound louder?

After you have done this exercise with one recording, pick another one and try it again. The processing that it employs may improve the sound of some recordings but may degrade the sound of others. Because the Sound Enhancer can result in positive or negative effects on a recording, it offers another critical listening exercise that we can use to determine differences in audio characteristics.

7.14 Analyzing Sound from Acoustic Sources

Live acoustic music performances can be instructive and enlightening in our development of critical listening skills. Except for people who play or study an acoustic instrument, I would estimate that most of the music that most people hear most of the time is mediated through electroacoustic transducers of some sort—that is, via loudspeakers or headphones. We may forget what an instrument sounds like acoustically, with its ability to project sound into all directions in a room or hall. At least one manufacturer of consumer audio systems used to encourage its research and development staff to attend concerts of acoustic music. This practice is incredibly important for developing a point of reference for tuning loudspeakers. The act of listening to sound quality, timbre, spatial characteristics, and dynamic range during a live music concert can fine-tune our skills for technical listening over loudspeakers.

It may seem counterintuitive to use such acoustic music performances for training in a field that relies on sound reproduction technology, but the sound radiation patterns of musical instruments are different from those of loudspeakers, and it is important to recalibrate the auditory system by listening actively to acoustic music. Acoustic music concerts, whether jazz, classical, contemporary acoustic, folk, or some other genre, allow audience members to hear the natural sound radiation patterns from each instrument and the natural reflection patterns and reverberation of the acoustic space. Sound emanates from each instrument into the room, theater, or hall and mixes with that from other instruments and voices. The spatial audio experience in a live space with acoustic music is much different from the experience of listening over speakers.

The next time you attend a concert of live acoustic music, try to listen to the sound with the ears of a mix engineer and think about the balance of sounds that you hear. If you somehow magically had faders that could rebalance the acoustic sounds, what would you change, if anything? Just as we can focus on the spatial layout (panning) and depth of a recording reproduced over loudspeakers, we can also examine these aspects in an acoustic setting. Try to localize the various members or sections of the ensemble that is performing. It may be easier to focus on the aural sensation with your eyes closed so that you can ignore what the sense of sight is reporting. Attempt to localize instruments on stage and think about the overall sound in terms of a "stereo image"—as if two loudspeakers were producing the sound and you are hearing phantom images between the speakers. The localization of sound sources may not be the same for all seats in the venue and may be influenced by early reflections from side walls in the performance space. If we were able to directly compare music being reproduced over a pair of loudspeakers to that being performed in a live acoustic space, the two sound images we perceive would be significantly different in terms of timbre, space, and dynamics. This kind of direct comparison is impossible since logistics make it difficult to move instantaneously from an audience seat during a performance to a seat in a recording control room to hear the same music played back over loudspeakers. Nevertheless, it is worth thinking about the loudspeaker listening experience and trying to imagine how it compares to a concert listening experience. Think about these questions to guide your listening:

- Try to estimate if the live music sounds wider overall or narrower than the typical width you might hear over stereo loudspeakers.
- Is the direct to reverberant ratio consistent with what we might hear in a recording?
- How does the timbre of the live music compare to what we hear over loudspeakers? If it is different, describe the difference.
- How well can you hear the quietest musical passages?
- How does the dynamic range compare?
- How does the sense of spaciousness and envelopment compare?

As audience members, we almost always sit much farther away from musical performers than microphones would typically be placed, and as such we are usually outside of the reverberation radius or critical distance—that is, the distance at which the direct and indirect sound are equal in level. In other words, most of the sound that we hear is indirect sound—reflections and reverberation—so it is therefore much more reverberant than what we hear on a recording. This level of reverberation would not likely be acceptable in a recording, but audience members find it enjoyable. Perhaps because music performers are visible in a live setting, the auditory system is more forgiving, or perhaps the visual cues help us engage with the music as audience members because we can see the movements of the performers in sync with the notes that are being played.

Ideally the reverberant field—the audience seating area—should be somewhat diffuse, meaning the indirect sound should be heard coming equally from all directions. In a real concert hall or other music performance space, this may not be the case, and it may be possible to localize the reverberation. If you find that you can localize a reverberation tail, try to focus on the width and spatial extent of it. Is it primarily located behind or does it also extend to the sides? Is it enveloping? Is there any reverberant sound coming from the front where the musicians are typically located?

We may also discern early reflections as a feature of any sound field. Early reflections usually arrive at a listener's ears within tens of milliseconds of a direct sound and are therefore usually imperceptible as discrete sounds. However, there are occasions when reflections can become focused due to a curved surface. Any curved wall will tend to focus reflections and cause them to add together and therefore increase in amplitude to a level greater than the direct sound. If the intensity of the reflected sound coming from one location is greater than the direct sound, a sound source may seem as though it is located at the point of reflection rather than at its location on stage. For example, Hill Auditorium, a 3,528-seat concert hall on the campus of the University of Michigan, has a large, curved wall that comes up from behind and to the sides of the stage and goes overhead to the back and sides of the auditorium. It is roughly parabolic in shape and, as you might imagine, it has some interesting acoustic focusing effects for music performed on stage, especially if you are sitting off-center in either the mezzanine (lower balcony) or the upper balcony. I have noticed that this strong focusing effect can make it seem like sound is emanating from somewhere on the curved wall as though from a loudspeaker, even when no sound reinforcement system is present. The effect is simply due to acoustic focusing by the parabolic shape of the hall.

In rectangular-shaped concert halls, early reflections from the side can help to broaden the perceived width of the sound image. Although we may not hear lateral reflections as discrete echoes, we can try to focus on the overall width. Try to focus also on how the direct sound blends and joins the sound coming from the sides and rear. Is the sound continuously enveloping all around, or are there breaks in the sound field, as there may be when listening to multichannel recordings?

Echoes, reflections, and reverberation are sometimes more audible when transient or percussive sounds are present. Sounds that have a sharp attack and short sustain and decay will allow indirect sound that comes immediately after it to be heard, because the direct sound will be silent and therefore will not mask the indirect sound. Each time you hear a live music performance, especially with no sound reinforcement, listen to the space in which the music is being played and see what you can learn about the space.

7.15 Bias in the Listening and Evaluation Process

Although we can certainly develop reliably consistent critical listening skills with enough practice and hard work, our perceptual systems are fallible, and we can still be fooled by auditory illusions and bias. One auditory illusion is called a Shepard tone. Named after cognitive scientist Roger Shepard, a Shepard tone sounds like a continuously rising or falling pitch that never seems to actually get higher or lower in pitch. Illusions such as this one seem to work whether we know what is going on with a signal or not. Even though I know how a Shepard tone is created, the illusion still holds each time I hear it.

Another factor that influences our perception of sound is the sound level or loudness since the frequency response of the human ear depends on listening level. This is quantified in the equal-loudness contours as discovered by Fletcher and Munson that we discussed in Chapter 2. If we want to compare two recordings or two different plugin settings, we need to ensure that the levels are matched as closely as possible so that the ear's frequency response is consistent. If a recording is played back at one level and then again at a different level, we tend to prefer the louder playback, even if the difference is as small as 1 or 2 dB. A tiny difference in level can be the sole reason why we think two recordings sound different.

But there is another aspect of the listening experience that can influence what we hear and the judgments we make about sound, and it is called *bias*. Because of our inherent bias, we can convince ourselves that we are hearing something that we are not. It can be easy to think we can hear differences when, in fact, there are none, especially when other factors are at play.

Nousaine (1997) wrote an article "Can You Trust Your Ears?" in which he outlined three categories of bias as they relate to listening:

- sensory bias,
- psychological bias, and
- social bias.

Sensory bias allows our perceptual systems to focus only on the most important events in our surroundings, so that our perceptual systems do not become overloaded and we can conserve energy. The best example of sensory bias is when we suddenly notice the sound of an air-handling system when it shuts off. Even if it is running in the background and is clearly audible prior to being shut off, the auditory system stops paying attention to it since it is a constant sound. We stop paying attention to any continuous sound until it changes in some way. Our auditory system, like our other perceptual systems, evolved to be most sensitive to changes in our environment. Similarly, you probably did not notice the feel of the clothing you are wearing until you read this sentence. We become acclimatized to our sensations. What this means in practical terms for audio and listening is that differences between two audio stimuli are most salient right at the point when we switch from one to the other. If you have ever switched to a different set of loudspeakers or headphones in the middle of a recording or mixing project, you know that the difference can be quite dramatic. But the longer we listen after the switch, the more we get used to the new set of monitors since we adapt to the timbre of the monitoring system (Toole 2026).

Psychological bias encompasses a few different types of bias, such as expectation bias and confirmation bias. In expectation bias, we may make up our minds about the sound quality of two stimuli based on information we know about the stimuli rather than on actual sonic differences in the stimuli. When we *expect* to hear changes in an audio signal, we often seem to hear them. Have you ever adjusted parameters on an equalizer and heard, or thought you heard, the timbre change, even though you realized that the equalizer had been bypassed the whole time and that the sound had not actually changed? I have. And apparently, I am not alone. Thinking one is boosting a frequency on an equalizer even though it is unknowingly bypassed and truly believing one hears the change is an example of expectation bias.

Tom Nousaine (1991) conducted listening tests in which participants reported hearing differences in sound quality 76% of the time when presented with two identical stimuli (A and B). In other words, listeners reported hearing differences between A and B even though there were no differences. When loudness differences were introduced in the experiment, listeners preferred the louder version of each pair.

Expectation bias also comes into play when comparing audio equipment such as loudspeakers. Floyd Toole and Sean Olive have done significant and important work to further our understanding of the relationship between subjective impressions of loudspeaker and headphones and the objective measurements such as frequency response. They found that when listeners know the make and model of loudspeakers in a listening test, they evaluate them differently than if they do not know the make and model (Toole and Olive 1994). In a *sighted* listening test, listeners know what they are listening to, and in a *blind* listening test, they do not know what they are listening to. Blind tests are more objective than sighted tests because confirmation bias is eliminated.

If a listener compares high sampling rate audio at 96 kHz, for example, to the same audio at a standard sampling rate of 44.1 kHz, while knowing which stimulus contains which audio signal, chances are good they will judge the 96 kHz version as sounding better because they think it *should* sound better. It is high-res audio after all, so it *must* be better. But as we discussed earlier in the chapter, blind listening tests have yet to confirm that anyone can reliably hear the difference between standard and high sampling rate audio with blind testing. As much as we try to convince ourselves that we can separate what we *know* about a piece of equipment or a signal and what we *hear* from it, we are not truly able to separate the two. If we know what we are listening to, we must assume that this knowledge will always influence what we hear. This is the reason double-blind tests are the gold standard for scientific research in any field.

Another type of bias is social bias, which plays out in listening sessions with a group of listeners. Group dynamics can shape what we think we hear when someone suggests some quality of sound for which we should listen. As others confirm that they hear the quality that has been suggested, we also begin to hear the same quality or at least believe that we hear it.

Celebrities and other high-profile individuals can shape our perceptions too. Advertisers across a wide range of products have been exploiting this phenomenon, known as the *endorsement heuristic*, for years. A heuristic allows us to make quick judgments about an unknown thing based on personal experience and information already known to us, such as an endorsement by a celebrity. In contrast, systematic thinking requires much more effort and background research than the much faster heuristic thinking to make a judgment. If a well-known musician or recording engineer endorses a particular piece of audio equipment or recording technique, we may rely on their endorsement rather than conducting a thorough evaluation of our own. Listening to a piece of equipment and reviewing its technical data prior to making a purchase takes time and energy, both of which are limited resources. Advertisers take advantage of this in all areas of life. If we are considering the purchase of a piece of equipment that is expensive and one that we will use regularly, it's probably best to take the time to evaluate it thoroughly.

As Toole and Olive's research has highlighted, one way to counter our inherent human biases is to make sure any listening tests we conduct are blind. If you want to conduct your own blind listening tests, one method we can use is an ABX test. The ABX test first was described by Munson and Gardner (1950) and is now available from a few different sources as a software tool.[3] An ABX test provides a way to compare two stimuli, such as audio coded at different bit rates or sample rates, or different analog-to-digital converters, for example. In an ABX test, two known audio stimuli are assigned to the labels A and B. The reference, labeled X, is randomly assigned by the ABX software utility to be the same as either A or B but it is not revealed to the listener which of A or B is the same as X. The listener's task is to audition each of the three stimuli (A, B, and X) one at a time and to indicate which of the two stimuli (A or B) matches the reference X.

When conducting any comparison between two audio signals, it is vital to change only one parameter at a time to avoid confounding multiple variables. For example, when comparing two different microphone models, we should set them up at the same time so that they record the same musical performance and place them as close as possible to each other. Different musical performances can produce differences in audio signals that are greater than the differences between microphones.

Overall level differences can also play a role in our perception. As an experiment, pick a recording and import it into a DAW. Create a second version of it on a new track and reduce the level of the copied track by only 1 or 2 dB. Now you have two versions of the same recording, and the only difference between them is level. Even though you know what the difference is between the two versions (it is not a blind test for you), compare them yourself and think about the differences you hear. Do you hear only a level difference, or do you hear differences in timbre, reverberation, dynamics? Because of expectation bias—you know what the difference is between the tracks—it will be difficult to make a fair judgment. A better idea is to ask some friends and colleagues to listen to the two versions and compare them back to back without any visual cues such as waveform or meters, and without telling them what the difference is. Ask them which one they prefer and what differences they hear, if any. This will be a blind test for them, and the results may be surprising for all, especially once you reveal what the difference really is. Level matching is crucial for listening, and this kind of exercise highlights the differences we think we hear with only a small level difference.

The next time you compare two pieces of equipment or audio signals, think about how bias may influence your judgment. Try to eliminate bias by making the comparison blind. With wrong or misleading information available about audio equipment performance, especially in consumer audio publications, in online sources, and in audio equipment reviews, along with the natural human tendency for bias, it can be difficult for us to separate audio myth from reality. With some awareness that bias plays a role in our listening, we can attempt to counter it and focus on what we hear rather than what we think we hear.

7.16 Summary

We can analyze sound whether we are listening over loudspeakers or headphones or if the sound we hear is purely acoustic. Almost any situation presents opportunities to listen intently and try to deconstruct and uncover characteristics and features of the sound image we hear. The more we listen to recordings and acoustic sounds with active engagement, the more sonic features we can pinpoint and focus on. With time and continued practice, our perception of auditory events sharpens, and we notice sonic characteristics that we may not have noticed previously. The more we develop these skills through active listening, the deeper our enjoyment of sound can become, but it does take dedicated practice over time. Likewise, as our listening skills become more focused and effective, we improve our efficiency and effectiveness in sound recording, production, composition, reinforcement, and product development. Technical ear training is essential for anyone involved in audio engineering and music production, and critical listening skills are well within the grasp of anyone who is willing to spend time being attentive to what he or she is hearing.

Here are some final words of advice: Listen to as many recordings as possible, even to genres that you may not choose when listening casually or for fun. Listen over a wide variety of headphones and loudspeaker systems. During each listening session, make notes about what you hear. Find out who engineered the recordings that you admire and find more recordings by the same engineers. Note the similarities and differences among various recordings by a given engineer, producer, or record label. Likewise, note the similarities and differences in recordings from one recording or mixing engineer to another. It can be helpful to find recordings of an artist whose recordings have been made by different engineers.

The most difficult activity to engage in while working on any audio project is continuous active listening. The only way to know how to make decisions about what gear to use, where to place microphones, and how to set parameters is by listening intently to every sound that emanates from one's monitors and headphones. By always listening actively, we can gain essential information to best serve the musical vision of any audio project. In sound recording and production, the human auditory system is the final judge of quality and artistic vision.

Notes

1 For more information about sound radiation patterns of musical instruments, see *Acoustics and the Performance of Music: Manual for Acousticians, Audio Engineers, Musicians, Architects and Musical Instrument Makers* (2009) by Jürgen Meyer. *Tonmeister Technology: Recording Environments, Sound Sources, and Microphone Techniques* (1989) by Michael Dickreiter is another good source, although it is now out of print.
2 Yo-Yo Ma, Stuart Duncan, Edgar Meyer, Chris Thile – Attaboy (Live). YouTube. www.youtube.com/watch?v=d-31e8Nlujw.
3 A few of the available ABX testers include:

- Lacinato WebABX/Shootouter, www.lacinato.com/cm/software/othersoft/webabx.
- ABX app, www.abxtests.com.
- ABX Comparator, www.foobar2000.org/components/view/foo_abx.
- ABXNULL S2 from the Tukan Studios plugin repository for REAPER, https://stash.reaper.fm/v/43504/TUKANPLUGINS.png.

Bibliography

Dickreiter, Michael. 1989. *Tonmeister Technology: Recording Environments, Sound Sources, and Microphone Techniques*. Translated by Stephen F. Temmer. Temmer Enterprises, Inc.

Doyle, Tom. 2015. "Classic Tracks: Cowboy Junkies 'Sweet Jane.'" *Sound on Sound*, October. www.soundonsound.com/techniques/classic-tracks-cowboy-junkies-sweet-jane.

European Broadcasting Union. 1997. *Assessment Methods for the Subjective Evaluation of the Quality of Sound Programme Material—Music*. Technical Document 3286-E. European Broadcasting Union. https://tech.ebu.ch/docs/tech/tech3286.pdf.

Ford, Natanya, Francis J. Rumsey, and Tim Nind. 2002. "Subjective Evaluation of Perceived Spatial Differences in Car Audio Systems Using a Graphical Assessment Language." *112th Convention of the Audio Engineering Society*, Preprint 5547.

Ford, Natanya, Francis J. Rumsey, and Tim Nind. 2003. "Evaluating Spatial Attributes of Reproduced Audio Events Using a Graphical Assessment Language: Understanding Differences in Listener Depictions." Paper presented at AES 24th International Conference on Multichannel Audio, Banff, Alberta, Canada. *AES 24th International Conference on Multichannel Audio*.

International Telecommunication Union. 2022. *Recommendation ITU-R BS.775–4: Multichannel Stereophonic Sound System with and without Accompanying Picture*. ITU.

Martin, Geoff. 2005. "A New Microphone Technique for Five-Channel Recording." *118th Convention of the Audio Engineering Society*, Preprint 6427.

Mason, Russell, Natanya Ford, Francis J. Rumsey, and Bart de Bruyn. 2000. "Verbal and Non-Verbal Elicitation Techniques in the Subjective Assessment of Spatial Sound Reproduction." *109th Convention of the Audio Engineering Society*, Preprint 5225.

Meyer, Jürgen. 2009. *Acoustics and the Performance of Music: Manual for Acousticians, Audio Engineers, Musicians, Architects and Musical Instrument Makers*. Fifth Edition. Translated by Uwe Hansen. Springer.

Montag, Matt. 2012. "Universal's Audible Watermark." January 10. www.mattmontag.com/music/universals-audible-watermark.

Montgomery, Christopher (Monty). 2012. "24/192 Music Downloads . . . and Why They Make No Sense." *Xiph.Org*, March 1. https://people.xiph.org/~xiphmont/demo/neil-young.html.

Munson, W. A., and Mark B. Gardner. 1950. "Standardizing Auditory Tests." *Journal of the Acoustical Society of America* 22 (5_Supplement): 675. https://doi.org/10.1121/1.1917190.

Nousaine, Tom. 1991. "Can You Trust Your Ears?" *91st Convention of the Audio Engineering Society*, Preprint 3177.

Nousaine, Tom. 1997. "Can You Trust Your Ears?" *Stereo Review*.

Toole, Floyd E. 2026. *Sound Reproduction: The Acoustics and Psychoacoustics of Loudspeakers, Rooms and Headphones*. 4th ed. With Sean Olive and Todd Welti. Routledge. https://doi.org/10.4324/9781003477495.

Toole, Floyd E., and Sean E. Olive. 1994. "Hearing Is Believing vs. Believing Is Hearing: Blind vs. Sighted Listening Tests, and Other Interesting Things." *97th Convention of the Audio Engineering Society*, Preprint 3894.

Usher, John. 2004. "Visualizing Auditory Spatial Imagery of Multi-Channel Audio." Paper presented at 116th Convention of the Audio Engineering Society, Berlin, Germany. *116th Convention of the Audio Engineering Society*.

Usher, John, and Wieslaw R. Woszczyk. 2003. "Design and Testing of a Graphical Mapping Tool for Analyzing Spatial Audio Scenes." Paper presented at AES 24th International Conference on Multichannel Audio, Banff, Alberta, Canada. *AES 24th International Conference on Multichannel Audio*.

Woszczyk, Wieslaw R. 1993. "Quality Assessment of Multichannel Sound Recordings." *Proceedings of the AES 12th International Conference on the Perception of Reproduced Sound*, 197–218.

Index

Note: Locators in *italic* indicate figures and in **bold** indicate tables.

For Product Safety Concerns and Information please contact our EU
representative GPSR@taylorandfrancis.com
Taylor & Francis Verlag GmbH, Kaufingerstraße 24, 80331 München, Germany

www.ingramcontent.com/pod-product-compliance
Ingram Content Group UK Ltd.
Pitfield, Milton Keynes, MK11 3LW, UK
UKHW050234200926
475702UK00020B/3028
* 9 7 8 1 0 3 2 6 0 9 3 3 1 *